HANDBOOK OF INDUSTRIAL ENERGY ANALYSIS

HANDBOOK OF INDUSTRIAL ENERGY ANALYSIS

I. BOUSTEAD and G. F. HANCOCK
The Open University,
Milton Keynes

ELLIS HORWOOD LIMITED
Publishers Chichester

Halsted Press: a division of
JOHN WILEY & SONS
New York - Chichester - Brisbane - Toronto

The publisher's colophon is reproduced from James Gillison's drawing of the ancient Market Cross, Chichester

First published in 1979 by
ELLIS HORWOOD LIMITED
Market Cross House, Cooper Street, Chichester, West Sussex, England

Distributors:

Australia, New Zealand, South-east Asia:
Jacaranda-Wiley Ltd., Jacaranda Press,
JOHN WILEY & SONS INC.,
G.P.O. Box 859, Brisbane, Queensland 40001, Australia.
Canada:
JOHN WILEY & SONS CANADA LIMITED
22 Worcester Road, Rexdale, Ontario, Canada.
Europe, Africa:
JOHN WILEY & SONS LIMITED
Baffins Lane, Chichester, West Sussex, England.
North and South America and the rest of the world:
HALSTED PRESS, a division of
JOHN WILEY & SONS
605 Third Avenue, New York, N.Y. 10016, U.S.A.

British Library Cataloguing in Publication Data

Boustead, I.
Handbook of industrial energy analysis/
[by] I. Boustead and G. F. Hancock.
1. Industry - Energy conservation
I. Title II. Hancock, G F
658.2'6 TJ163.3 78-40636

ISBN 0-85312-064-1 (Ellis Horwood, Publishers)
ISBN 0-470-26492-6 (Halsted Press)

Typeset in Press Roman by Ellis Horwood Ltd.
Printed in England by Biddles of Guildford.

Table of Contents

Authors' Preface

The calculation of primary energy required to manufacture an artifact or product from raw materials in the ground is a relatively recent innovation, even though similar calculations for specific production plants and machines have been the practice of engineers for many years. However energy analysis, as this subject has become known, is becoming increasingly important in view of increasing fuel costs, the prospect of dwindling supplies of both fuels and raw materials and a greater awareness of the environmental impact of industrial operations. As a result, both Governments and commercial organisations who have to make decisions in this area are calling increasingly for more detailed and accurate information on energy matters. In the space of a decade, energy analysis has progressed from a minor academic research interest to an essential tool of management.

In the early days of energy analysis, individual analysts developed techniques to suit their own purposes. To a considerable extent the techniques were tailored to suit the information available at the time; frequently statistical data of uncertain accuracy. As a result doubt remains in some industrial circles as to the value of energy analysis, and this is compounded by the widely differing published energy requirements to be found in the literature for apparently similar operations. Many of the inconsistencies arise from the particular assumptions taken when calculating the energy requirement and from the application of different analytical techniques. However it is frequently difficult if not impossible to detect the reasons for the discrepancies from the published literature. What is needed is a textbook which develops logically the ideas needed to perform a satisfactory energy analysis, which highlights the limitations of the methods used and which explains the assumptions that must be made. This is the purpose of the present book.

The book is divided into two parts. In Part 1 Chapters 1 through 8 develop the principles underlying energy analysis with an emphasis on process energy analysis, the most useful technique for industrial purposes. Chapters 9 through 12 give some detailed examples of how these techniques may be applied in specific industrial cases and demonstrate the amount of detail that can be incorporated into the analysis. At the same time they introduce in a simple way the potential application of the subject to energy conservation. Part 2 of the book has been included because one of the major difficulties encountered by anyone attempting to perform even the simplest analysis lies in obtaining the necessary data base. There is in fact a large body of published information on energy requirements of industrial processes but it is scattered throughout the literature. Discovering it and then converting it all to the same base so that the results are comparable is a very time consuming task and so Part 2 presents a wide selection of data for many different operations collected together and converted to a standard form wherever possible. We therefore hope that this will provide a useful tool for those wishing to carry out their own analysis.

At the end of the book, the reader should be able to perform a detailed energy analysis of any plant – providing of course that he has the time to do so.

The book is aimed primarily at managers, both technical and non-technical, who may be called on to provide answers to energy problems, or asked to interpret published data. For this reason we have limited the mathematical treatment and restricted the introduction of scientific content to that needed to develop the relevant arguments. We have aimed to provide a reasonable commentry on the fundamental scientific theory necessary to understand the basis of the many calculations used in energy analysis. The book must not therefore be regarded as a textbook on fuel science, for example, for which there are many more detailed texts already available. As a result, some areas will inevitably appear to be superficially treated by their respective specialists. However, we felt it important that the development of ideas used in energy analysis should take precedence over academic completeness.

Apart from the potential industrial readership, we also think that the development presented here would be suitable as an introduction to the subject for undergraduates or graduate students in energy analysis and environmental sciences.

Finally we must acknowledge the help we have received over the years by being able to talk to many people in industry and academic institutions; these discussions have formulated many of the ideas contained in the book. We would also like to thank Ellis Horwood Ltd., the publishers, for their help in the preparation of the book and to Pat Brittin and Avril Jenner who helped with the preparation of the manuscript. We hope that the text is free of errors, but if any are discovered, we would apologise and the blame for them must fall on ourselves alone.

I. BOUSTEAD & G. F. HANCOCK
The Open University
1978

Part 1

METHODS AND APPLICATIONS

Chapter 1

Energy analysis and thermodynamics

1.1 ENERGY AND ENERGY ANALYSIS

Energy may be defined in scientific terms as the ability to do work and energy analysis is a technique for examining the way in which energy sources are harnessed to perform useful functions. This book is concerned with the energy analysis of industrial operations. Although it is primarily devoted to current industrial practices, it also provides the basis for a consideration of future developments and changes and introduces techniques for doing this.

The philosophy underlying energy analysis is not new. For over a hundred years engineers have taken a close interest in the energy consumption of the machines they have designed and built. An electrical engineer can readily describe the energy requirements of an electric motor under various conditions of loading, for example, and a mechanical engineer can describe the behaviour of a steam engine with great precision. Moreover, this interest of engineers in the energy consumption of industrial operations has not been confined to smaller machines. As far back as the mid-nineteenth century, chemical engineers were developing whole plants constructed to minimise energy use. One example of this is the Solvay process for manufacturing sodium carbonate from common salt. This plant requires a supply of steam, but the steam was generated at a considerably higher pressure than that needed by the plant. The excess pressure was used to drive an electrical generator thus providing the electricity needed in the plant. The steam output from the generator was then available to satisfy the steam requirement of the process. This original design was so effective that the process remains fundamentally unchanged to this day. Historically, engineers have been interested only in the performance, in energy terms, of the machines and plant directly associated with a production process. In recent years however, energy analysis has been expanded to encompass not only the actual manufacturing processes but also the ancilliary operations and services that are needed to support present-day industry. The impetus for this expansion was the result of many factors but three in particular stand out.

First, the 1960's were marked by an increasing awareness of environmental considerations by the population as a whole. One such consideration was the large and increasing release into the atmosphere of heat energy and fumes. Attention was focused dramatically onto some aspects of this by localised climatological effects such as those experienced in Los Angeles and Tokyo. Atmospheric pollution from the internal combustion engine and other sources brought about the tough legislation passed by the American Government relating to exhaust emission from automobiles. The 1960's also saw the rise of pressure groups as a force to be reckoned with in society, so requiring manufacturers to be more concerned with the impact of their products and processes on the environment.

A second factor resulting in a demand for a more detailed knowledge of energy use arose from a gradual appreciation of the finite quantities of conventional fossil fuels available from the earth. This problem was highlighted in some alarming, and at times alarmist, predictions of the date at which fuel supplies would be exhausted. These predictions were given additional weight as some oil wells became exhausted [47, 68, 91]. Neither of these reasons, of themselves, really resulted in any major changes in industrial practices but they did create a climate of opinion that was extremely sensitive to the third reason, namely the rapidly increasing cost of energy. So large were the cost increases that energy was fast becoming a significant factor in the overall cost of manufactured products, even in operations where it had hitherto been a relatively small contributor.

It remains true that in many manufacturing industries the machines and operations directly used in the production processes represent the largest consumers of energy. However, although the ancilliary operations may individually consume only small quantities of energy, they can also be responsible for large total energy consumptions because of the large numbers of such operations that are frequently needed in a production complex. Operations are called 'ancilliary' when they do not contribute directly to the manufacture of the product under consideration. They would include, for example, the heating and lighting of factory buildings, the heating and lighting of administrative areas associated with the factory, the energy to operate pollution control machinery, and so on.

The significance of the energy consumption in ancilliary operations is best illustrated by a simple example that will be considered in more detail later, namely the manufacture of metal cans. A typical can-making factory is supplied with sheets of metal. These sheets are lacquered, printed and cut into blanks. The blanks are then bent to form the can body and the side seam is soldered. For a typical can-making plant, the energy associated with each can for carrying out these operations is of the order of 0.23 MJ. However, the heating and lighting of the plant, which would certainly be regarded as essential by the workers but which nevertheless contributes nothing directly to the manufacturing process, requires an additional 0.08 MJ/can. Thus of the total energy of 0.31 MJ/can, only 74% is attributable directly to the production of the can bodies [16] and the remaining 26% is due to heating and lighting the plant.

The need to consider the energy consumption of ancilliary operations (that is those which do not contribute directly to the actual production process) and the capital energy associated with the manufacture of the plant itself has led to the concept of a **total production energy** or **gross energy requirement** (see Section 3.6). This is the energy associated with *all* of the operations needed to support the production of a commodity or the provision of a service. One of the main aims of this book is to show how such total production energies can be calculated. At first sight this appears to be a relatively easy aim to satisfy, but examination of some apparently simple operations soon shows that they are considerably more complex than originally expected.

Some of the difficulties can be readily appreciated by considering a very simple non-industrial operation, namely that of driving a car to the supermarket. Suppose that it is necessary to calculate the energy associated with such a journey. The main consumption of energy is as the fuel required to make the car move. However, in order to provide this fuel further energy must have been expended in the oil refinery to produce it from crude oil and yet more to deliver it to the petrol station from which it was bought by the car driver. In addition, the car itself had to be manufactured and this involves an energy expenditure not only in the construction of the car, but also in the back-up industries that supply the parts and raw materials to the car industry. Furthermore, during the drive use is made of a road, whose construction and maintenance also requires the expenditure of energy. This list could be extended to include all of the many additional operations needed for someone to drive from home to the supermarket. Associated with each of these operations there is an energy consumption and some proportion of this energy consumption is attributable to the single journey under consideration. For some of the factors listed the energy attributable to the journey will be small, but for others it will be much more significant. A fundamental problem lies in deciding which of these operations should be included in a calculation of the total energy for the drive. The decision as to which operation to include requires that some formal procedure be established. The setting-up of this procedure and the reasoning behind it demands detailed explanation and this is dealt with in later Chapters of this book.

Clearly, energy analysis of complete systems is not as simple as it appears at first sight. Much of the early work done in this area necessarily involved making arbitrary choices as to the inclusion or exclusion of certain operations, and such choices frequently give rise to uncertainties in the interpretation of published values for system energies. It is however hoped that the procedures developed in this book will help to eliminate such uncertainties in the future.

1.2 THE LAW OF CONSERVATION OF ENERGY

In common with most abstract concepts, 'energy' does not readily lend itself to a simple pictorial description. Scientists and engineers develop a 'feel' for the concept which has more to do with the familiarity gained by continued use in many applications rather than with a mere appreciation of the basic definition given in the previous Section. In each of these applications some facet of energy becomes apparent and as a result energy is thought to be governed by two fundamental scientific laws often referred to as the first two laws of thermodynamics.

The **first law of thermodynamics**, also called the **law of conservation of energy**, states that **energy can be neither created nor destroyed but can only be changed from one form to another**. Such a statement implies that energy is available in a number of forms and so it should be possible to construct a list of these energy forms. One such list might be that given in Table 1.1, which also gives an example of each type together with the magnitude of the energy involved in the example. It is important to recognise that such a list is not definitive. For example, heat energy can be related to the motion of the atoms and molecules in a material and so is strictly a form of kinetic energy. Similarly sound, which has not been included in the list, is also due to motion induced in the atoms or molecules of the medium through which it is travelling and so is also a form of kinetic energy. The inclusion or exclusion of sound from such a list is arbitrary and the list can be extended or shortened depending upon the amount of detail required and upon the purpose for which the list is drawn up.

Table 1.1

Different forms of energy together with examples of each form

Energy form	Example	Maximum energy present in the example MJ
Kinetic (motion)	A mass of 1 kg moving with a speed of 30 m.p.h.	0.00009
Potential (position)	1kg of water at a height of 500 m above reference level	0.005
Heat	1 kg of water at 373K above a reference temperature of 293K	0.34
Electricity	A current of 1 A flowing for 1 hour across a potential of 230 V	0.83
Light	Sunlight falling on an area of 1 square metre of earth's surface per hour at midday on a bright sunny day	3.4
Chemical	1 kg of oil burnt in air to produce carbon dioxide and water	45
Nuclear	Fission of 1 kg of uranium-235	80,000,000

However, using such a list as in Table 1.1 it is not difficult to construct routes by which one form of energy can be converted to another. In some instances the transformation is accomplished in a single step as, for example, when chemical energy in the form of coal or oil is converted to heat and light energy by burning in air.

In others the overall change involves a number of transitions from one energy form to another. For example, the conversion of the potential energy of water in a mountain reservoir to electricity by a hydro-electric scheme requires that the potential energy of the water be converted first into kinetic energy by causing the water to fall through a pipe from the resevoir to the generating station. The kinetic energy of the water is then used to impart kinetic energy to the rotor arm of the generator and a proportion of this rotor kinetic energy is converted to electricity. Some examples of the ways in which energy can be converted from one form to another are shown in Table 1.2.

Notice that not all of the conversions involve a single step. The hydro-electric process for converting potential energy to electrical energy passes through an intermediate energy form, that is kinetic energy. Some of the spaces in Table 1.2 have been left blank because there is no simple single operation that will carry out the conversion. However, schemes can be devised which will allow the conversion between energy forms indicated by these blanks. For example, potential energy could be converted to light by using the potential energy to produce electricity in a hydro-electric scheme and then using this electricity to operate an electric light bulb.

Table 1.2 also illustrates one other important feature useful in considering the practical use of energy, and this is the concept of **energy concentration**. Study of the conversion routes shown in Table 1.2 soon shows that some of the conversion mechanisms will take place more readily than others. For example, oil or coal will readily burn in air so converting chemical energy to heat and light. In contrast, the conversion of electricity into nuclear energy can only be accomplished by the use of particle accelerators, which are usually only found in nuclear research laboratories. Such machines require enormous quantities of electricity to produce relatively infrequent events. Therefore some property must exist in energy forms which promotes or inhibits the ready conversion to other energy forms. This property can be thought of as the **energy concentration**, and can be regarded as **the maximum amount of energy that can be extracted from a given amount of material in a specified state**.

Returning to Table 1.1, it can be seen that for every 1 kg of water in a reservoir 500 m above some reference level an energy of 0.005 MJ could, in principle, be extracted if the water were allowed to fall to the reference level. In contrast, for every 1 kg of uranium-235 that took part in a fission reaction a maximum energy of 80,000,000 MJ could be obtained. Clearly the energy available from 1 kg of fissionable uranium-235 is much greater than that available from 1 kg of water at a height of 500 m. Hence it is reasonable to say that

Table 1.2

Examples of operations for converting energy from one form to another

TO / FROM	Kinetic	Potential	Heat	Light	Electricity	Chemical	Nuclear
Kinetic	Gearbox	Roller-coaster	Friction	Axe hitting a stone (sparks)	Generator		
Potential	Waterfall	Bouncing ball			Hydro-electric plant		
Heat	Turbine	High pressure steam boiler	Heat-exchanger		Thermocouple		Fusion reactor
Light	Light vane		Greenhouse	Phosphor	Photo-electric cell	Photo-synthesis	
Electricity	Electric motor	Electric pump	Electric fire	Light bulb		Electrolysis	Particle accelerator
Chemical	Rocket		Burning	Burning	Battery	Chemical processing	
Nuclear	Particles from radioactive decay		Fission	Cerenkov radiation	Nuclear reactor		

the nuclear energy concentration in 1 kg of uranium-235 is 16,000 million times greater than the potential energy concentration of 1 kg of water at a height of 500 m. This concept is important because conversions which involve transformation of energy from a concentrated form to a less concentrated form are relatively easily carried out while transformations in the reverse direction are not. Interestingly, this observation has been used as the basis of a statement of the second law of thermodynamics which will be considered in the next Section [105].

As Table 1.1 demonstrates, sources of chemical and nuclear energy exhibit significantly higher concentrations of energy than the other forms, and so it comes as no great surprise to find that these represent the two principal forms of energy now in use. The predominant use of chemical energy, harnessed as it is by a much simpler technology than nuclear energy, perhaps reflects historical development. A comparison of the relative concentrations of energy in these two forms indicates the relatively enormous quantities of energy that can, in principle, be derived from nuclear sources. Generally materials in which the energy concentration is high and from which some or all of this energy can be readily liberated are used as fuels. Industry relies upon these materials for its sources of energy.

Table 1.1 also illustrates one more feature concerning energy: the need to quote or to imply some reference level from which measurements are to be taken. This need arises because there is no naturally occurring absolute zero of energy and so the best that can be done is to consider the energy change between one state and another. In the case of the potential energy of water, the height of the water relative to some base line must be stated. Similarly, in the case of heat energy the reference state is defined as the thermal conditions existing in the surrounding room. In other examples the reference level is not stated explicity but it is assumed to be that which would exist if the relevant change had not occurred. For instance, in the case of kinetic energy the energy is calculated by reference to the state when the mass is not moving.

As this brief discussion should make clear, industrial processing relies upon the conversion of highly concentrated energy forms known as fuels into forms of lower concentration, This gradual 'downgrading' of energy during use is a fundamental property of nature, and is summed up by the second law of thermodynamics.

1.3 THE SECOND LAW OF THERMODYNAMICS

The second law of thermodynamics is probably one of the most fundamental laws of the physical sciences and yet it cannot be proved by direct experiment. However, its validity has been established because it has been used successfully to predict a large number of phenomena that can be directly verified. This law is of interest in the present work because it provides an ultimate set of conditions which always apply to processes involving the transformation of energy from one form to another, and which govern the ways in which energy can be used in industrial processing. While the first law of thermodynamics is concerned with energy itself, the second law is concerned with energy **changes** and the constraints that exist during these changes. The intention here is not to present a complete and quantitative account of the second law but rather to examine qualitatively some of the implications of this law for industry.

At this point it would be very convenient if the law could be stated in a simple and unambiguous manner which would encompass all of its implications. Regrettably this cannot be done. There are a number of different statements of the second law as consultation of the standard textbooks on the subject can quickly confirm. All of the different statements lead ultimately to the same conclusions, but their diversity reflects different ways of looking at the physical world. The statement used here is not that in common use as a basis for the development of the mathematical theory of thermodynamics, but it forms an extremely useful starting point for a descriptive treatment of the implications of the law.

This statement is based on the work of the American scientist G. N. Lewis, and it recognises that certain changes occur in the physical world without the interference of man. These changes are said to be **spontaneous.**

Typical examples are salt dissolving in water, hot metal cooling when allowed to stand in air and iron rusting in air. It is important to note that when a process is said to be spontaneous, the specific conditions under which this spontaneous reaction occurs must be stated. When reading the list of examples above, the automatic tendency is to think of the conditions existing in a room or in some open space on the earth's surface where, within certain limits, parameters such as temperature, pressure and humidity are defined very approximately. However, conditions under which a so-called spontaneous reaction ceases to be so are not difficult to envisage. For example, a kilogramme of metal at 350K is certainly hot compared with a typical room at a temperature of 290K and so the metal would cool down. However, this same piece of metal in an oven at 600K will in fact heat up rather than cool down. So although spontaneous reactions can be discussed, it should be remembered that spontaneity implies behaviour under specific conditions.

Bearing in mind the need to specify conditions, one important property of a spontaneous reaction is that it is accompanied by a release of energy. The hot metal in the example above will give out heat energy to its surroundings during cooling and so, in principle, this heat energy could be collected and made to perform a useful function. Perhaps less obviously, heat is released during the dissolution of a salt in water. If a thermometer was inserted into 1 kg of water during the process and if the container was carefully lagged, the investigator would find that there is a small temperature rise indicating that energy was being released, although in a much smaller quantity than in the case of cooling the hot metal. At the other extreme the reaction of hydrocarbons with oxygen (burning) is certainly spontaneous, and the energy release is very large indeed.

These observations are important because the source of all usable energy lies in the occurrence of spontaneous reactions and the whole of industrial processing relies upon the engineer's ability to utilise the energy released during a spontaneous reaction to effect some desired change, for example to manufacture a product, to heat a room, to pump water, and so on. This energy is normally referred to as **free energy** in thermodynamics, indicating that it is available for use. The term does not imply that the energy is free in a financial sense.

The need to supply energy to bring about the desired changes arises from the corollary of these ideas, because most industrial processes aim to reverse spontaneous effects. Industry may want to obtain solid salt from a solution, to heat a cold metal until it is hot or to convert iron oxide (rust) into iron. In each of these operations, the intention is to return to the state that existed before the spontaneous reaction occurred. For example, by evaporating the water from the salt solution and condensing it, the salt and water can be separated. By removing the oxygen from the iron oxide, the iron can be recovered from the iron oxide, and by placing the cold metal in a hot oven, its temperature can be raised. However, although the effect of the spontaneous reaction has been reversed in each of these operations, it has only been at a price. In all three examples, energy has had to be supplied to cause the reverse change.

It is worth noting that, in common with other metal oxides, iron oxide can be dissociated to metal and oxygen in principle merely by heating. However, because the temperatures required for the dissociation of metal oxides are usually excessively high, it is impractical to produce the metal in this manner. Consequently the oxide is often reacted with another element which has a greater affinity for oxygen than the required metal. The added element therefore removes the oxygen from the metal oxide by combining with it preferentially, producing the required metal and the oxide of the added element. Such reactions usually occur at much lower temperatures than those required for dissociation.

Not all of the free energy required to reverse a spontaneous change need come from the burning of chemical fuels or from the use of electricity. Most chemical processes involve changing reactants from one chemical compound to another and some of these reactions absorb energy whereas others release energy in a manner similar to that occurring during the burning of coal and oil. In many industrial processes that involve chemical reactions, more than one reaction often occurs in the same reaction vessel. If one reaction involves the liberation of energy, this energy may be used to support a parallel reaction which is absorbing energy. Under these conditions, energy that is required by the overall process and that must be supplied from external sources such as the burning of chemical fuels, will be less than would be needed if the energy-absorbing reaction were taking place in isolation.

One simple practical example illustrates this point. The large scale production of aluminium has grown out of the discovery in 1886 by Hall and Heroult that molten cryolite, Na_3AlF_6, is a good solvent for aluminium oxide, Al_2O_3, the naturally occurring form of aluminium, and that the resulting solution can be electrolysed to give aluminium without appreciably decomposing the solvent. Electrolysis takes place at a temperature of about 1250K and causes the reaction shown by Equation (1.1).

$$\frac{2}{3} Al_2O_3 = \frac{4}{3} Al + O_2. \tag{1.1}$$

The free energy change during this reaction is +0.84 MJ, where the positive sign, by convention, indicates that the energy is absorbed. At first sight the whole of this energy might seem to have to be supplied as electrical energy. However, carbon anodes are used in the electrolytic cell and the oxygen evolved at the anode reacts with the carbon to produce carbon dioxide according to Equation (1.2).

$$C + O_2 = CO_2. \tag{1.2}$$

However, the free energy change during this reaction is −0.40 MJ, where by convention the negative sign indicates that energy is liberated. The overall reaction inside the electrolytic cell is therefore the sum of Equations (1.1) and (1.2) and is shown in Equation (1.3).

$$\frac{2}{3} Al_2O_3 + C = \frac{4}{3} Al + CO_2. \tag{1.3}$$

The free energy change for this composite reaction is therefore 0.84 − 0.40 = 0.44 MJ. The net result is that by burning away the carbon anodes the external energy input is reduced by almost 50% compared with that which would be required if the anodes were made of some other inert material [42].

We are now in a position to understand a statement of the second law of thermodynamics based on Lewis's statement. **Every spontaneous reaction results in a release of energy and the reversal of such a process requires a supply of energy.**

This law implies that under conditions when a particular reaction occurs spontaneously, the reverse reaction will not occur spontaneously. If a desired reaction is not spontaneous under a given set of conditions, then the conditions must be changed until it becomes spontaneous and this change of conditions requires a supply of energy. When the law is stated in this way, the goal of industry can be regarded as **the manipulation of conditions until a desired change in a material will occur spontaneously.** Manipulation of conditions in practice may involve just a change of physical conditions, for instance providing a high temperature to promote the spontaneous heat flow needed to heat a piece of metal. Alternatively, it may require a change in the physical conditions together with a change in the chemical environment, such as adding chemicals to promote a desired sequence of chemical reactions.

The idea of spontaneous change can be pursued further to give an insight into those factors which govern energy use in a practical sense. All industrial operations rely upon the use of machines. Here the word 'machine' is used in a very broad sense, to include not only objects such as mechanical machines, but components such as electrical conductors, furnaces and nuclear reactors. However, for simplicity consider a mechanical machine because it will produce readily observed effects. A suitable example is a flywheel on an axle that is supported in some way. Now suppose that the wheel is set in motion by spinning it with the hand, thus imparting kinetic energy to it, and that once moving it is not interfered with by man in any way. In time the wheel will come to rest. No matter what steps are taken to reduce friction in the axle bearing, or air resistance, the inevitable result is that the wheel will always come to rest. The spontaneous product is therefore a stationary wheel. Our statement of the second law of thermodynamics implies that the intital kinetic energy of the wheel has been released in some way: it has been spontaneously converted into heat energy in the bearings and possibly kinetic energy of the molecules in the surrounding air. If this flywheel formed part of a larger machine intended to run at a constant speed, these spontaneous loss processes would have to be reversed and as stated in the second law this could only be done by supplying energy to the machine.

Spontaneous energy loss mechanisms such as these are not confined to mechanical machines. Electrical energy is lost when an electric current is passed through a conductor because of the resistive effects of the conductor, and this electrical energy loss manifests itself as heat energy raising the temperature of the conductor. Similarly, heat losses occur in furnaces because of heat conduction through the furnace walls. In both examples the existance of an energy loss mechanism can be proved by cutting off the supply of energy to the machine, which then reverts spontaneously to a state different to that experienced when the energy supply is connected. The furnace will cool and the electric current will no longer flow.

The energy losses occurring in all these examples result from natural spontaneous phenomena such as friction, electrical resistance and thermal conduction. Although the magnitude of such losses can be reduced by careful design and operation of the machines, they cannot be completely eliminated and this leads to the oft-quoted statement that no machine with an efficiency of 100% is possible. These losses also rule out the possibility of perpetual motion machines.

The importance of the second law of thermodynamics to industrial practices can now be appreciated more fully. Essentially we live in a world in which the products of industry have to be used under the conditions existing at the surface of the earth, that is a temperature of about 290K, a pressure of 1 atmosphere, and so on. We desire many products that do not occur naturally under these conditions. In other words, there are no spontaneous reactions that will allow the desired products to form. Therefore localised regions of space must be provided where materials can be taken and where conditions will allow the desired changes to occur spontaneously. We call these localised regions of space factories and the required conditions are maintained using machines such as furnaces and moulding machines. After manufacture, the products from these factories have to be returned to the everyday world outside the factory for use.

An outside observer can examine the materials before they enter the factory and can describe the change that he wishes to be brought about within the factory. From a knowledge of the scientific laws governing the change he wishes to effect, the observer can calculate the energy input required.

Consider the simple example of heating a block of metal so that it melts and flows into a mould to produce a given shape. This process requires that two physical effects be induced, first raising the temperature of the metal to its melting temperature and then changing its state from solid to liquid.

The energy required to raise the temperature is the product of the mass, the specific heat capacity and the temperature change, where the specific heat capacity is defined as the energy required to raise the temperature of a unit mass of the material by 1K and is a property of the material. The energy needed to cause melting once the metal has reached its melting temperature is given by the product of the mass to be melted and the specific heat of fusion, which is defined as the energy to melt a unit mass of material. Table 1.3 gives values of these parameters for a selection of metals.

Using the data in Table 1.3, if a moulding operation were to be carried out on a sample of aluminium of mass 0.5 kg at an initial temperature of 290K, the energy required would be:

$$
\begin{aligned}
&(\text{Energy to heat to melting temperature}) + (\text{Energy to melt}) \\
&= (0.5 \times 913 \times (932 - 290)) + (0.5 \times 397000)\,\text{J} \\
&= 293073 + 198500\ \text{J} \\
&= 491573\ \text{J} \\
&= 0.49\ \text{MJ}.
\end{aligned}
$$

This energy would be released into the atmosphere by conduction, convection and radiation as the moulded shape cooled back to room temperature and would represent the **minimum** energy required to produce the specified change.

This example is relatively trivial in industrial terms but it serves to illustrate the point that a minimum energy value can be calculated for any change. Similarly, if the desired change involves a number of chemical reactions then the

Table 1.3
Specific heat capacities, specific heats of fusion and melting temperatures of some metals.

Metal	Specific heat capacity	Specific heat of fusion	Melting temperature
	$J\ kg^{-1}K^{-1}$	$MJ\ kg^{-1}$	K
Aluminium	913	0.397	932
Antimony	209	0.161	903
Beryllium	1824	1.356	1551
Bismuth	126	0.052	544
Cadmium	230	0.055	594
Chromium	460	0.331	2163
Cobalt	418	0.276	1763
Copper	385	0.205	1356
Gold	130	0.063	1336
Iron	452	0.272	1803
Lead	130	0.230	600
Magnesium	1017	0.368	923
Manganese	477	0.267	1493
Molybdenum	251	0.289	2895
Nickel	444	0.297	1725
Platinum	134	0.100	2043
Potassium	753	0.061	336
Silicon	711	1.799	1700
Silver	238	0.111	1234
Sodium	1226	0.113	371
Tin	209	0.059	505
Titanium	523	0.418	2073
Tungsten	134	0.192	3660
Zinc	389	0.113	692

energy change for each reaction occurring must be known in detail before the minimum energy required can be calculated. The energy value produced by such calculations is known as the **minimum thermodynamic energy requirement** for the reaction.

To return to the example of making an aluminium moulding, although an observer *inside* the factory could also calculate a minimum energy requirement for the process as 0.49 MJ, he is also conscious that in creating, within the factory, the localised conditions for a spontaneous increase in temperature of the aluminium, he will also cause other spontaneous changes to occur. For example, to melt the aluminium he must provide a furnace at a temperature of 932K. Because this machine (furnace) is surrounded by air at a much lower temperature, spontaneous loss of heat energy will occur by conduction, convection and radiation. Although the operator can design his machine to reduce spontaneous processes other than the desired change, he knows that because of the second law of thermodynamics he will never be able to eliminate these other processes and the associated energy losses. He must therefore resign himself to the fact that he will always have to supply more than the predicted 0.49 MJ of energy to his furnace in order to end up with the desired product. He therefore recognises a **practical energy requirement**, which will always be greater than the minimum thermodynamic energy requirement.

Before leaving the role played by free energy initiating a reaction, let us consider further two of the examples discussed earlier, the melting of aluminium to form a casting and the reduction of iron oxide to iron. In moulding, all the energy required to effect the change is first supplied to the metal and afterwards is given back to the surroundings as the moulding cools. In the extraction of iron, the energy requirement of 7.35 MJ per kg of iron is the minimum thermodynamic energy needed to convert iron oxide to iron [131]. This requirement is constant however the reaction is performed and is *not* returned to the surroundings unless the 1 kg of iron reoxidises completely (rusts) to the oxide. The 1 kg of iron may therefore be regarded as having 7.35 MJ of energy 'tied-up' within it and this has potential as a fuel, a fact that will be raised again later.

1.4 THERMODYNAMICS AND THE REAL INDUSTRIAL WORLD

As the brief discussion of thermodynamic principles in Sections 1.2 and 1.3 makes clear, two energy values can be associated with any industrial process. The first represents the theoretical minimum energy change that must occur between the starting materials under a specified set of conditions and the finished products measured under the same set of conditions. The second energy value will be the energy actually used by the machines and processes which carry out the reaction and will always be greater than the theoretical minimum. From thermodynamic considerations, no process will consume less than the predicted theoretical minimum and this represents the energy requirement of that process in a perfect world without phenomena such as friction and heat losses.

Turning to the industrial operator, the thermodynamically predicted energy value provides him with a yardstick against which he can measure the energy efficiency of his process. If the minimum theoretical energy requirement is E_t and the practical energy requirement is E_p, it is possible to define an energy efficiency η, as:

$$\eta = \frac{\text{Minimum theoretical energy}}{\text{total energy used}} = \frac{E_t}{E_p} \qquad (1.4)$$

The difference between these two energy values, $(E_p - E_t)$ represents the energy losses of the process. Many research laboratories can and do devise experimental arrangements where E_p very closely approaches E_t so that the efficiency defined in Equation (1.4) approaches unity. However, these experimental arrangements are seldom constrained by non-technical factors such as cost, which are often far more important in the industrial scene. As a result, it comes as no surprise to find that industrial energy efficiencies are considerably lower than experimentally attainable values.

Two examples illustrate this point well. Thermodynamics predicts that the reduction of iron oxide to iron requires an energy of 7.35 MJ per kilogramme of iron [131], yet a blast furnace requires of the order of 20 MJ per kilogramme of iron produced. Similarly, the theoretical energy needed to produce 1 kg of silica glass from its raw materials is of the order of 2 MJ [48, 78] whereas the practical melting energy in industry is of the order of 10 to 15 MJ/kg [15]. Clearly to account for these discrepancies there must be some very significant differences between industrial practice and the theoretical conditions assumed in thermodynamics. Separate sets of reasons exist for these differences in each process but it is possible to separate the reasons very roughly into three groups. First there is the inadequacy of the theoretical model needed for the calculation of the minimum thermodynamic energy. Secondly there are practical considerations, and lastly there is the commercial nature of industrial plants. Let us examine each of these reasons in turn.

1.4.1 Inadequacies of theoretical models

In most modelling exercises, the real world situation must be simplified in order to produce a problem amenable to solution by the available theoretical techniques. Thermodynamics is no different in this respect. Many of the reactions encountered in industrial practice are extremely complex. For example, the production of iron does not simply involve the removal of oxygen from the oxide ore. The ore is not a pure oxide but a mixture of iron oxides together with impurities that could be deleterious to later processing. Additional materials are therefore added to the blast furnace charge forming a slag which helps to eliminate many of these impurities from the iron produced and which also affects the reaction itself. Hence the blast furnace reaction is much more complicated than that represented by the simple Equation (1.5) showing the reduction of iron oxide to iron [14].

$$Fe_2O_3 = 2Fe + 1\tfrac{1}{2}O_2. \quad (1.5)$$

The use of such an equation to predict the energy requirement of the process will therefore be an oversimplification of a much more complex set of reactions. Similarly, the production of silica glass involves more than simply melting together appropriate quantities of the main ingredients, sand, limestone and sodium carbonate. Small quantities of additives are included to change colour and to remove gas bubbles and these affect the simple reaction process. With sufficient time and effort the thermodynamics of these complex reactions can certainly be worked out in detail, but it is doubtful if any significant benefits would accrue from such knowledge. In many respects the relatively simple thermodynamic descriptions give an approximate indication of the theoretical energy requirements, but it must always be remembered that they are often an oversimplification of industrial practice.

At this stage one very important limitation on the use of Equation (1.4) in defining an energy efficiency for an industrial process should be pointed out. Equation (1.4) assumes that a value for the thermodynamic free energy change for a reaction or process can be derived in all cases. As noted previously this is usually possible for most physical and chemical changes bearing in mind the need to simplify the system. However, for certain industrial operations it is difficult if not impossible to derive such an energy efficiency value satisfactorily; transport is one example. Consequently Equation (1.4) is severely limited in its application to overall industrial processes.

1.4.2 The practical nature of industry

Obviously a complete list of all the practical reasons for energy losses in all industrial processes cannot be provided but some examples will demonstrate the nature of the problems and the very real practical limitations on their solution.

One of the simplest and most serious sources of energy loss has already been mentioned in Section 1.3; this is friction. Friction is a force existing between two surfaces and which inhibits the relative motion of the two surfaces. Almost all industrial processes use machines with moving parts and no matter how well maintained and lubricated these machines may be, they always exhibit frictional forces which can only be overcome by the expenditure of energy. Indeed some machines do very little more than overcome frictional forces. A lorry for example expends almost all of the traction energy derived from its fuel in counteracting the external friction between the tyres and the road and between the vehicle and the air moving past it (viscous drag) as well as internal friction between the moving parts of the engine, gearbox, and transmission. Other machines present less extreme examples of frictional energy loss but such losses are none-the-less significant and extremely costly to industry in both energy and economic terms.

Frictional losses in machines can be reduced by careful design, construction, maintenance and lubrication of components but there is a limit to the amount of money that can be spent to this end. As with factors other than energy, where industry is capital oriented, a balance must be struck between the cost of preventative measures and inefficiency resulting from frictional losses.

Heat processes present a second area in which significant energy losses occur because of practical considerations. One of the main problems encountered in all thermal processes is ensuring that heat is transferred efficiently from the source of heat energy to the material to be heated. The many types of steam boilers and furnaces in existence illustrate how designs have been varied to give the best possible heat transfer in particular situations. For example, we find that the blast furnace heats a mass essentially from within the bulk of the material by mixing the fuel into the furnace charge, whereas the ingredients in a glass furnace are heated either by flames directed onto the surface of the furnace contents or by the passage of an electrical discharge through the bulk.

Yet another complication arises from the physical state of the materials to be heated. In glass making the furnace charge is in powder form and conducts heat poorly in this condition. Although heat conduction becomes much more efficient as the charge fuses, there are significant heat losses at first because high temperatures must be maintained to counteract the poor thermal conductivity of the ingredients. Similar problems occur in almost every thermal process.

In addition there is the continual problem of heat losses through the walls of the furnaces or reactors. Considerable additional capital expenditure could reduce such losses, but once again an economic balance must be struck between the initial capital cost and the ongoing cost of the losses.

One area associated with thermal processes which frequently offers a sensible investment for capital is in the recovery of energy from hot flue gases. Recirculation plant enables the distribution of such gases to preheating ovens which in turn reduce the thermal load in subsequent processes (see for example Figure 2.3).

1.4.3 The commercial nature of industry

Discussion of the thermodynamic basis of industrial processes has been concerned with establishing the conditions favourable to the occurrence of a particular reaction. No consideration has been given to the speed at which the reaction would then proceed. This omission neglects one crucial difference between the experimental researcher and an industrial plant operator, who must not only make the reaction occur but must complete it as quickly as possible to take best economic advantage of both men and machines. However, trying to increase the rate of a reaction frequently leads to the use of conditions which tend to increase energy losses.

A simple example will illustrate this. Suppose that a chemical compound, A, can be converted to some other chemical compound B, simply by heating. If a quantity of A is sealed in a reaction vessel maintained at a temperature T_1 and the amount of the product B is monitored as a function of time, the conversion

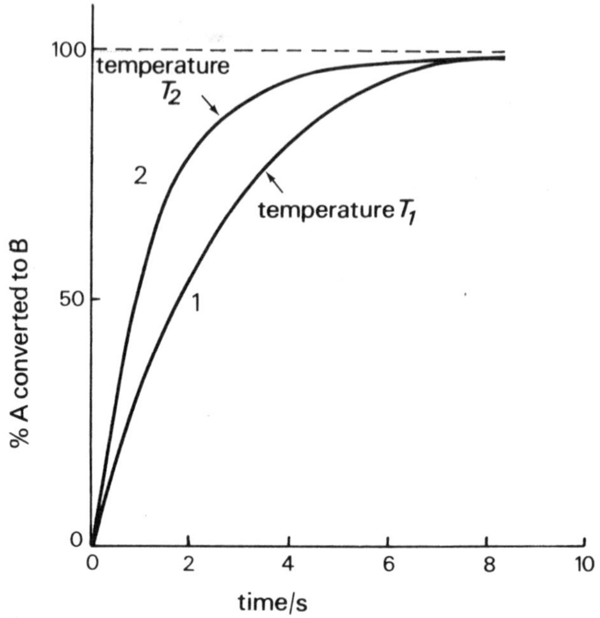

Figure 1.1 Simple rate diagram showing the conversion of hypothetical reactant A to product B under two different sets of thermal conditions 1 and 2. See the text for a detailed explanation of the graphs.

of A to B could be represented graphically as shown in Figure 1.1, Curve 1. After about 8 seconds almost all of the input materials have been converted to product, but the first 50% of the conversion occurred during the first 2 seconds, whereas the remaining 50% required a further 6 seconds. Accordingly, it might make sound commercial sense to stop the reaction after 2 seconds, separate the product B from the unconverted compound A, and return the unconverted residue to the beginning of the process. The effective conversion time would be halved but only at the cost of installing a separation plant. Alternatively the reaction might be speeded up by the use of a higher reaction temperature T_2, to produce Curve 2 of Figure 1.1. At this temperature almost 100% conversion would be achieved in about 4 seconds with 50% conversion after only 1 second. As before, this reaction could be stopped after 1 second, the reactant and product separated and any unconverted reactant fed back to the beginning of the process. On the other hand, the improved reaction rate achieved at this higher temperature implies that additional energy might be lost because of the more severe thermal conditions needed. Operating conditions and procedures are not usually chosen with energy as the sole criterion. Indeed energy is only one contributory factor in the total financial cost of a product. In reality, industrial processes are significantly affected by factors such as speed of throughput and plant costs which do not enter into thermodynamic discussions.

Curve 1 in Figure 1.1 raises one further consideration. If an operator allows the reaction only 2 seconds before stopping it and separating the reactants and products, he may find it cheaper in economic terms to discard the unconverted reactant. This may seem absurd when the quantities in question amount to 50% of the input, but in a real situation with smaller quantities they might well be discarded. The importance of this to the energy analyst is that if he were calculating a total production energy, he would have to take account of the energy needed to manufacture both the reactant discarded and the reactant actually consumed. The *efficiency* with which industry uses materials is closely linked to energy analysis and this highlights the integration of energy analysis with resource analysis generally.

1.5 THE PURPOSE OF ENERGY ANALYSIS

Energy analysis has a role to play at three levels, global, national and company, and we can consider its purpose by examining problems from each of them.

One of the most fundamental global problems and one widely discussed in the late 1960's is the possible environmental impact of the release of unlimited quantities of energy as heat into the earth's atmosphere. This heat is dissipated by radiation into outer space, a process dependent upon the temperature of the earth; the greater the temperature, the greater the radiation rate. If increased industrialisation and urbanisation should release more heat into the atmosphere, the temperature of the earth would be raised until the rate of radiation balanced the rate at which heat was released. The exact inter-relation between radiation rate, R, and temperature, T, is not known with any certainty, although Stefan's Law is frequently assumed to apply, in which case:

$$R \propto T^4. \tag{1.6}$$

The detailed implications of a rise in the earth's surface temperature are unknown but significant climatological changes are expected. Campbell [24] for example has argued that the combination of heat release and the accompanying atmospheric contamination with waste products from burning fuels could lead to a temperature rise of 1K. Whilst this does not appear significant at first sight, the resultant melting of the polar ice could lead to a rise of 50 m in the sea level.

An alternative view is that if heat release were greater than 1% of the energy received from the sun, similar effects would be produced [25]. Whilst the average heat release over the whole surface of the world is well below this value, certain localised areas are already well in excess of it. For example, London is thought to release heat equivalent to about 5% of the solar energy input into the same area. So, while the average amount of heat released over the whole earth's surface may pose no immediate threat, large scale industrialisation in the, as yet, under-developed countries coupled with the adoption of more energy-intensive technologies as raw materials become scarcer could alter the situation dramatically.

We need to know just how close this critical situation is in reality and to consider ways in which industrial practices can be tempered to prevent any serious climatological and ecological disturbances. Surprisingly, our current knowledge of global energy release and of the environmental impact of different industries is still limited, and energy analysis has a part to play in this work.

The second global problem is the availability of fuels. By far the greatest sources of energy at present are the fossil fuels oil and coal. They exist in strictly limited quantities which will inevitably be exhausted in time. Despite the insistence of the first law of thermodynamics that energy cannot be destroyed, the second law implies that although the energy may still be there in the global system, once used it will be in some way 'downgraded' and no longer sufficiently concentrated to be usable. Thus once a fuel has been used, a significant proportion of its energy content is lost forever as far as industry is concerned.

The present state of technology seems to indicate that nuclear fuels will be the next generation of energy sources and these undoubtedly offer appreciable quantities of energy. There are limits to the availability of nuclear fuels such as uranium however. First, rich supplies of uranium ore are already becoming scarce, even though the total amount of uranium in the earth's crust is very large indeed. However, in common with most metals, the extraction of uranium from its ores becomes increasingly more energy-intensive as the quality of ore becomes poorer. The effect of ore grade on the energy required for extraction is best illustrated with copper, because detailed calculations have been made for this metal. This is shown in Figure 1.2 and a similar curve is expected for uranium. Eventually the energy required to extract uranium from its ore will become greater than the energy that can be derived from it.

Secondly, the nuclear power industry itself consumes a large amount of energy indirectly in the form of materials required for the construction and maintainence of the power plant and distribution facilities. Moreover, increasingly large supplies of energy will be needed to treat and store spent fuel. It is suggested that once the size of the industry grows above a certain level, it could well end up consuming more energy than it produces [34]. Finally, workable nuclear fuels no less than fossil fuels will someday be exhausted, and the sources of energy then available are still in the realms of speculative technology. A useful discussion of future energy sources is given by McMullen, Morgan & Murray [93].

Accordingly it would be sensible to make the prudent use of existing fuels a priority and energy analysis can play a part in predicting potential conservation measures. Moreover, as new sources of energy are proposed, energy analysis can evaluate their potential. For example, one important consideration in the use of solar energy is whether or not more electricity can be generated in the lifetime of a solar cell than was expended in its manufacture and installation.

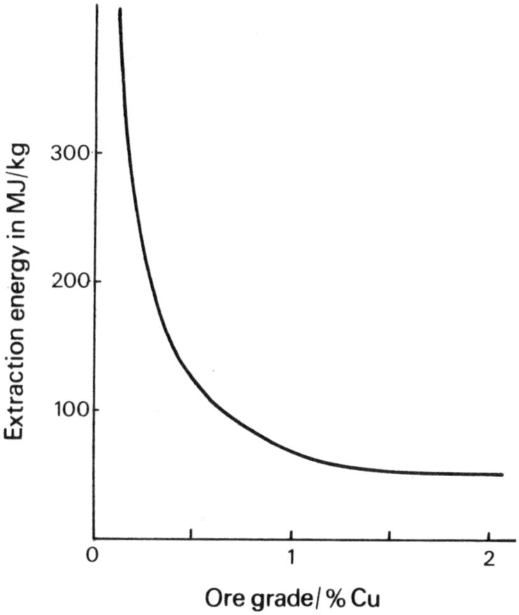

Figure 1.2 The extraction energy of a metal depends upon the grade of ore available. For copper the extraction energy varies as shown. Thermodynamically extraction energy is expected to vary exponentially with ore grade.

The problem of the availability of fuels becomes apparent in a somewhat different form at a national level. Given an adequate source of supply, the availability of a primary fuel relies upon the ability and willingness to work it. Trade union, political and economic pressures can lead to a scarcity of one fuel type or another. In addition, the availability of the principal secondary fuel, electricity, depends upon the willingness of government to invest in the necessary generation and distribution facilities. There are already indications that the unlimited expansion of this facility is being seriously reconsidered in view of the cost and relatively low efficiency of energy conversion and distribution. Considerations such as these imply that any industry heavily dependent upon any one fuel type will be vulnerable to interruption in supply from whatever cause. Energy analysis can identify these sectors of industry and search for ways to make them less vulnerable to fluctuations in availability.

This application of the analysis also provides useful information in mapping national energy flows essential to forecasts of future demands. Most countries import some or all of their fuel and this is reflected as a drain on national economic resources. The effects of the recent large increases in oil prices can still be seen in most national incomes. Any savings in the use of imported fuels, large or small, are to be welcomed and energy analysis can indicate the high energy consuming areas, formulate energy saving measures and evaluate the size of any potential savings.

At company level, fuels contribute to the cost of the finished product along with other factors such as labour, raw materials and capital expenditure. The principal aim of any industry is to improve efficiency of operation whether in terms of raw materials use, labour requirement or energy consumption because all of these factors affect the cost of the product. Energy analysis is an essential aid to such company functions.

As we have seen there are in general two values of energy associated with any operation; a theoretical value which represents the absolute minimum energy needed to produce a change under 'ideal' circumstances and a practical value representing the energy actually used. Energy analysis is primarily interested in this latter value and so can be regarded as providing the basic data needed for a description of industrial processes in the real world. Energy analysis also provides the only meaningful value for those operations where a theoretical value for the energy requirement is difficult or impossible to calculate.

Moreover, energy analysis goes further than this. Industry cannot be regarded as a collection of isolated operations functioning independently of one another. It is a complex network of operations so closely inter-linked that the production of seemingly simple artifacts requires the services of a number of operators who are often under the control of different organisations and in different countries. Thus the total energy used in producing some commodity or providing some service is not found only in the specific operation of interest but also in a whole sequence of associated operations. While evaluation of the energy associated with such a production scheme is not part of the engineers brief, the energy analyst makes use of data that would also be used by an engineer.

As an example of the importance of considering a group of operations as a whole, consider the manufacture of a metal can. In recent years there have been a number of arguments put forward suggesting that the metal can wastes resources in general and energy in particular. In Section 1.1 we quoted the example of the metal can manufacturing process to illustrate the importance of ancilliary operations within a factory and stated a value of 0.31 MJ/can as the total energy consumed directly in a typical can-making plant. Now it would be unrealistic to regard this energy as the total energy associated with a metal can because it is only the energy required to produce a can from the raw material, tin-plate. If all of the processes needed to produce sufficient tin-plate for one metal can are considered, these involve a direct energy requirement of about 1.5 MJ, which clearly represents a very significant additional energy contribution. This example illustrates the point that more than the energy consumed in the factory making the product must be considered when attempting to describe the total energy associated with a product. We will refer to this issue time and again throughout the book.

Clearly all the relevant contributors to the total energy requirement of a process or a product should be defined and considered. The virtue of the detailed type of analysis described later lies in applying the technique so that rational judgements can be made regarding the boundaries of the system being studied and the areas of high energy consumption identified. This enables the analyst to suggest ways of modifying production schemes to improve their energy efficiency.

As we shall see in Chapter 2, the total energy of a system may be influenced by the *way* in which the contributing operations are linked together. To illustrate this point, Figure 1.3 shows a method of distributing milk. An energy can be calculated for each of the operations represented by a box, but the energy needed to drive the whole system is not simply the sum of the energies of the individual boxes because two feedback loops exist. For example, the reclamation of the broken glass (cullet) from the disposal end of the system requires additional energy. However, recycling this cullet reduces the requirement for virgin raw materials and so the energy needed to produce them forms a smaller fraction of the overall energy requirement of the bottle.

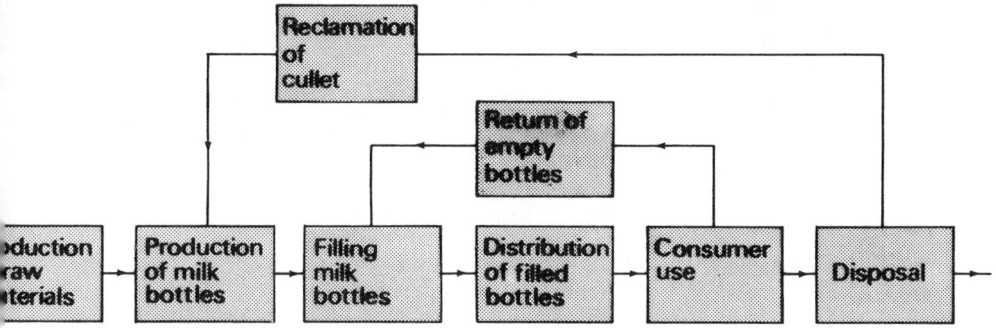

Figure 1.3 Flow diagram showing the principal operations involved in the delivery of milk in returnable glass bottles. Also shown is a recycling loop that would exist if some or all of the glass in the bottles were reclaimed when the bottles were no longer serviceable as containers.

First and foremost then, energy analysis is concerned with accurate descriptions of present industrial energy consumption in the same way as an economic analysis requires the results of the accountants work in its initial stages. This descriptive function is one of the principal preoccupations of energy analysts at the present time because relatively little basic data is available, particularly at a detailed factory level. This descriptive function is not however the sole aim of energy analysis. It is essentially a practical subject and descriptions of processes provide the essential information needed to compare different practices, to examine new and untried systems or modifications and to evaluate systems already existing either at company or national level. Of course, factors other than energy consumption are important in each of these areas of potential interest.

Economic considerations are probably dominant in the minds of most industrialists, but recent events have highlighted the dependence of modern industry on adequate and stable fuel supplies. There is an increasing need for some means of monitoring the energy requirements of industrial processes.

To some extent studies can be based on money as a unit of calculation, but this is a unit that continually changes its value for reasons not related to industrial considerations. In contrast, energy units can be used as an accounting currency because they represent an unchanging absolute quantity not susceptible to outside influences. We are not suggesting that energy analysis should replace economic considerations or that the information derived from such analyses is of more value than other information available to companies. However, we do believe that energy considerations play an important role in industrial decision-making and should not be ignored.

The future importance of energy analysis lies in the use that industrial operators can make of the results not just for the benefit of individual companies but for the benefit of the population as a whole. As we have seen, the type of application depends upon the scale of the system being considered and ranges from the global down to the individual. Many of the potential applications of energy analysis were recognised before the data for their detailed consideration were available. Hence an examination of potential applications of the technique tends to reflect problems that have existed for some time because they have stimulated the development of the technique.

Chapter 2

Theoretical industrial systems

2.1 THE GLOBAL INDUSTRIAL SYSTEM

The previous Chapter introduced some of the basic concepts governing the way in which energy is used and the concepts developed assume that the energy needed in a particular industrial operation can be measured. As indicated earlier, this task can be more complicated than appears at first sight and so our purpose in this Chapter is to develop some ideas which will allow industrial energies to be evaluated.

Let us begin by considering the purpose of industrial operations. Irrespective of motive, all industrial operations are concerned with processing materials in some way. Some changes induced in materials are strikingly obvious; when a metal is extracted from its ore or a polymer is produced form crude oil there is no doubt about the nature of the change. However some processes which may seem to have little to do with industrial operations induce far more subtle changes which can be just as important. A laundry does not change the physical or chemical properties of clothes but prolongs their lifetime because soiled garments need not be replaced. Similarly, transport operations change the geographical locations of materials even though the physical and chemical nature of the materials is again unchanged. We can therefore make the simplifying assumption that **all industrial operations are concerned in some way with the processing of materials.**

An examination of the wide range of operations in use makes it immediately apparent that the fuel and materials requirements of any process can be traced back to some previous processing stage. The laundry may boil its water using fuel oil derived from an oil refinery while the oil refinery in turn derives its supplies of crude oil from an oil well.

In fact these requirements can be traced back until a point is reached at which the supplies are actually derived from the earth so completing the chain of supply. Similarly the fate of the materials *outputs* from any operation can be followed through further operations until they too, after varying periods of use, are returned back into the earth.

In this argument the word 'earth' is used in a broad sense to include not just the land but also the seas and the atmosphere. **So the 'natural' boundary of**

industry is defined by the points at which materials are extracted from or deposited back into the earth.

This view of industry as existing to process materials derived from and eventually returned to the earth brings its own problems. How can this simple concept be used as an aid to understanding the complexity of the real industrial world with its vast number of inter-dependent and inter-related operations? At first sight any attempt to analyse the behaviour of this intricate network seems to be faced with insurmountable obstacles.

However, industry's behaviour can be described to a considerable extent by regarding it as a **system**. In essence a system can be regarded as a collection of operations or objects which perform some recognisable function. The systems approach to problem solving has been widely adopted in the physical and biological sciences as well as in management **because it permits analysis of the behaviour of complex organisations such as the nervous system without demanding a detailed knowledge of the way in which the component parts behave.**

If the systems concept is to be applied to industry we therefore need to be able to identify a suitable collection of operations that perform some specified function.

We have already shown that there is a 'natural' boundary to any industrial system defined by the points at which materials are extracted from or returned to the earth. Without attempting to select any specific group of operations from the total industrial network at this stage, we can use this 'natural' boundary to examine the behaviour of a system composed of all the industrial operations in the world – we shall refer to this as the **global industrial system**.

This system can be represented very simply by Figure 2.1. The box labelled 'global industry' represents the system containing all industries operating on the earth at the time of this description. The input to this system is made up of materials of mass M_i extracted from the earth and there are two outputs. One represents materials of mass M_o which are returned to the earth, while an energy E_o, released during the processing as waste heat, accounts for the other.

This simple model highlights some very important properties of the global industrial system which are also possessed by individual industrial operations, as we shall see later. The earlier definition of a general system requires this global system to exhibit some recognisable function. Figure 2.1 identifies the sole function of this system as **the processing of a supply of potentially useful raw materials into a stream of useless waste materials**. This may seem a sobering reflection of the purpose of industry but it is an inevitable consequence of the second law of thermodynamics if we set the boundaries at the earth.

It should however be remembered that there is a stage at which the processed materials exist for some useful lifetime, though this lies within the system boundary in Figure 2.1, and so does not form a part of the present description of the overall system. However, it will be seen later that it is sometimes necessary to identify components within the system in order to relate the quantitative description to easily measurable parameters (Chapters 11 and 12). This system is of course subject to the laws of physical science in common with all other physical phenomena. Thus the **law of conservation of mass**, which states that mass can be neither created nor destroyed, must apply so that:

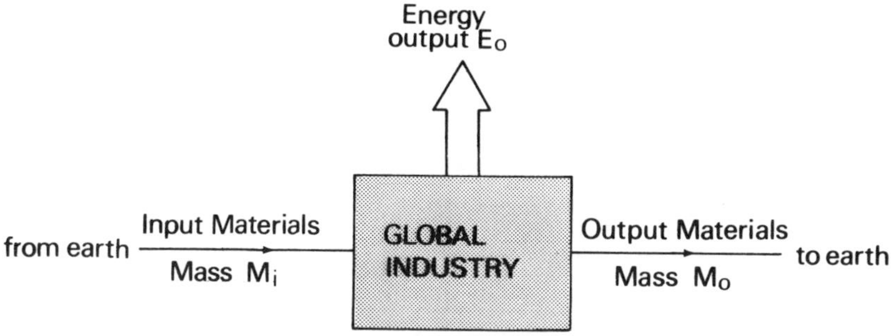

Figure 2.1 The global industrial system.

$$M_i = M_o. \tag{2.1}$$

This inviolable law applies not only to the overall global system but also to any component operation within it.

An application of this law forms a useful check when drawing up a detailed flow diagram to model a sequence of processing operations because it will usually reveal any omissions of inputs or outputs of materials. However, there may sometimes be apparent deviations from this law in continuous processes. For example, in any production process involving a number of machines used in a sequence, input materials will accumulate within the system if one of the machines breaks down and the input to the overall sequence is maintained. Restoring the defective machine to working order eliminates this build-up of mass so that the output mass would be greater than the input mass for a short time. Averaged over the whole time of the disturbance, the law is still obeyed even though an outside observer would see an apparent violation during the departure from normal running. Similar apparent violations can be detected in factories holding significant buffer stocks of raw materials or part-finished components and this practical point should be borne in mind when assembling data for energy analysis.

Figure 2.1 also illustrates an apparent violation of the first law of thermodynamics (conservation of energy) because energy is produced without any seemingly being supplied. However this is not true because all energy used in industry at present is derived from *materials* entering the system. Chemical fuels such as coal and oil, and nuclear fuels such as uranium are all materials inputs, while the energy required for hydro-electric schemes derives from potential energy possessed by the water in the reservoir supplying the power station. Consequently a proportion of the materials inputs, M_i, are in fact energy inputs. This factor can be taken into account if Figure 2.1 is modified to produce Figure 2.2. Here the two types of inputs have been separated. The one labelled 'raw materials inputs' represents those inputs of mass that remain as matter throughout the processing. The other, which is labelled 'fuels inputs', represents those materials taken in for the purpose of releasing energy from them.

Because Figure 2.1 and 2.2 are equivalent, it follows that:

$$M_i = M_m + M_f. \tag{2.3}$$

We now seem faced by a paradox, because if the masses entering and leaving the system are the same, even though some of the inputs are supplying energy to the system, then energy is apparently being obtained from an unchanging mass. This paradox can be explained by the property of matter first postulated by Einstein [49]. He supposed that mass and energy are interchangeable forms of the same physical quantity and expressed this interchangeability by the equation:

$$E = m.c^2, \tag{2.3}$$

where E is the energy associated with a mass m and c is the speed of light ($= 2.99 \times 10^8$ m.s^{-1}). If matter is changed in some way that makes it release 1 MJ of energy, then the decrease in mass accompanying this liberation of energy will be 1.12×10^{-11} kg, a change that is too small to be detected by all but the most sophisticated experiments. Thus the law of conservation of mass is not strictly adhered to when energy is derived from materials, but the minute changes predicted by Einstein's equation are undetected on an industrial scale and so for all practical purposes it is reasonable to assume that the law of conservation of mass still applies.

It is tempting to suggest that this departure from the law of conservation of mass can be used to differentiate between the materials inputs and the fuels inputs in Figure 2.2. Because the fuels inputs will be used to produce energy, there will definitely be a departure from mass conservation, while we may reasonably expect the raw materials inputs to obey the law because they remain as materials even though their physical and chemical form may be changed during the processing. Unfortunately, the second law of thermodynamics states that the changes in physical and chemical form of these materials will be accompanied by energy changes as well and so their mass will alter according to the Einstein equation. This makes any such differentiation impossible.

Returning to the more mundane aspects of Figure 2.1, everyday experience dictates that what any entrepreneur needs first is a potential **consumer** for his proposed product. In everyday language the word 'consumer' is usually applied to the housewife or some other person who ultimately buys the finished artifact.

However the global system of Figure 2.1 includes a large number of operations whose prospective consumers in fact operate further stages of processing. This is a very much broader definition of the word consumer because there is in fact no difference in purpose between the metal can manufacturer who 'consumes' tin-plate for use in his factory and the housewife who 'consumes' metal cans in transporting and storing food or beverages.

Thus Figure 2.1 will include all consumers, including the housewife, as operators within the global system. The need to include operators such as the metal can manufacturer within the industrial system is perhaps obvious, but the

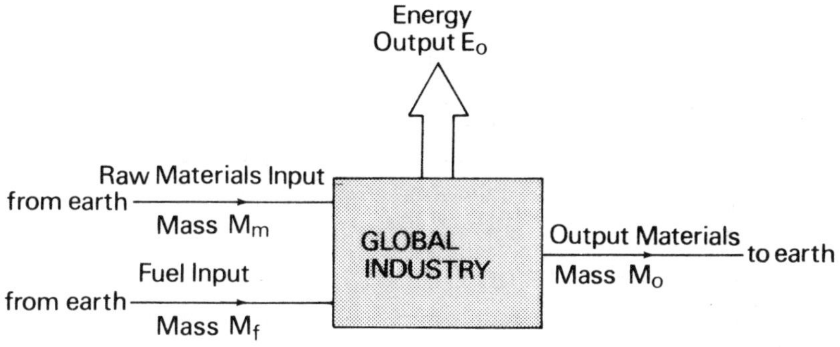

Figure 2.2 The global industrial system with the inputs separated into materials and fuels.

inclusion of the housewife as an industrial operator may be less so. However, further processing may occur after the materials have left the housewife's hands. For example, it might be thought worthwhile to reclaim plastic packaging from household refuse. The amount of processing needed would depend crucially upon factors such as the cleanliness of the waste and the ease with which the plastics could be separated from the rest of the domestic refuse and the housewife can exert a significant control in both of these cases. Hence she is an essential link in many industrial processing chains and must be included in such systems.

The representation of global industry by Figure 2.1 is essentially static in that it can only be used to describe the behaviour of this system at one moment. This does not limit the use of such a model seriously because industrial change does not happen overnight. However, this limitation makes it important to recognise that the descriptions of input materials as 'potentially useful raw materials' and of output materials as 'useless waste materials', are time dependent. Hence if Figure 2.1 were used to describe global industry as it exists today, all of the materials inputs could in principle be identified quantitatively. However, if a new process making use of one of the present-day waste outputs were developed, this would be incorporated into the 'global industry' box in future descriptions.

Its input would then no longer be included in the waste output of the global operation, although the waste arising from this new operation would replace it. Future descriptions of global industry therefore must inevitably exhibit a different set of materials and energy inputs and outputs compared with the present-day description.

This simple description of global industry shows how a very large and complex network of operations can be reduced to a form that can be manipulated and examined in very simple terms. However, we could only do this by assuming that a hypothetical interface can be drawn around all of the operations of interest and that inputs and outputs to the sytem are restricted to those quantities which cross this interface. We refer to it as the **system boundary**. The system boundary for global industry is relatively easy to determine because we can assume that all inputs spring from the earth and all outputs return to the earth. Choosing boundaries for the smaller component systems raises some difficult problems, as we shall see later, but once the boundaries have been drawn, the behaviour of the enclosed system is identical to the global system.

2.2 INPUTS AND OUTPUTS OF INDUSTRIAL SYSTEMS

Taken by itself the global system described in Section 2.1 is of little immediate practical use in detailed energy analysis of industrial processing, but it helps to introduce a number of very important concepts by providing a useful and readily understandable example. The overall operational block labelled 'global industry' in Figures 2.1 and 2.2 contains a large number of inter-related and inter-dependent operations which form the total global industrial network. Energy analysis examines the behaviour of these operations individually or in selected groups and to describe them in practice it can use the relatively simple description of the global system with only minor modifications.

A description of the behaviour of any system demands that the inputs and outputs of that system should be defined. Therefore the ability both to identify the inputs and outputs and to represent them pictorially using a method such as that demonstrated in Figures 2.1 and 2.2 is essential. However simplicity has been achieved in Figure 2.1 only at the expense of a loss of detail and Figure 2.2 shows how such a representation can be improved to permit the inclusion of greater detail. The aim of this Section is to show how these representations can be improved still further and be applied to detailed individual operations. This development entails a more careful consideration of both materials and energy inputs and outputs.

Consider first the materials inputs. The distinction already made between raw materials inputs and fuels inputs is fundamental, in both a philosophical sense and a practical physical sense. Usually the reaction which releases energy from the fuels is physically separated from the processing of the desired materials, both to contain it and to avoid any possibility of the materials products being contaminated. When the flow diagram showing the movement of mass between a group of operations is drawn, the masses shown represent the flow of raw materials used as fuels.

The law of conservation of mass can then be applied rigidly because both input and output masses can be monitored in practice. In contrast, the input masses of fuels can be measured but the output masses of the residues are usually unknown.

Notice that on some occasions the distinction between raw materials and fuels is rather blurred. For example, the manufacture of steel from pig-iron essentially involves melting iron and blowing oxygen or air through the molten mass. This oxidises some of the carbon present in the input pig-iron to carbon dioxide gas so that the steel contains less carbon than the pig-iron. The heat generated by the carbon oxidation reaction helps to maintain the reactor contents at a high temperature, yet this carbon was included in the mass of the input pig-iron and is a characteristic component of pig-iron. Moreover the oxidation of carbon is not the only reaction occurring because the presence of iron and oxygen together promotes the reaction.

$$2Fe + 1\tfrac{1}{2}O_2 = Fe_2O_3. \qquad (2.4)$$

This is a spontaneous reaction of the sort discussed in Section 1.3 and is accompanied by a liberation of energy. This energy release also helps to support the high temperature of the reaction vessel and so the small amount of iron that is burnt to iron oxide in the reaction vessel is strictly speaking acting as a fuel. Nonetheless such inputs are more conveniently regarded as *raw materials* inputs rather than fuel inputs because of the small masses involved in the fuel reaction. Corrections for effects such as these can be made later as the need arises.

In fact there is less interest in the *mass* of fuels such as oil or coal than in the *energy* that can be derived from them and while the industrial plant operator may purchase his fuels by mass, this is only a convenient measure for use by the merchant. Because energy analysis is concerned with the energy delivered to an operation, an analyst finds it more convenient to replace the mass of the fuels, M_f in Figure 2.2, by an equivalent amount of energy, E_f. Another point to remember is that energy may one day be supplied to the global system shown in Figure 2.2 from sources which do not involve the input of materials. For example, sunlight falling onto the earth might be converted into electricity. Although the conversion machinery such as the solar cells and storage facilities would then fall within the system boundary as defined earlier, the sunlight crossing the system boundary would be one form of energy input not involving the transport of mass. Similarly, consideration of the energy inputs to individual operations within the global system shows that most operations take in at least a part of their total energy requirement as electricity, a form of energy that is also delivered without the movement of mass.

Both solar energy and electrical energy can be expressed quantitatively as an energy value E_e and can be included as energy inputs to the processing operation. To retain simplicity in the model it is most convenient to add together the energy inputs from materials and non-materials sources to give an overall energy input E_i so that:

$$E_i = E_f + E_e. \tag{2.5}$$

Although the two types of energy represented by the parameters E_f and E_e have been combined, there are essential differences between them which will become apparent later. Suffice it to note at present that naturally occurring materials which are fed into industrial processes as energy sources are often referred to as **primary fuels**. In contrast, electricity and other fuels that are fed into industrial processes in a form different to the one which originally provided the energy supply from the earth are often referred to as **secondary fuels** (see Chapter 6).

The energy output from any industrial process is most simply represented by a single parameter E_o, as in Figures 2.1 and 2.2, but it can often be represented as the sum of two main components. The first of these arises as a result of the second law of thermodynamics, which predicts that some energy will always be lost as waste heat no matter how good the design of the equipment used. The heating of moving parts in a machine because of friction is typical of such losses. This energy cannot easily be reclaimed for industrial use and so represents an inbuilt energy loss. The second component of the energy output covers those outputs from which energy could be recovered. Hot flue gases are a good example of this type of energy loss. Such outputs frequently arise because industrial plants may have to be operated in a manner not automatically implying the most efficient use of energy (see Section 1.4).

Because the concentrations of energy encountered in such outputs are often relatively high, practical schemes can in many cases be devised to reclaim some of this otherwise wasted energy. It is quite common practice to pre-heat the materials inputs to many high temperature operations by passing them through the hot flue gases from the reactor. The thermal cracking of naphtha shown in Figure 2.3 provides one example of this technique. Naphtha is a complex mixture of hydrocarbons derived from crude oil and must be converted into a simpler mixture of specific composition before it can be used in the petrochemical industry. This is achieved by mixing the naphtha with steam and heating to a temperature of about 1100K. The simplest furnace would be a tube leading the mixture through the furnace chamber, but in practice the input naphtha is pre-heated by passing it through a tube that runs through the flue gases of the cracking furnace. The steam is preheated in a similar manner, added to the naphtha and the mixture again passed through a tube heated by the furnace flue gases before being fed into the furnace chamber itself, where the reaction takes place. Figure 2.3 shows the route taken by the input materials in this example.

If the energy output due to thermodynamic losses is E_t and the energy losses due to the practices adopted in industry is E_p, perhaps representing a thermodynamically inefficient operation adopted for economic reasons, then the total energy output E_o will be:

$$E_o = E_t + E_p, \tag{2.6}$$

where a proportion of E_p may be reclaimable.

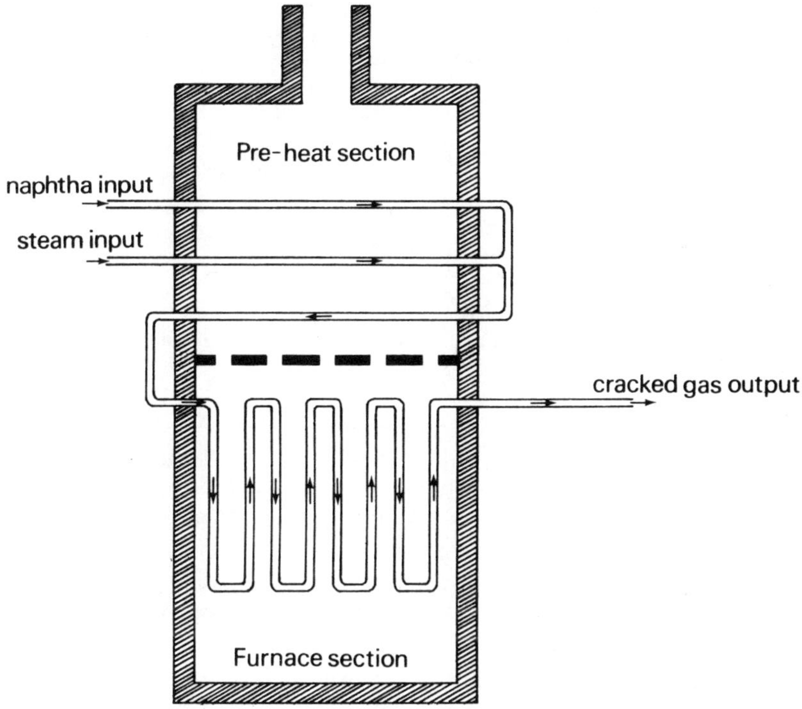

Figure 2.3 Schematic diagram of a simple thermal cracker for naphtha showing the way in which input naphtha and steam may be pre-heated with the flue gases from the cracking furnace before themselves passing into the furnace section of the plant.

Perhaps the major modification to the model embodied in Figures 2.1 and 2.2 needed before it can be applied to individual industrial operations arises from the nature of the materials output. The first and most obvious change is that these outputs are not of course waste materials but the desired products of the operation. Thus M_o in Figures 2.1 and 2.2 will represent the masses of the products, assuming that no waste materials are produced. Collecting together the ideas presented so far, an individual industrial operation can then be represented by Figure 2.4. A comparison of Figures 2.4 and 2.2 shows that there are only minor differences between the two representations.

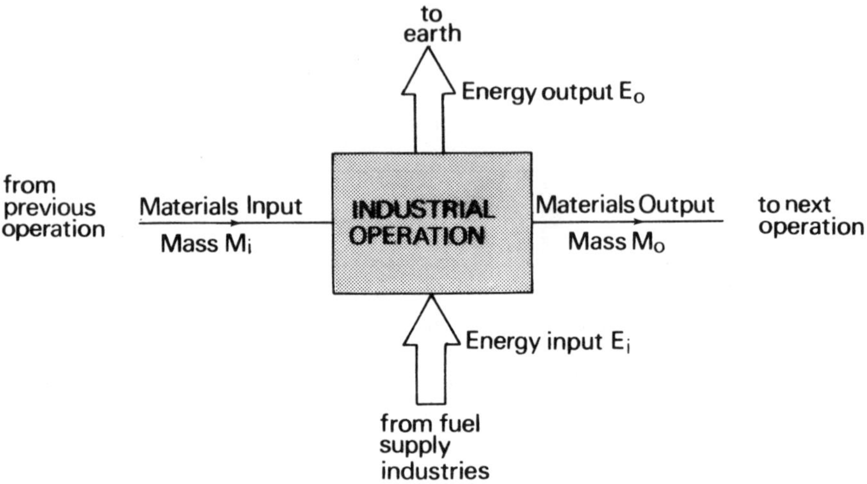

Figure 2.4 Schematic representation of the materials and energy flows in a simple industrial system.

However, Figure 2.4 is somewhat idealised, particular with regard to the output materials. Few operations will proceed without forming any non-saleable waste materials and so the representation should distinguish between these and the saleable or useable part of the output. Furthermore, many industrial processes result in the formation of more than one product. For example, the electrolysis of brine to produce chlorine gas also produces hydrogen gas and sodium hydroxide (caustic soda), both of which are saleable products. The simple model shown in Figure 2.4 must accordingly be modified to allow for waste materials and multiple products, resulting in Figure 2.5, which represents the production of waste of mass m_w and two products of masses m_1 and m_2.

As before, the law of conservation of mass applies and so:

$$M_i = m_1 + m_2 + m_w, \tag{2.7}$$

and comparing Figures 2.5 and 2.4;

$$M_o = m_1 + m_2 + m_w. \tag{2.8}$$

The efficiency with which materials are converted to saleable or useable products is an important factor in all materials processing and one which influences the energy efficiency of a process significantly, as we shall see later.

This parameter can be summed up in a simple mathematical form by defining a materials conversion efficiency, C, as

$$C = \frac{\text{Mass of saleable or useable products in output}}{\text{Total input mass}}. \qquad (2.9)$$

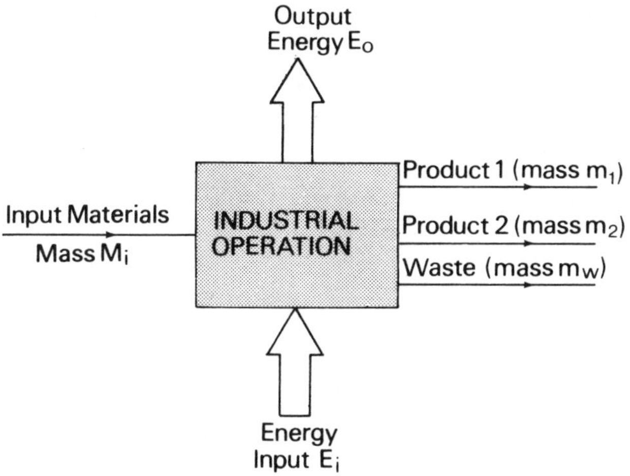

Figure 2.5 Schematic representation of the materials and energy flows in an industrial operation which produces more than one product as well as waste products.

Thus from Figure 2.5, the conversion efficiency in terms of all the saleable products, C_p, can be written down as:

$$C_p = \frac{m_1 + m_2}{M_i}$$

$$= \frac{m_1 + m_2}{m_1 + m_2 + m_w}. \qquad (2.10)$$

A simple numerical example will show the way in which numbers can be assigned to this conversion efficiency. Sodium metal is produced by the electrolysis of fused sodium chloride and Figure 2.6 shows a typical system with the input and output masses that might be used to represent it. Note that of the two outputs both sodium and chlorine are saleable, while the waste output represents the materials losses of the process from all sources. Comparison of the total input and output masses shows that the law of conservation of mass is obeyed. The conversion efficiency, C_p, for this process can be calculated by adding the output masses of the sodium and the chlorine and dividing by the total input mass as shown by Equation (2.11):

$$C_p = \frac{0.39 + 0.59}{1.00}. \qquad (2.11)$$

This process could therefore be said to have a 97% conversion efficiency in terms of materials used.

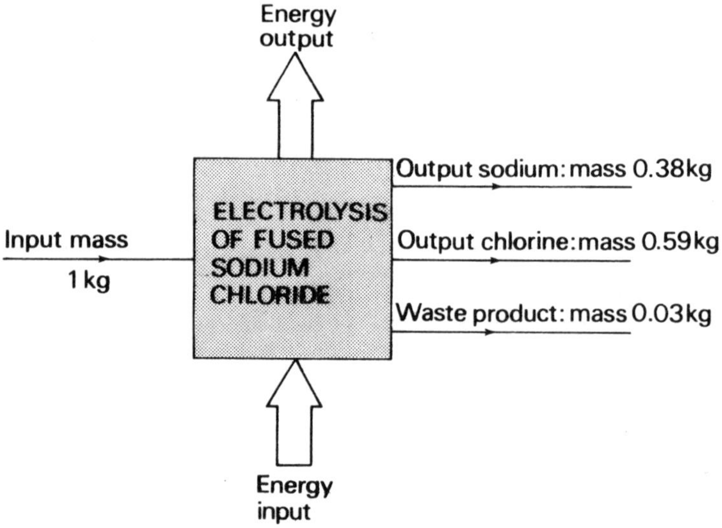

Figure 2.6 The electrolysis of fused sodium chloride as an example of a simple industrial system showing the materials flows.

Although the simple definition of conversion efficiency given in Equation (2.9) seems straightforward, it is well to be aware of a number of variations that are quite commonly used. One such variation arises when there are several products, as shown in Figure 2.5. Frequently interest may be concentrated upon only one of the products and it is often convenient to consider the yield of this one product alone. Thus a conversion efficiency, C_1, may be defined for product 1 as shown in Equation (2.12):

$$C_1 = \frac{m_1}{m_1 + m_2 + m_w}. \qquad (2.12)$$

Applying this definition to the production of sodium in Figures 2.6, a **partial** conversion efficiency, C_{Na}, for sodium can be calculated as:

$$C_{Na} = \frac{0.38}{1.00} = 0.38. \qquad (2.13)$$

Similarly, a 'partial' conversion efficiency, C_{Cl}, can also be calculated for chlorine thus:

$$C_{Cl} = \frac{0.59}{1.00} = 0.59. \qquad (2.14)$$

Note that the sum of these 'partial' conversion efficiencies is 0.97, the same as that determined using Equation (2.10) for the overall process. When such partial efficiencies are quoted, the basis of the numerical value must be clearly stated.

Another variation arises from the way in which the inputs are specified. Consider the materials taken into an iron ore concentration plant. The input mass to such a plant essentially consists of two components, the iron oxide, which is wanted, and the gangue which is not. Such an input could be specified in three ways; as the total mass of iron oxide alone, as the total mass of iron oxide plus gangue, or as the mass of elemental iron present, neglecting the mass of the gangue and combined oxygen in the input. Depending upon the method of specifying the input, a different conversion efficiency would be obtained. Therefore the inputs must be specified as closely as the outputs if ambiguity is to be avoided.

2.3 ENERGY ANALYSIS AND SYSTEMS

The method of describing a single industrial operation in terms of its inputs and outputs developed in the previous Section is satisfactory assuming that the energy analyst knows the system to be treated. If faced with the problems of describing the energy requirement of a blast furnace, he would be able to provide the description based on the ideas of Section 2.2 because the system has already been defined for him.

However, the questions that an energy analyst is called upon to answer are rarely of this simple type. Instead the problem is frequently posed in the form of questions such as, 'What is the energy needed to produce 1 kg of steel?' The important difference between this type of question and the description of industrial processing outlined earlier is that it is couched in terms of a **product** whereas the earlier description is related to an industrial **operation**. So the first task of any analyst faced with such a demand is to convert this product-oriented question into an operationally oriented one. This involves choosing a system boundary which encompasses all the operations which the analyst regards as necessary to describe the process needed to manufacture the product.

This will identify a whole series of individual operations, or **sub-systems**, and the analyst must evaluate the materials and energy inputs and outputs for each of these sub-systems in the manner discussed in Section 2.2.

Once the inputs and outputs of the individual sub-system have been evaluated the analyst must combine these numbers to obtain a description of the whole system. Therefore as a general rule energy analysis follows three quite distinct stages; **identification of the operations of interest, evaluation of the materials and energy inputs and outputs associated with each of these operations or sub-systems, and combination of these individual materials and energy requirements to give a set of inputs and outputs for the overall system**. The remainder of this Chapter considers how each of these three stages can be carried out. As we shall see, the procedure is iterative in that the original choice of system boundaries may be modified in the light of the numerical data collected and the information available.

2.4 CHOOSING A SYSTEM

The choice of system and the identification of its component sub-systems presents the energy analyst with his most difficult problem. It also gives rise to the greatest difficulty in interpreting published values for energy requirements. A simple example can illustrate this. Suppose that you were faced with the question, 'What is the energy required to produce glass bottles for a beverage filler?', which system would you use? Figure 2.7 shows 29 of the operations that might be considered necessary to carry out the function specified in the question. These operations are linked in the diagram by lines representing the flow of *materials* between them. Although there is a transport step between each pair of linked operations in Figure 2.7 and the energy contribution of these operations must also be included in the energy required to operate the system shown, they have been omitted in Figure 2.7 for the sake of clarity. Note that all of the inputs on the left-hand side of the Figure are derived from the earth. Only one output representing the main product has been shown on the right-hand side of the Figure, although there will also be other products, some waste and some useful, from the other operations in the system. Not all of the materials flows within this simplified system have been shown. For example, operation 28, the production of oil fuels, will be linked to almost all the other operations. Also omitted are those operations needed to construct the buildings and machinery involved in each of the sub-systems shown.

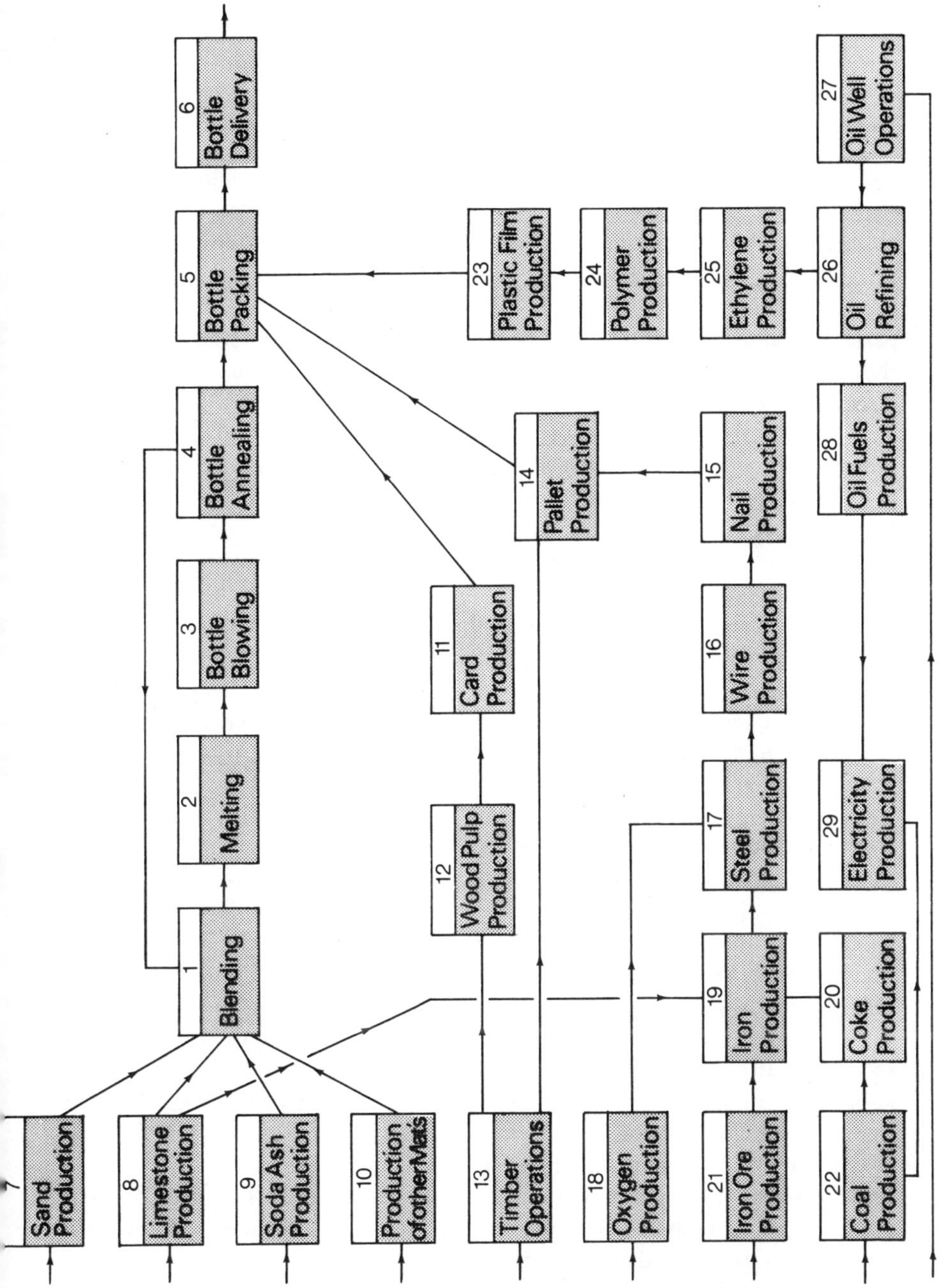

Figure 2.7 Simplified network of operations for the production and delivery of glass bottles to a filler. Note that all of the flows within the network are not shown; electricity production for example will feed all of the other operations.

Clearly the system shown in Figure 2.7 can only be regarded as a single, simplified, possible system which could be used to calculate the energy requirement and not as the definitive system that must be used. Some critics would argue that it contains too many operations because several are relatively insignificant contributors to the function under consideration. An adequate description could probably be obtained by using a system composed only of operations 1 to 10 plus the transport operations occurring between them. Others would maintain that leaving out of account operations such as those involved in the construction of the buildings and machinery needed represents a significant and serious omission. Since all objectors to Figure 2.7 can produce perfectly valid arguments to justify their positions, the inevitable conclusion is that there is no such thing as a 'correct' system which can be used to provide answers to the type of product-oriented question mentioned earlier. Indeed, experience has shown that no single system that can be used to describe any industrial process will be agreeable to all, because different systems must be defined to satisfy the requirements of the specific users of the information. For example, a plant operator may only be interested in the energy content of the fuels actually consumed by his plant, whereas company policy decisions require information concerning not only the energy to run the plant but also that needed to construct the plant in the first place and maintain it in working order thereafter. Similarly, the analyst may be obliged to define his system to match the available information. Consequently, although the energy for a system can be evaluated and stated, the precise system to which this energy relates must also be specified as an inseparable part of the answer.

This disagreement over the choice of systems boundary has serious implications for energy analysis and the nature of these implications can best be illustrated by a very simple numerical example. Let us suppose that we wish to calculate the energy required to generate steam using a reservoir of water. There are three possible operations involved in this process; the pumping of water, the operation of the boiler and the lighting of the boiler house. These three operations, together with hypothetical materials and energy throughputs, are shown in Figure 2.8. Using these three operations, four possible systems that could be chosen for the production of steam are shown by the systems A, B, C and D in Figure 2.8 (a) to (d). Each system will have steam as output. The energies per kilogramme of steam for each of the four systems are shown in Table 2.1. Note that the overall energy of the system depends upon the system chosen. This is a conclusion of fundamental importance and one which cannot be overstated. **There is no such thing as a 'correct' or 'absolute' value for the energy needed to produce a kilogramme of any commodity. The values obtained depend critically upon the systems boundary chosen.**

To ensure that any quoted energy value can be immediately interpreted in terms of a specific system, we need to approach all industrial systems in the same way. Despite the internal complexity of many systems, there are some

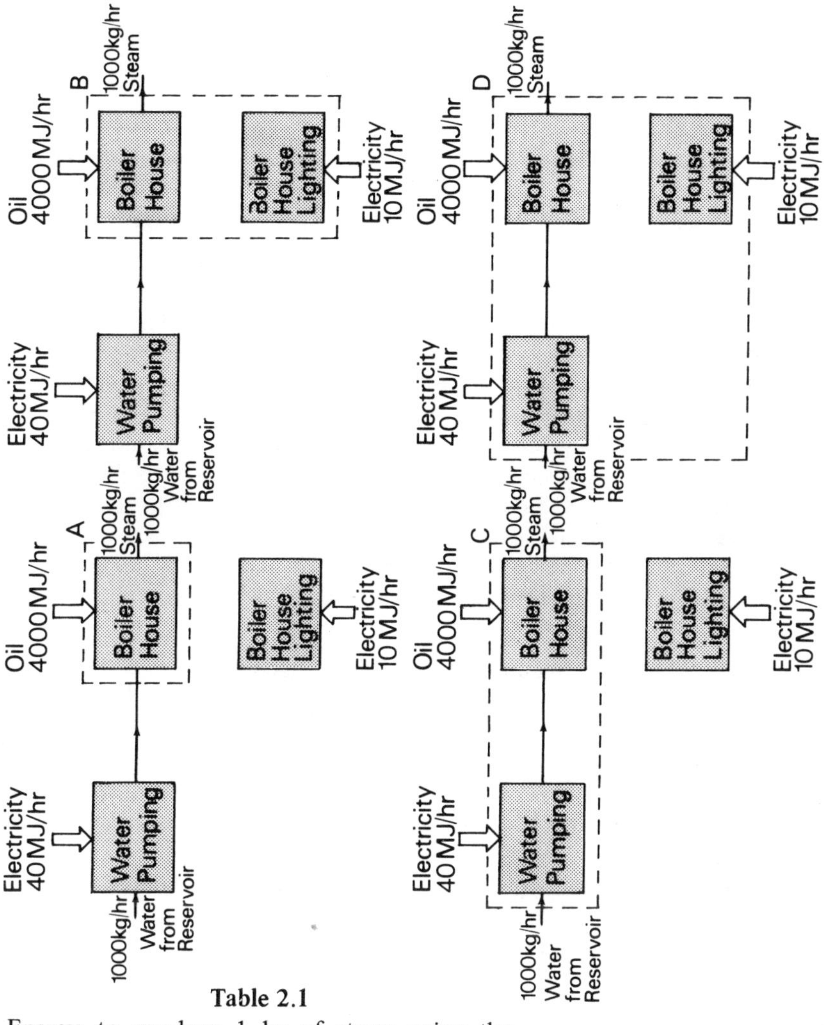

Table 2.1

Energy to produce 1 kg of steam using the systems defined in Figure 2.8

System	Energy/MJ
A	4.00
B	4.01
C	4.04
D	4.05

Figure 2.8 An example of how four different systems *A*, *B*, *C* and *D* may be defined for the same operation – here the production of steam from water in a reservoir.

properties common to all that can be isolated and used to establish a standard procedure. These common properties can be stated as three propositions which provide the basis of rational energy analysis.

Proposition 1 Any industrial process can be represented by a system whose aims are to maintain a throughput of materials and to modify the state of these materials. This modification may be physical or chemical or it may simply be a change in the geographical location, as in transport operations.

Proposition 2 The system describing any industrial process can be defined by the inputs and outputs of materials and energy and by the function that the system performs.

Proposition 3 All industrial processes, no matter how complex, can be subdivided into a sequence of operations linked by a flow of materials such that the useful output mass from one sub-system forms the input mass to the succeeding sub-system. Conversely, a collection of suitable operations can be assembled into one interlinked sequence to satisfy a specific function using mass flow as the linking parameter.

The propositions are stated in terms of mass because this is frequently the most easily measured flow parameter in industrial processing and in general all other flow parameters can be related to it.

Propositions 1 and 3 provide the basis for drawing up flow diagrams such as that shown in Figure 2.7. Proposition 2, however, is the essential guide to defining the system chosen. The inputs and outputs, particularly of materials, can usually be readily identified and described in a quantitative manner, but the **function** of the system is more difficult to state with any precision. A flow diagram such as Figure 2.7 is then invaluable because it shows precisely those operations that are to be included within the system boundary.

A problem arises if no limits are set by the questioner because the ultimate limit to any processing operation is the extraction of the raw materials from the earth at one extreme and their disposal back into the earth at the other as discussed earlier.

The preparation of a flow diagram to represent a production system then raises the problem of how far back the analyst should trace a particular production sequence. If in doubt, it is wiser to construct a system so that these 'natural' limits are included, at least within the main production sequence. For example, in Figure 2.7 the main production sequence is represented by the operations 1 to 10 and in this instance the system energy should be calculated to include all operations from the extraction of the raw materials needed for glass-making through to the delivery of the finished container to the filler. This last operation is chosen because it is the limit specified in the original question.

In terms of the need to include operations that feed materials to the main chain (such as the feeds of packaging materials to operation 5 in Figure 2.7) there is no absolute criterion that can always be applied at the outset. The initial system must be made as complete as possible. It can then be simplified if the collection and evaluation of data indicate that certain inputs are insignificant. For example, if the evaluation of the first few components of any side branch feeding the main production sequence yields energy values that are small compared with the energy of this sequence and outside the chosen limits of accuracy, then further evaluation of the energy due to this side branch can be discontinued. The major disadvantage of this approach is the need to perform a large number of calculations which will ultimately be discarded. However, it does mean that sub-systems are only discarded on the basis of rational calculations. It follows that the system boundary is only finally fixed during the evaluation of the system energy and not at the outset. Although there appears to be something inherently unsatisfying in the approach, it merely reflects the essentially practical nature of energy analysis.

Remembering that no absolute value can be fixed for the energy requirement of a process, this approach yields an energy value for a specific defined system enabling *meaningful* comparisons to be made with other analyses.

One further practical limitation should be borne in mind when preparing a detailed flow diagram to describe a system, and this relates to the form in which data is available in practice. The purpose of the flow diagram is to identify the sub-systems of the main system. Subsequent analysis aims at evaluating the inputs and outputs for each of these sub-systems. However, in practice information concerning the inputs and outputs of various operations is only available in the form in which it is monitored within a factory. The sub-system must therefore be matched to those industrial operations for which the data are available.

The inclusion of superfluous sub-systems may be of interest from the point of view of understanding the technology of the process, but it produces redundant information as far as energy analysis is concerned. Similarly, available information is lost in the aggregation process if the detail in the flow diagram is inadequate.

Figure 2.9 (a), representing the principal operations occurring in a glass container manufacturer's premises, gives an example of this. Assuming that the energy supply to every operation within the factory is monitored separately, then the degree of detail shown would be necessary and would allow a close examination of the relative importance of each operation within the production scheme. However, monitoring of individual operations within the factory is seldom carried out and the only description available for inclusion in a network such as Figure 2.7 would be the single sub-system of Figure 2.9 (b).

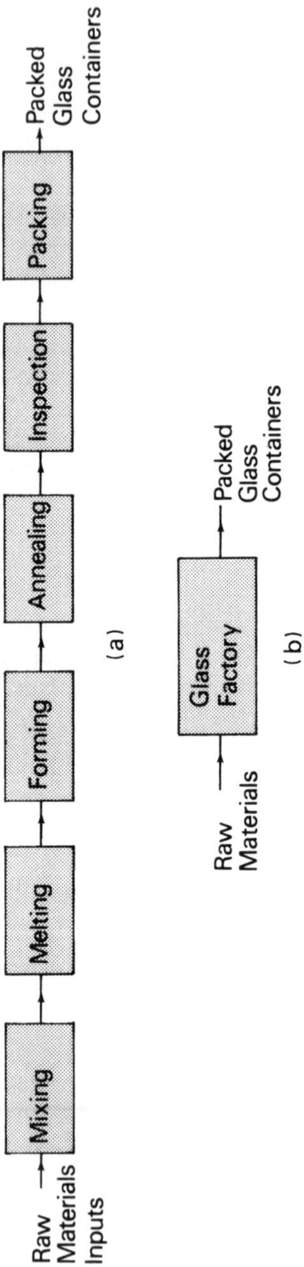

Figure 2.9 (a) The principal operations taking place within a glass container manufacturing plant. (b) In practice the detail shown in scheme (a) may be redundant if the information available describes only system (b).

2.5 ASSIGNING ENERGY VALUES TO SUB-SYSTEMS

The net product of the first stage of the energy analysis is therefore a description of the complete system together with an identification of the component sub-systems. Each sub-system will exhibit inputs and outputs of materials and energy as shown by Figure 2.5 and some unambiguous energy value must be assigned to the operations represented by each of these sub-systems.

Assume initially that the necessary data are available from the relevant industrial operator and assume also that no waste products are formed in any of the sub-systems. This latter assumption is unrealistic but may be corrected later. Each sub-system can therefore be represented by Figure 2.4. At this stage, assume also that the whole of the energy output, E_o, is dissipated as waste heat. The total energy requirement of the operation is E_i, which is of course dependent upon M_o, the mass of material being processed. This dependence of energy requirement upon throughput can be removed by calculating the energy per unit mass throughput. This normalises the energy in terms of mass throughput. A system energy E_s per unit output can then be defined by Equation (2.15).

$$E_s = \frac{E_i}{M_o}. \tag{2.15}$$

In general the energy requirement is always normalised with respect to some physical measure of the system output. Mass is the most obvious, but it could be volume in the case of liquids or numbers of components.

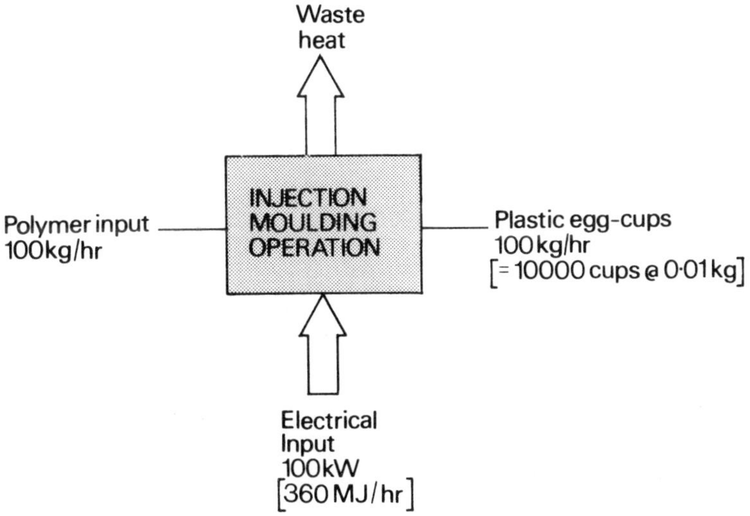

Figure 2.10 Simple injection moulding operation represented as a system. Detailed calculations based on these values are given in the text.

This model is adequate to describe the performance of very simple industrial operations. For example, suppose that an injection moulding machine is converting raw polymer resin into plastic egg-cups. The system diagram representing such an operation is shown in Figure 2.10, the total energy input is 360 MJ/hr and the total output of materials is 100 kg/hr. None of the energy output is used for any other purpose and so the system energy, E_s, can be derived from Equation (2.15) as:

$$E_s = \frac{360}{100} = 3.6 \text{ MJ/kg}. \qquad (2.16)$$

If each egg-cup is assumed to weigh 0.01 kg, then the system energy could also be expressed as 0.36 MJ/egg-cup.

Figure 2.5 describes the system when the production of waste materials and the formation of more than one product is considered. Because the system energy is concerned with making useable or saleable product, waste products are usually regarded as a nuisance and most industries attempt to minimise them. However, system energy is affected by waste production and this is taken into account by defining a system energy, E_s', in terms of the saleable or useable output rather than the total output. Thus E_s' for the system shown in Figure 2.5 will be as shown by Equation (2.17).

$$E_s' = \frac{E_1}{m_1 + m_2}. \qquad (2.17)$$

Note that E_s' will be higher than the corresponding system energy E_s calculated on the basis of the total output and that E_s' will be sensitive to the mass of waste materials produced, that is to the efficiency with which the process converts input materials to saleable products. The system energy is therefore related to the materials conversion efficiency, C, as defined in Equation (2.9). So for Figure 2.5:

$$C = \frac{m_1 + m_2}{M_i}, \qquad (2.18)$$

or

$$m_1 + m_2 = C.M_i. \qquad (2.19)$$

Replacing $(m_1 + m_2)$ in Equation (2.17) by $C.M_i$ from Equation (2.19) gives a system energy E_s' as:

$$E_s' = \frac{E_i}{CM_i}. \qquad (2.20)$$

Combining Equations (2.15) and (2.20) establishes a simple relationship between E_s and E_s' as:

$$E_s' = \frac{E_s}{C}. \tag{2.21}$$

It is clear then that the value of the system energy E_s' will vary from a minimum of E_s when no waste products are formed to very large values indeed as C approached zero, that is when large amounts of waste are produced. This dependence of E_s' on conversion efficiency is shown graphically in Figure 2.11. The energy per unit product doubles as the conversion efficiency falls from 100% to 50%. At lower values of conversion efficiency, a dramatic increase in the energy per unit product can be clearly seen.

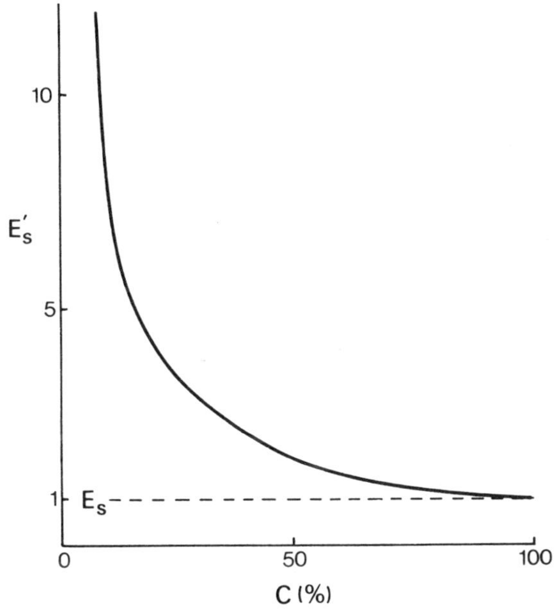

Figure 2.11 Graph showing the dependence of the energy of a system, E_s', upon the materials conversion efficiency of the system.

The effect of waste production upon system energy can be demonstrated using the plastics moulding example given earlier. Let us assume that with the same inputs as shown in Figure 2.10 there is a 5% loss of saleable product due to waste production that is thrown away. This situation is shown in Figure 2.12. For every 100 kg of input there is now only 95 kg of saleable output and the corresponding energy per unit output is now 360/95 = 3.79 MJ/kg as compared with the value of 3.6 MJ/kg occurring when there is no waste. Not only has the energy increased, but it has increased by more than the 5% change in output materials – in fact it has risen by 5.3%, as might have been expected from Figure 2.11.

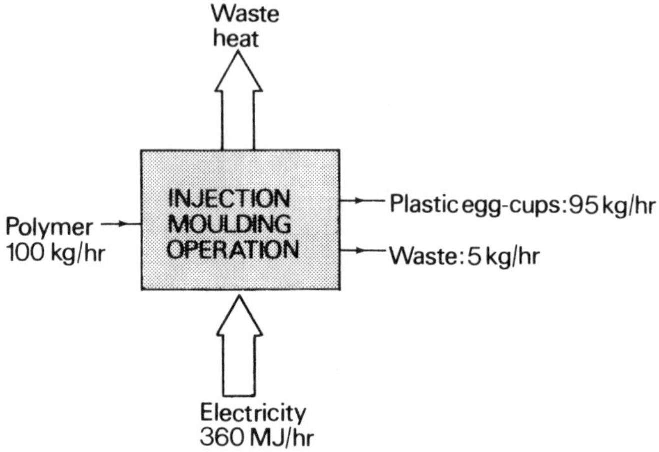

Figure 2.12 The simple injection moulding operation of Figure 2.10 when waste products are formed during moulding.

Moreover, when a waste output occurs within an operational system there is no way to improve the energy performance *of that system* other than by eliminating the waste. For example, the recycling of the waste in the injection moulding example might be thought to lead to improved energy performance. Figure 2.13 shows the system that would occur if the operator of the moulding machine were to recover the waste manually and feed it back to the machine. Using this means the recycling step itself does not incur the expenditure of further energy, as it would if the waste were to be returned to the moulding machine by another machine. The steady state requirement of virgin polymer is now reduced to 95 kg/hr and together with the 5 kg/hr of recycled plastic this maintains a throughput of 100 kg/hr. However, the saleable output is still only 95 kg/hr, as it was when no recycling was practised, and so the energy per unit product is still 3.79 MJ/kg. Initially this suggests that there is no good reason for even contemplating recycling the waste, since it does not appear to lead to any energy saving. However, while it is true that recycling does not reduce the energy consumption of the moulding operation itself, it does reduce the requirement for virgin polymer from 100 kg/hr to 95 kg/hr. Typically the energy required to produce a polymer will be of the order of 100 MJ/kg. Thus calculating a combined production energy, that is polymer production plus moulding, the energy consumed without recycling will be $(100 + 3.79) = 103.79$ MJ/kg, whereas the energy required when all the waste is recycled as indicated by Figure

2.13 will be (95 + 3.79) = 98.79 MJ/kg. There is therefore an *overall* energy saving arising because of the reduced input mass per unit output of product when recycling is taken into account. However, we are now examining a larger system than one just encompassing the moulding machine. Recycling problems are considered further in Chapter 11.

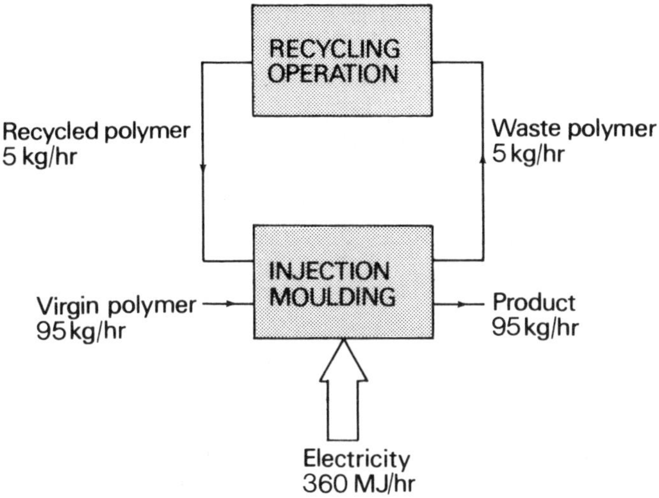

Figure 2.13 Injection moulding with a feedback loop which reclaims any waste products.

We shall return to the use of feedback loops such as that used in Figure 2.13 when we discuss methods of reducing energy consumption in more complex systems, but the above example demonstrates clearly that efficiency of materials conversion is of crucial importance in determining the energy attributable to a product from any process.

2.6 COMBINING THE ENERGIES OF SUB-SYSTEMS

Adopting the procedures outlined in the previous Section, an energy per unit mass of output can be derived for each of the sub-systems comprising the system under examination. The final stage of the analysis is the combination of the energies associated with each of the sub-systems into a single value which represents the energy for the whole system.

When a sub-system is considered as a part of the complete system, the energy consumed within the *sub-system* so as to yield unit mass of product from the overall system can be written as $\mathcal{E}$. $\mathcal{E}$ will be the product of the energy per unit mass, E, for that sub-system and the mass, m, of the output from that sub-system.

Thus:

$$\mathcal{E} = m.E. \tag{2.22}$$

The difference between $\mathcal{E}$ and E can be demonstrated by examining a system consisting of say three sub-systems labelled 1, 2 and 3 arranged as shown in Figure 2.14. Suppose that the materials conversion efficiency of each of the sub-systems is only 50%. To obtain a product output mass M_o from the overall system, the output from sub-system 3 must be M_o and because of the 50% conversion efficiency in this sub-system the input to block 3 must be $2M_o$. This input is the output of sub-system 2 and because of the 50% conversion efficiency this again requires a materials input twice as large, that is $4M_o$. This represents the output from sub-system 1. Putting $M_o = 1$ (kg), the outputs from each of the sub-systems will be 1 kg from block 3, 2 kg from block 2 and 4 kg from block 1. If the energy consumed per unit mass of output from each of these sub-systems is E_1 from sub-system 1 and so on, then the energy actually consumed in each of these blocks when they are linked as shown to yield unit output from the overall system, that is 1 kg from sub-system 3, will be:

$$\mathcal{E}_1 = 4.E_1, \tag{2.23}$$
$$\mathcal{E}_2 = 2.E_2, \tag{2.24}$$
$$\mathcal{E}_3 = E_3. \tag{2.25}$$

Clearly the quantitative materials flow through a system must be known in detail before the energy consumed by its sub-system can be established and this underlines the importance of establishing an accurate flow diagram at the outset of the analysis.

When the values of $\mathcal{E}$, the energy consumed by each sub-system, have been established, these values must be combined to give the overall system energy. This overall energy is not always the simple sum of all the values of the sub-system because it depends upon the way in which the individual sub-systems are linked together. Of course, a given set of industrial operations may be combined in many ways in practice and obviously all possible arrangements cannot be covered here. However, three types of sequence occur frequently and we have described these. Most processing sequences can be reduced to groups of operations of these types and so they provide a useful set of building blocks for use in more complex sequences.

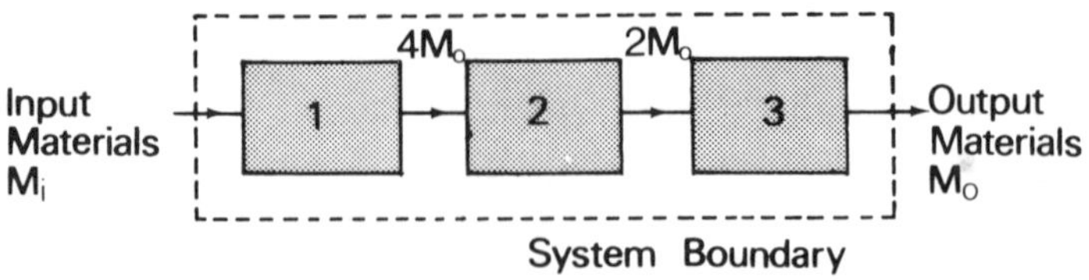

Figure 2.14 Simple linear sequence of operations showing the materials when there is a 50% loss of materials at each processing stage.

2.6.1 Multiple feeds

One of the most common arrangements found in practice is that shown in Figure 2.15, where a number of operations feed materials to a single process. Typical examples are the feeding of iron ore, coke and limestone to a blast furnace or the feeding of sand, limestone, soda ash and many minor ingredients to a glass furnace, as shown by operations 1, 7, 8, 9 and 10 in Figure 2.7. Following the convention developed above, values of $\mathcal{E}_i$ can be assigned to the energy actually consumed in the i'th sub-system in order that a unit mass of product will emerge from sub-system 4 and hence from the overall system. The system energy per unit output, E_s, will therefore be given by Equation (2.26).

$$E_s = \mathcal{E}_1 + \mathcal{E}_2 + \mathcal{E}_3 + \mathcal{E}_4. \qquad (2.26)$$

This is the equation for the overall system energy in its simplest form, but it can be written another way. Putting m_i as the mass of material flowing from the i'th sub-system and E_i as the energy per unit mass of output from the same, the system energy can be written as shown in Equation (2.27):

$$E_s.m_4 = E_1.m_1 + E_2.m_2 + E_3.m_3 + E_4.m_4, \qquad (2.27)$$

or

$$E_s = E_1.\frac{m_1}{m_4} + E_2.\frac{m_2}{m_4} + E_3.\frac{m_3}{m_4} + E_4. \qquad (2.28)$$

Note that the coefficients of the energy terms E_1, E_2 and E_3 represent the fraction of the overall total output mass contributed by the mass fed from each sub-system. Thus m_1/m_4 is the fraction of m_4 contributed by sub-system 1 to the final product and for a given process this fraction is a constant determined by the materials requirements of operation 4.

The effect of the conversion efficiences of the various sub-systems on the total system energy in Figure 2.15 should be considered. Since sub-systems 1, 2 and 3 are all directly linked to the final sub-system 4, any change in the conversion efficiency of operation 4 will have a proportional effect on the materials requirements from operations 1, 2 and 3. If the conversion efficiency of operation 4 were to be halved, operations 1, 2 and 3 would all have to double their supply of materials to it. On the other hand, operation 1 is independent of operations 2 and 3 and so any change in its conversion efficiency will have no effect upon them, nor indeed upon operation 4, because the effect of a change in conversion efficiency is only felt by those operations further back down the chain of supply. Similarly, operation 2 is independent of operations 1 and 3 and operation 3 is independent of operations 1 and 2.

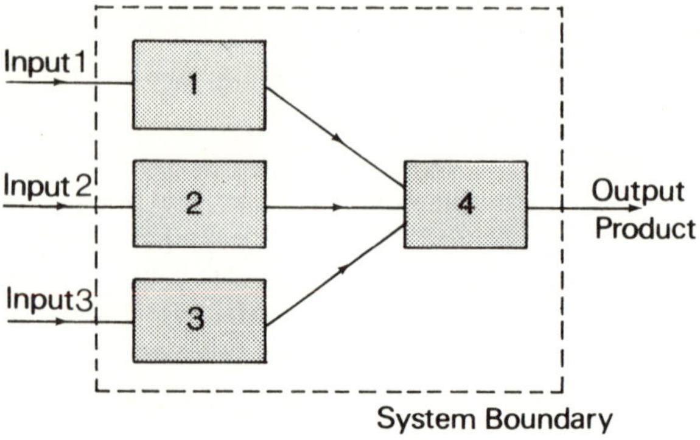

Figure 2.15 Production system composed of multiple feed sub-systems.

The mutual independence of operations 1, 2 and 3 and their dependence on sub-system 4 can also be shown mathematically by calculating the system energy E_s in terms of the energy per unit output for each sub-system, E_i, and the conversion efficiency, C_i, for each sub-system. Writing M_i as the mass input to the i'th sub-system and C_i as the conversion efficiency of the i'th sub-system, the following equations can be immediately written down:

$$m_1 = C_1.M_1, \tag{2.29}$$

$$m_2 = C_2.M_2, \tag{2.30}$$

$$m_3 = C_3.M_3, \tag{2.31}$$

$$m_4 = C_4.M_4, \tag{2.32}$$

$$M_4 = C_1.M_1 + C_2.M_2 + C_3.M_3. \tag{2.33}$$

Combining these equations with Equation (2.2) to eliminate m_1, m_2, m_3 and m_4 gives a system energy as shown in Equation (2.34):

$$E_s = \frac{E_1C_1M_1 + E_2C_2M_2 + E_3C_3M_3}{C_4(C_1M_1 + C_2M_2 + C_3M_3)} + E_4. \tag{2.34}$$

However, the factor $$\frac{C_1M_1}{C_1M_1 + C_2M_2 + C_3M_3}$$

is simply the fraction of the final product mass supplied by sub-system 1 and is a constant for any specific process; it can therefore be written as f_1. Similarly,

$$f_2 = \frac{C_2M_2}{C_1M_1 + C_2M_2 + C_3M_3}, \qquad (2.35)$$

$$f_3 = \frac{C_3M_3}{C_1M_1 + C_2M_2 + C_3M_3}. \qquad (2.36)$$

Thus the system energy from Equation (2.34) reduces to:

$$E_s = E_1.\frac{f_1}{C_4} + E_2.\frac{f_2}{C_4} + E_3.\frac{f_3}{C_4} + E_4. \qquad (2.37)$$

As expected, the system energy depends only on the conversion efficiency of sub-system 4.

2.6.2 Linear sequences

Many industrial processes involve linear sequences of operations such as that shown in Figure 2.16. Writing $\mathcal{E}_i$ as the energy associated with the i'th sub-system, then for a unit output of mass from the overall system the overall system energy, E_s, will be given by Equation (2.38).

$$E_s = \mathcal{E}_1 + \mathcal{E}_2 + \mathcal{E}_3 + \mathcal{E}_4. \qquad (2.38)$$

This form of equation is exactly the same as that derived in Equation (2.26) for the multiple feed example. However, in practice there is one very important difference. Unlike the multiple feed example, this operational sequence contains no independent production elements. If the conversion efficiency of any sub-system changes, the effect of this change will pass back along the chain of supply. The difference between Equations (2.38) and (2.26) becomes more apparent when the system energy is calculated in terms of the conversion efficiency for each operation, C. Using the earlier notation, it can readily be shown that the system energy E_s in terms of the energy per unit output from each sub-system, E_i, is as given in Equation (2.39).

$$E_s = \frac{E_1}{C_2C_3C_4} + \frac{E_2}{C_3C_4} + \frac{E_3}{C_4} + E_4. \qquad (2.39)$$

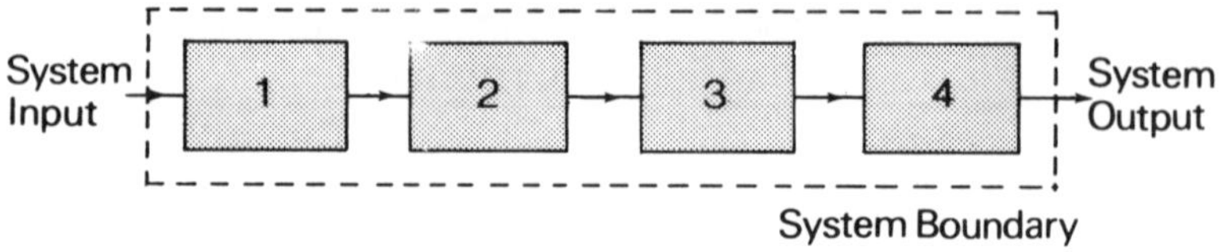

Figure 2.16 Production system composed of a simple linear sequence of sub-systems.

When this is compared with the corresponding Equation (2.37), it becomes apparent that the term associated with the energy of any sub-system in Equation (2.39) contains a reference to the conversion efficiency of each of the later sub-systems. Thus the energy associated with sub-system 1 involves the energy per unit mass of output from that sub-system, E_i, as well as the conversion efficiencies C_2, C_3 and C_4 for the three subsequent sub-systems.

2.6.3 Feedback loops

The third type of situation regularly encountered in practice is the feedback of materials from one part of a production sequence to an earlier stage. This normally occurs when waste materials produced during manufacture are recycled within the plant as in the hypothetical case of the manufacture of plastic egg-cups. Other typical practical examples are waste glass (cullet), which is produced inside a glass factory and fed back to the furnace, or spoilt metal castings, which are produced in a casting shop and returned to the melting furnace to reclaim the metal.

A typical sequence is shown in Figure 2.17, where operations 4 and 2 of a linear sequence of operations labelled 1 to 5 are linked with a feedback loop via an additional operation 6. Suppose that of the mass entering operation 4, a fraction f is returned via the feedback loop and the remaining fraction (1-f) carries on to operation 5. As before, assume that an energy E_i per unit mass of output is consumed by the i'th sub-system.

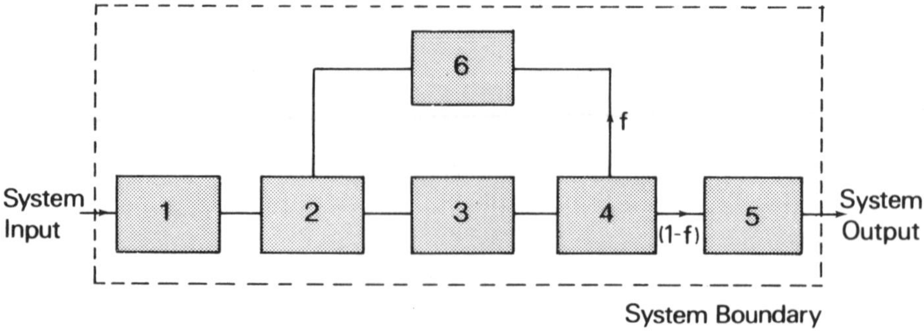

Figure 2.17 Production system composed of a simple linear sequence of sub-systems but incorporating a single feedback loop.

The operation of this system can be most easily considered if the effect of a feed of mass m into sub-system 1 is examined and it is assumed there are no materials losses throughout. The whole of this mass will therefore pass along the linear chain of operations 1 to 4 with an energy consumption of:

$$E_1 m + E_2 m + E_3 m + E_4 m. \qquad (2.40)$$

At block 4, a mass of $(1-f)m$ will pass onto block 5, leaving the system and consuming an energy:

$$E_5(1-f)m. \qquad (2.41)$$

The remainder of the mass, that is mf, will pass around the feedback loop through operations 6, 2, 3 and 4, consuming an energy:

$$mf(E_6 + E_2 + E_3 + E_4). \qquad (2.42)$$

Again at block 4 a fraction $(1-f)$ of this mass, that is $(1-f)mf$, will continue on to block 5, where it will consume an energy:

$$E_5(1-f)mf, \qquad (2.43)$$

and the remainder of the mass, that is mf^2, will again pass around the feedback loop to consume a further energy of:

$$mf^2(E_6 + E_2 + E_3 + E_4). \qquad (2.44)$$

This circulating procedure will carry on until eventually, at infinite time, all of the mass will have emerged from block 5. The total energy expended in achieving this state will be the sum of all the energy contributions represented by expressions such as (2.40) to (2.44) and the total sum can be easily simplified to give a system energy per unit mass of throughput as:

$$E_s = (E_1 + E_2 + E_3 + E_4 + E_5) + \frac{f}{1-f}(E_2 + E_3 + E_4 + E_6). \qquad (2.45)$$

Note that the first group of terms represents the energy consumption of the simple linear sequence involving operations 1 to 5 while the second group of terms involves all of the operations comprising the closed loop. Thus when $f = 0$, corresponding to the absence of the feedback loop, the coefficient $f/(1-f) = 0$ and Equation (2.45) becomes the same as that for the linear sequence shown by Equation (2.38). Similarly, as f increases from zero, the second group of terms increases in importance until at high values (that is as f approaches unity) it may dominate the value of E_s. This underlines the earlier point that although recycling of materials results in the conservation of materials, it may also increase energy consumption. A more detailed discussion of the application of Equation (2.45) is given in Chapter 11.

Chapter 3
Real industrial systems

3.1 COMPLEXITY WITHIN INDUSTRIAL SYSTEMS

Chapter 2 examined in outline the types of analysis that can be performed on industrial systems once the precise system has been defined and Figure 2.8 highlighted the different results obtained for the energy of systems which apparently perform the same function but for which different system boundaries are defined. Clearly the choice of system boundary is of crucial importance in energy analysis and the present Chapter examines this aspect in more detail to enable rational and sensible choices of system boundary to be made for real systems.

The overwhelming problem facing the analyst is the complexity of industrial operations. The obvious physical complexity of industrial plant is clear on seeing any petrochemical plant or iron and steel works. However the *total* network needed to support any particular operation is even more complex, a fact which may be difficult to appreciate because the machines involved are seldom in the same geographical location.

Such complexity is best illustrated by a simple example. Figure 2.8 showed the three main operations needed to produce steam from a water supply and one of these operations was water pumping. The bare essentials for this operation are a water pump and a supply of electricity and Figure 3.1 shows the simplified network needed to supply these basic requirements starting with raw materials in the ground. For simplicity all transport operations have been omitted from the diagram. With the exception of the water pumping operation itself, all other elements of the network are large industrial agglomerates displaying an internal complexity of the type already mentioned.

Figure 3.1 shows only some of the links existing between the different elements of the network. It is clear that there is an interchange of products between a number of manufacturing units and this is a recurring problem in energy analysis. The steel industry for example supplies a proportion of its output to the electricity industry which in turn feeds electricity to the steel industry. Indeed, producers of electricity and steel as well as oil feed their outputs to almost every other sector of industry. Not surprisingly therefore, the energy to produce steel is influenced by the energy to produce electricity, which is in turn influenced by the energy needed to produce the steel used in the construction of the plant, and so on. This interdependence of production energy requirements is examined later because evaluation of production energies requires an understanding of the interdependence relationship.

In practice, the links between the different operations in Figure 3.1 are either continuous or intermittent. The supply of electricity to the water pump is continuous while the pump is running. In contrast, the link between the production of water pumps and the water pumping operation is intermittent because a pump, once installed, will operate for a significant period of time before it needs replacement. These two different types of flow in the network can be reduced to a single type by the simple expedient of smoothing out the intermittent flows so that only continuous ones need be considered. If for example a water pump has to be replaced every 1000 days, it is convenient to regard this as a continuous flow of 0.001 pumps per day. Thus calculation of the energy associated with water pumping requires that one thousandth of the energy to manufacture a water pump be attributed to the daily pumping energy.

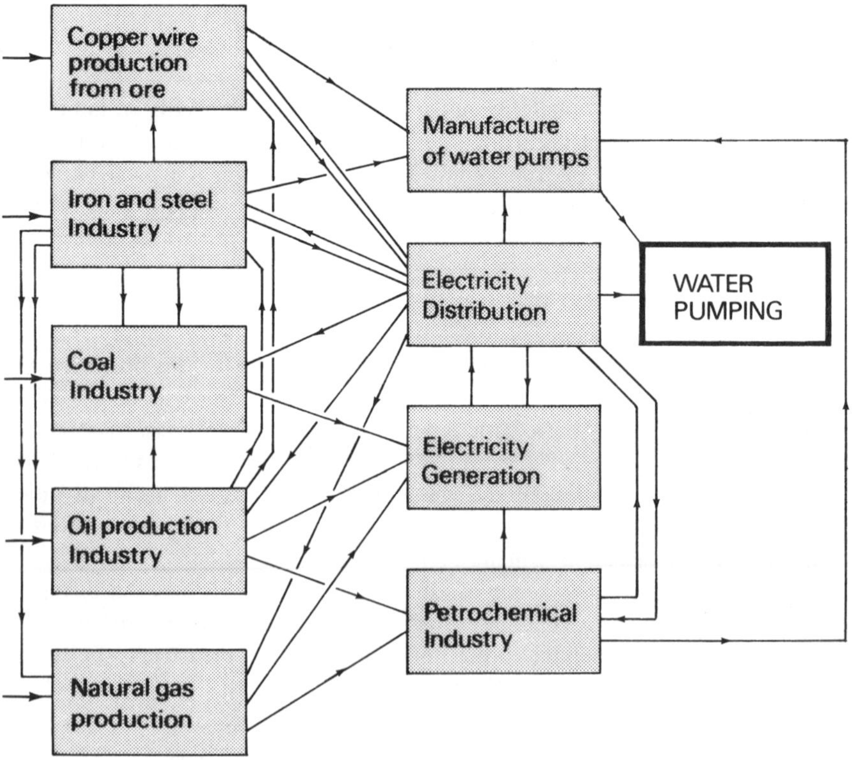

Figure 3.1 Simplified network showing the provision of facilities needed to carry out water pumping.

The type of network shown in Figure 3.1 is typical of those encountered in industry. In the form shown it is not readily amenable to numerical analysis of the type discussed in Chapter 2 because the operational blocks encompass whole sectors of interlinked manufacturing industry. It is therefore necessary to develop techniques which allow it to be simplified to a manageable system or set of systems without introducing errors that make the model no longer descriptive of the real world situation.

3.2 THE MODULAR APPROACH TO REAL SYSTEMS

An analysis technique can be developed using the proposition stated in Chapter 2, namely that all industrial processes, no matter how complex, can be subdivided into a sequence of operations linked by a flow of materials. This proposition is merely a recognition of the fact that within processing industries materials are changed from their input state to their output form by passing through a sequence of operations within a factory. The important feature is that materials flows and not energy link the processes. Conservation of mass exists within a factory and so the input mass must equal the mass of the products plus the mass of any waste materials produced. In contrast, energy is dissipated at each processing stage and output waste energy is seldom, if ever, measured.

Part of the complexity of Figure 3.1 arises because it contains both materials and energy flows. A first attempt at simplification would therefore be to eliminate all but materials flows. However, as already mentioned, most of the blocks shown can hardly be regarded as single operations; rather they represent significant sectors of industry as a whole. These blocks must therefore be broken down into sequences of simpler and more manageable production units. Figure 3.2 for example shows such a breakdown for the production of copper wire using only the main materials flows. Remember that at each processing step there may be a materials loss and there may also be some internal recycling of waste materials from one stage to some previous processing stage. Figure 3.2 also includes the transport steps that were omitted from Figure 3.1. It is important to remember that such steps will occur between every adjacent operation although they will not all be of the same type. The movement of ore to the concentration plant may be by lorry or train but the movement within the factory might be by conveyor belt.

Analysis of a complex processing system into a sequence of discrete sub-systems posses a number of advantages. First, each sub-system is a small system similar to those discussed in Chapter 2 and may be treated theoretically using the type of manipulation outlined earlier. Secondly, there will be an energy feed to each individual sub-system to effect the desired change in the materials throughput, and this energy requirement will be independent of the adjacent sub-systems. It may therefore be caluclated without reference to them. The main advantage of this is that the energy data used in any one sub-system can be improved and

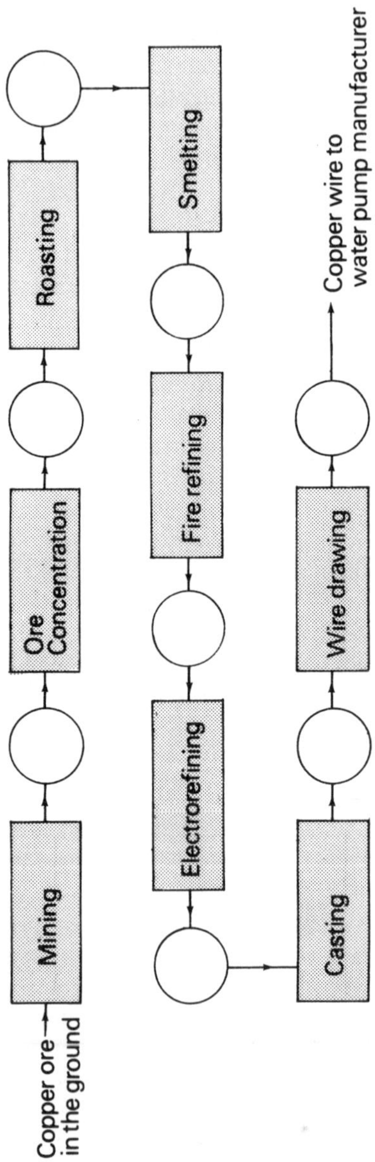

Figure 3.2 Possible sequence of operations involved in the production of copper wire from copper ore in the ground. Circles denote possible transport operations.

up-dated as fresh information becomes available. Furthermore, it is possible to completely replace one sub-system in a sequence by another without disturbing the rest of the sequence, providing the relevant input and output materials are matched. This allows examination of the effect of modifications to production processes. A third advantage of this type of analysis is that, because it describes the flow of materials in the network, quantitative data should in principle be available to the anlayst. Moreover, since the energy supplies to each sub-system are separate inputs, there can be no confusion over whether the indicated flows are materials or energy. Hence the energy supply industries feeding each sub-system are treated as separate entities and can be eliminated from a pictorial description of the system. Also, as will be seen later, it is much more convenient to separate materials processing operations from fuel processing operations.

Breaking down complex operations into sequences of simple operations is often referred to as **process analysis** and is widely used because it is often possible to measure accurately the energy requirements of simple operations. The component sub-systems are occasionally referred to as **production modules** [87].

3.3 RELATIVE CONTRIBUTIONS TO A SYSTEM

It should be clear from examination of a sequence of operations such as that shown in Figure 3.2 that the energy requirement per unit output from each sub-system will be different. Some operations such as smelting are obviously highly energy-intensive whereas others, such as transportation, require much lower energies. Because of this, it is very tempting to simplify such sequences by neglecting those operations for which the energy contribution is exceedingly small. Such a procedure is acceptable if these omissions introduce no significant errors into the overall system energy. However, some care is needed because the contribution of any sub-system to the overall system energy depends not only on the energy per unit output from that sub-system but upon the actual mass output from that sub-system needed to maintain a unit output from the system as a whole as discussed in Section 2.6.

Hence one fundamental requirement of flow diagrams such as Figure 3.2 is that the magnitude of the materials flows must be included. Alternatively, the materials conversion efficiency of each sub-system must be known. Without this information it is not possible to assign an energy requirement to each sub-system within the overall system.

For most industrial operations it can be extremely difficult to predict in advance which of the sub-systems in a processing sequence will make a negligible contribution to the overall system energy. However, a number of special instances exist where the relative values of adjacent sub-systems can be estimated with some certainty and these give rise to the concept of **orders** or **levels of analysis**.

3.4 ORDERS OF ANALYSIS

The concept of 'orders of analysis' is useful in limiting the number of preliminary calculations needed to establish the significant contributors to a system energy. To illustrate this, consider the example of the production of oil fuels. The system is essentially that of extracting oil from the ground, transporting it to the refinery and there processing it to give the desired products. The system is therefore essentially one with a materials input of crude oil and a materials output of finished oil products. The nature of the oil industry is such that essentially all of the materials inputs are converted to saleable products, and although a small proportion of the input may be used internally, assume for the sake of simplicity that there are no materials losses during processing.

For the purpose of the present description, assume also that the whole of the oil industry is itself fuelled by oil. To sustain a throughput of materials in the oil system it is therefore necessary to supply additional oil which provides energy for all the operations within the system. This system can be represented as shown in Figure 3.3, where a throughput of oil of mass M is sustained by supplying an additional mass m. This mass m can also be represented by $f.M$, where f is some suitable multiplier. In this example, we would expect f to be less than unity because the purpose of the system is to produce more oil products than are consumed.

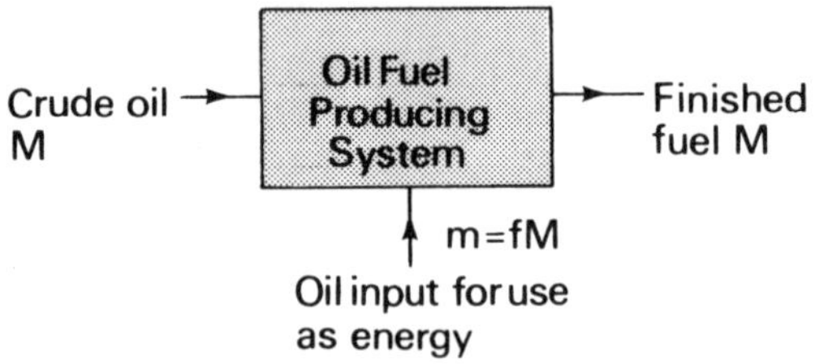

Figure 3.3 Simple oil-fuel producing system.

However the oil products of mass $f.M$ must have been produced at some earlier time using a production system similar to that shown in Figure 3.3. If the total energy associated with the production of oil fuels is to be calculated, the energy needed to produce this oil used for processing must be included.

In Figure 3.4, block A represents the system shown in Figure 3.3 but block B is now included to allow for the production of the oil of mass $f.M$ fed into block A for use as fuel. Block B will be similar to block A except that it will be on a smaller scale because the output $f.M$ will be less than M. For block B, the throughput of oil will be $f.M$ as shown, but in the same way as additional fuel had to be supplied to block A in order to process the oil throughput of that block, precisely the same is true of block B. By analogy with block A, block B

requires an oil input of $f^2.M$. By a similar argument, block C represents the oil-producing facility needed to feed block B and block D is needed to supply block C. Clearly this sequence of processes can be extended indefinitely.

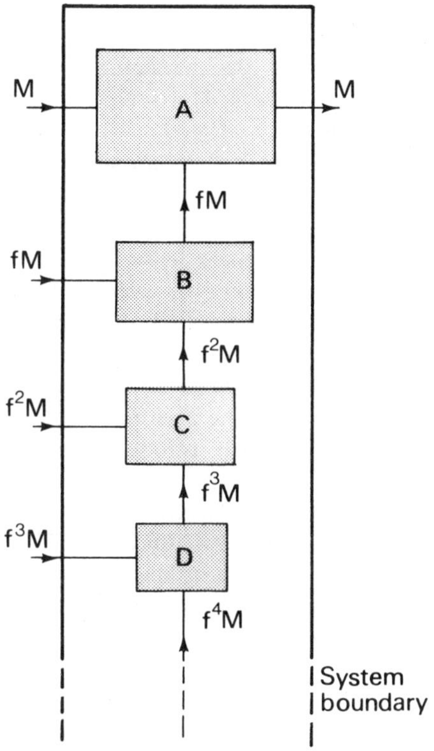

Figure 3.4 Oil-fuel producing system such that all inputs are derived from raw materials in the ground.

The complete fuel production system includes blocks A, B, C, D, and so on, and the system boundary is shown by the heavy line. All the inputs shown on the left hand side of the diagram are raw materials in the ground and give rise to the single output shown on the right hand side of Figure 3.4. The total input of crude oil will there be:

$$M + f.M + f^2.M + f^3.M + \dots\dots \qquad (3.1)$$

This is a geometrical series and in this instance since f is less than unity, the sum will be:

$$M.\frac{1}{1 - f}. \qquad (3.2)$$

If a materials conversion efficiency, C, were calculated for this process using Equation (2.9), it gives:

$$C = 1 - f \tag{3.3}$$

An interesting sideline here is that because we are dealing with oil which is predominantly used as a fuel, that is a source of energy, the value for C also represents the energy efficiency of production of oil fuels. This is because the energy supplied and consumed will be proportional to the masses of the oil fuels flowing in the system. We shall be returning to this concept later in the discussion of fuels (Chapter 6).

The example described above was deliberately chosen so that the sum of the inputs given in expression 3.1 could be calculated exactly. In general however such an exact summation is not possible. Instead the contribution of each of the blocks labelled A, B, C and D, to the overall total must be evaluated separately. It is therefore helpful to have some guidance as to how far down the chain of operations such calculations must be performed before the contributions to the overall total become negligible. This introduces the idea of the **order** or **level of analysis.**

For example, in Figure 3.4 we might choose to consider only block A when an input of $(M + f.M)$ will lead to an output of M. This is regarded as first order analysis following the nomenclature of Maddox [87] and, from Equation (2.9), will give a conversion efficiency of C_1, where:

$$C_1 = \frac{1}{1 + f}. \tag{3.4}$$

For the production of oil fuels, a reasonable value for f is 0.2. From Equation (3.4), first order analysis leads to a value of 0.83 for C_1, compared with the exact value of 0.80 calculated from Equation (3.3). By considering only the first order analysis the calculated conversion efficiency lies within 4% of the exact value.

For a system including blocks A and B the imput would be $(M + f.M + f^2.M)$ and the output would be M. This can be regarded as a second order analysis and would give a conversion efficiency C_2 of:

$$C_2 = \frac{1}{1 + f + f^2}. \tag{3.5}$$

This will take the value 0.806 when $f = 0.2$. Hence second order analysis gives a value lying within 0.75% of the exact value. By introducing the third order operations the value comes yet closer to the exact value, but the improvement is much smaller than that obtained by moving from first to second order. In fact, by considering third order analysis the value is about 0.63% in error.

It is clear then that in this example, a second order analysis would be adequate to yield a value that is in error by less than 1% of the true value. Moreover, improvements in values produced by successive orders become progressively less and unless there is some pressing need to achieve extremely accurate results, it is seldom necessary to consider long processing chains such as that in Figure 3.4.

In general second order analysis is adequate to provide sufficiently accurate results in fuel processing systems. For materials processing systems however, first order analysis will include only the main processing operation, which requires a feed of materials and fuels from second order operations. The materials inputs will comprise the inputs needed to construct both the plant and the machinery needed to provide the facilities required by the processing operation. The contribution of these operations to the system energy is usually so small that second order analysis will be adequate. However, in such a materials processing system the fuel producing operation corresponding to block *A* in Figure 3.4 will lie in the second order, as shown in Figure 3.5 for water pumping, and *not* in the first order. Since errors of less than 1% in fuel production energies require analysis of operations lying at least one order higher than the primary production system, it is usually necessary to consider a third order analysis for fuels.

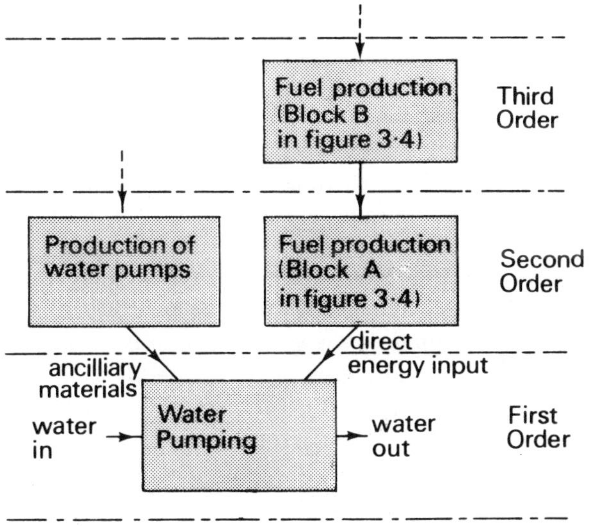

Figure 3.5 Water pumping operations showing the ordering of the different operations.

This requirement is well illustrated by the example of water pumping shown in Figure 3.5. First order analysis considers only the energy directly consumed by the pump. Second order analysis includes the energy required to produce this electrical energy and also that needed to produce the water pump in the first place. Third order effects would include contributions from operations such as the manufacture of machines needed to produce water pumps, which would undoubtedly be negligible. However, the third order also contains those operations needed to produce the fuels required by the second order operation providing the direct energy to the water pump. As demonstrated earlier, this may represent up to 4% of the system energy and so must be included in the analysis.

The importance of this concept of orders of analysis is that it allows any production system to be expanded outwards to include all relevant operations no matter how far removed from the principal operation. Without imposing any limits all systems would incorporate almost every industrial operation in existence and the system would be quite unmanageable. It is therefore useful to be able to decide which level or order of analysis is to be carried out, because this provides a useful guide to follow when deciding whether or not to include a particular operation or set of operations within a system. Most analysts adopt the concept of orders or levels of analysis, although variations exist in the detailed nomenclature. (Compare for example References [69] and [87].) Alternative nomenclatures are discussed in Chapter 8.

3.5 COMPLETENESS OF A SYSTEM

Widely different values have been published for the energy required to manufacture most commodities, Part 2 of the book gives a selection of such energies. Much of the controversy and discussion arising from an examination of the results of energy analysis is due to comparisons of results from essentially non-comparable systems. In large measure this arises from a desire to discuss products rather than processes and to assume that systems are equivalent if the outputs are the same. Such is not the case. A system is only properly defined when the outputs, inputs and function of the system are defined. Thus of the systems capable of producing steel, the inputs could be pig-iron, sponge iron, scrap or even a combination of these. The processing which is a part of the system function could in turn be by means of one of three main types of furnace and the system could be defined to encompass varying degrees of order in the analysis. Not surprisingly therefore, systems defined to include different statements of these parameters will yield different results for the system energy, even though all produce steel as the principal product.

It must be re-emphasised that there is no such thing as a 'right' or 'correct' system for describing any operation. Providing that the definition of the system is proper and complete, then that system is perfectly good within the defined function. However, the purpose of energy analysis is to provide as complete a description of the energy requirements of an operation as possible and so although there may be no 'correct' system, it is certainly true that some systems are more 'complete' than others. For example, the oil processing industry discussed in the previous Section is more completely described if the second order analysis is carried out than if only the first order system is considered. A system will therefore be totally complete only when an infinite number of orders are included.

On the other hand, the natural boundary imposed on any individual processing sequence will be the points at which raw materials are derived from the earth and so *within any one order of analysis* the outer limits are defined by extraction of the raw materials feeds from the earth.

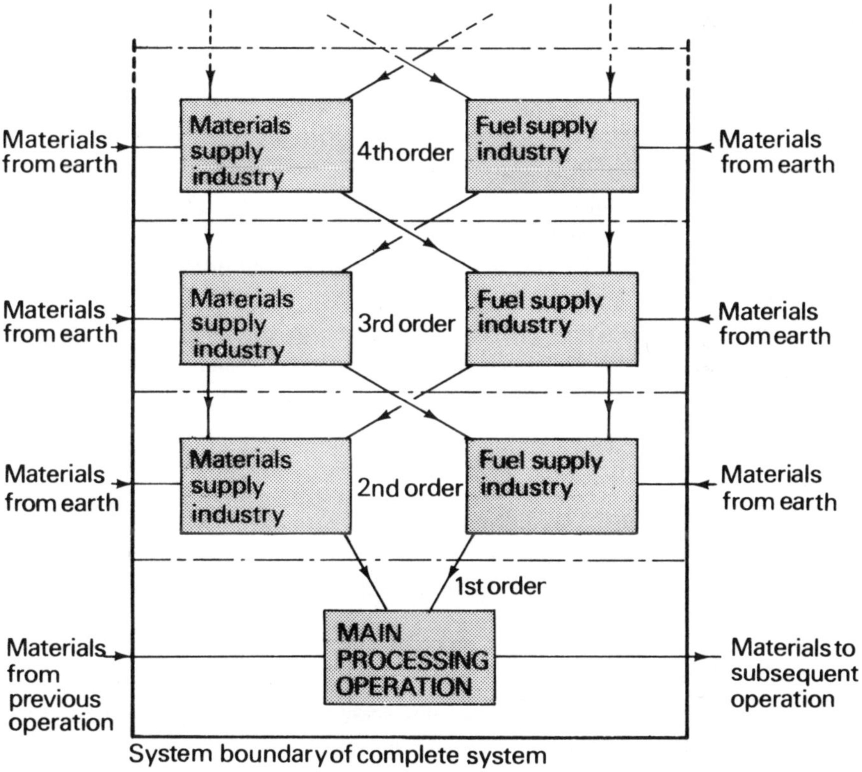

Figure 3.6 Representation of the complete system associated with any industrial operation.

To sustain any processing operation there is therefore a need to supply raw materials and fuels to that system. Moreover, these supplies of fuels and raw materials will always be derived from operations in the next highest order, as in Figure 3.6. The main processing operation is that which changes the materials inputs into the defined outputs, and the fuels and materials requirements for this system constitute the first order analysis. The requirements by this main processing operation for fuels and materials are derived from industries lying in the second order of the analysis, but here the inputs in the complete system are raw materials from the earth. Similarly these industries are fed with materials and fuels from the relevant industries in the next highest order as shown. An important feature is that inputs at all levels other than first order are from the earth. If the levels of analysis were contined indefinitely, then we would define the system as complete and the boundary of this complete system would be as shown in Figure 3.6.

The representation of the complete system shown in Figure 3.6 appears to be simple. However, although the various sub-systems are arranged in a more logical order than say Figure 3.1, it is in fact no simpler in terms of analysis because no sub-systems have in fact been deleted. This is readily apparent from Figure 3.7, which shows a system composed of two linked operations *A* and *B*. Into each of these sub-systems are feeds of materials and fuels as shown.

It is assumed that these second order operations can be traced back to the earth with only two sub-systems so that the second order will contain eight sub-systems. Each of these second order operations requires a feed of materials and fuels from the third order and so there must be 32 third order operations, again assuming that there are only two sub-systems needed to trace these back to the earth. In reality it is likely that there will be more than one materials and one energy feed to each sub-system in the next lower order and it is also likely that each materials feed will need more than two operations to trace its manufacture back to raw materials in the earth.

So, to carry out a third order analysis of the simple two-operation system illustrated in Figure 3.7 would require analysis of 42 sub-systems and in reality the number would probably be much higher. Quite obviously then, even though the system may be arranged in a more logical order, some means is still needed to simplify it while retaining the essential detail.

3.6 CHOOSING A SYSTEM IN PRACTICE

Simplification of complete systems such as that in Figure 3.6 begins by recognising that the sub-systems present in each order are of two quite distinct types; materials producers and fuel producers. Furthermore, although the number of potential materials producing sub-systems is extremely large because of the great variety of materials inputs that may be needed by any industrial operation, the number of fuels producing sub-systems will be small.

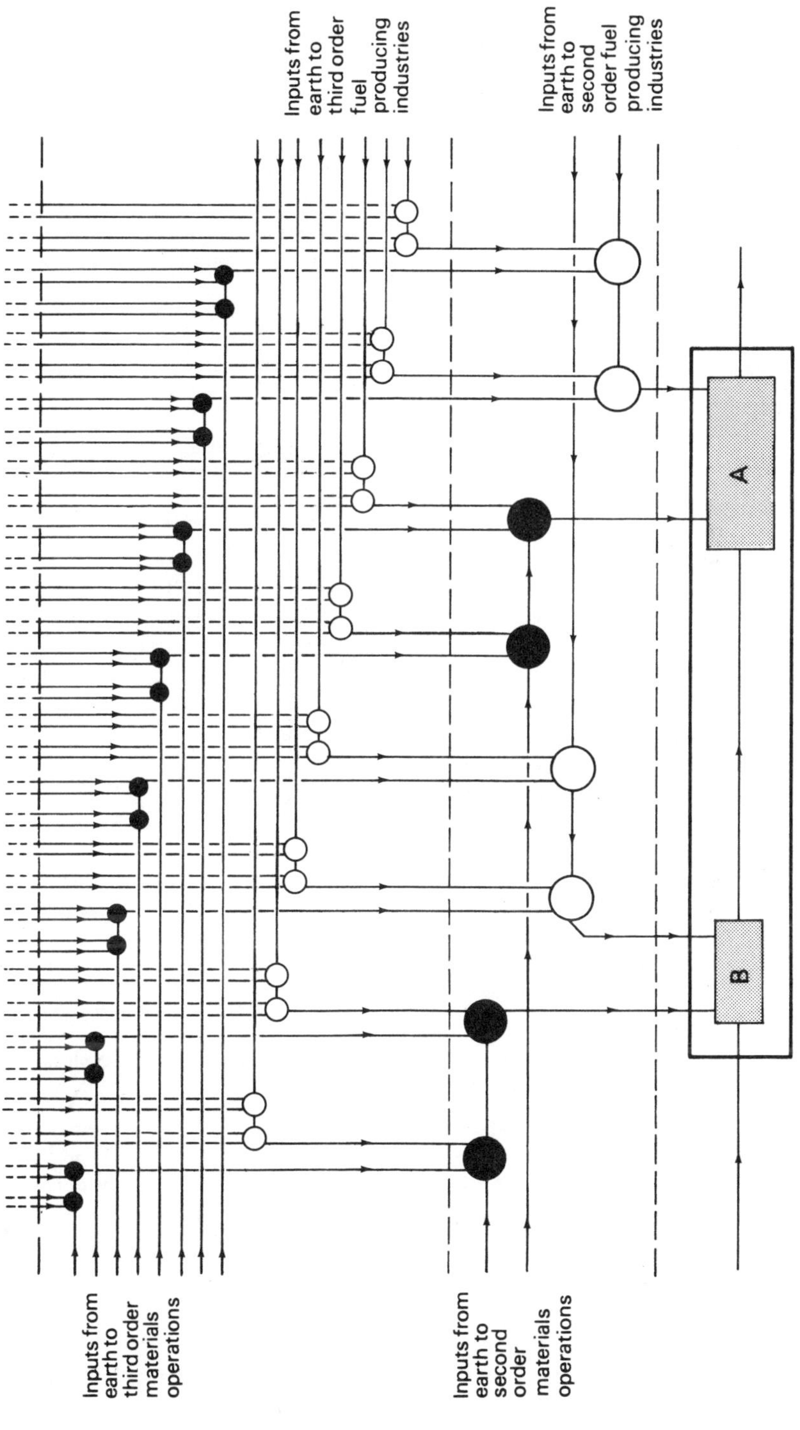

Figure 3.7 Complete system for a two operation process. Even though it has been assumed that any processing sequence consists of only two operations it is clear that the number of sub-systems rapidly increases in the higher orders. Open circles denote fuel handling sub-systems; black circles indicate materials handling sub-systems.

At the present time, the only fuels in widespread use by industrialised countries are coal, coke, oil, natural gas, manufactured gas and electricity. Because of their small number it becomes worthwhile to evaluate the energy requirements of the complete systems needed to produce these six fuels since one or more of these sub-systems will be used in each order of analysis for all other systems. The individual sub-systems referring to fuel production in Figure 3.6 can therefore be replaced by the single sub-system shown in Figure 3.8 and labelled 'fuel supply industry' and the details of this sub-system can be evaluated on a 'once and for all' basis before beginning any analysis.

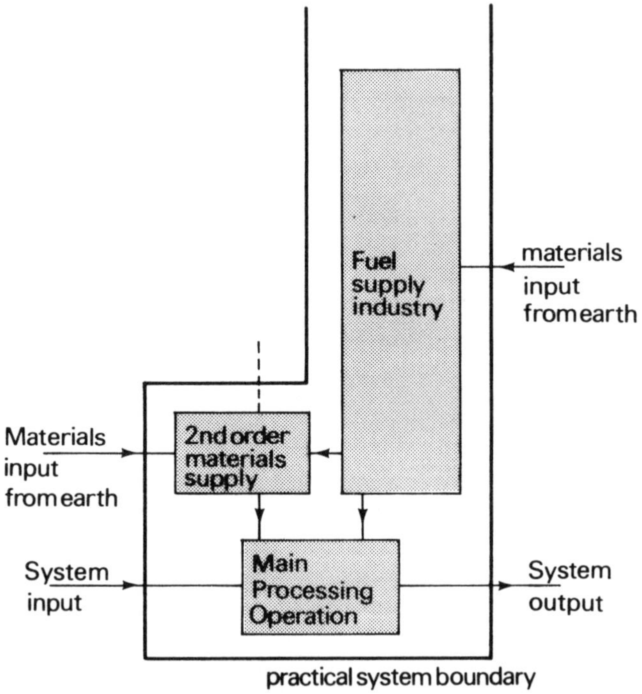

Figure 3.8 A commonly used practical system. Compare this with the complete system of Figure 3.6.

Of the materials supply operations shown in Figure 3.6, each will contain sequences of operations similar to those shown in Figure 3.2 for copper wire production and there will be a separate chain of processes for every commodity demanded by the operation it is feeding. Unlike the fuel supply industries, there will be no standard collection of supply chains guaranteed to cover the requirements most likely to be met in practice although some, such as the production of steel, will often recur. The number of such production sequences must therefore be reduced by some other means. To do this, the relative importance of the various orders of materials supply must be assessed, but this can only be done sensibly with hindsight. One contribution to the main supply system will be the production of the machines and buildings comprising the main production unit. This component of energy corresponds in the economic sphere to the capital cost of the plant and so is frequently referred to as the **capital energy**.

This contribution has been variously estimated from published values, but typically it is expected to be about 5% of the total system energy. If the demand of materials from the third order is of the same magnitude, then their contribution to the total system energy will only be about 0.25%. In other words, if the third and higher order materials supply operations are ignored, we can expect the value for the total system energy to be in error by only about 0.25%. For most practical purposes such a level of accuracy is quite acceptable and it is therefore common to ignore materials supplies above second order.

It should be pointed out that even when this simplification has been made the second order may still contain a large number of production sequences for which a detailed and accurate evaluation could be excessive. Many published results ignore the contribution from these operations, in which case the system energy will of course be in error by about 5%. Often, however, the principal contributions from this sub-system are evaluated using published data of uncertain accuracy so that the overall system error lies somewhere between 0.25 and 5%.

If the second order materials supply system is included in the overall system and all higher orders are excluded, a typical practical system boundary will be as shown in Figure 3.8.

The development of the argument leading to the establishment of the practical system shown in Figure 3.8 draws on two quite distinct ways in which the energy contributions to a particular system energy arise. These may be thought of in terms of the two ways in which the materials flows within any system may be traced back to the earth.

The first way in which this tracing back is carried out is based on the idea of 'order of analysis' and involves a **vertical** tracing back of the progressively decreasing contributions arising from higher and higher orders. As demonstrated in the example of oil fuel production, there is rarely any great improvement in accuracy to be achieved by considering orders higher than the second in fuel producing systems, or beyond third order when a fuel producing sub-system lies in the second order. In general, the contribution to the total system energy from an operation within any one order will always be greater than the energy contributions from the operations to which it is linked in the next higher order. Therefore if it can be shown in a particular analysis that the energy arising from third order sub-systems is negligible, it immediately follows that the fourth and higher order analyses will also be negligible and so need not be considered further.

The second type of tracing back is the **horizontal** expansion of sub-systems *within any one order* until the inputs are derived from the earth. An example of this is shown by Figure 3.2. The essential difference between the operations in a horizontal sequence and those in a vertical sequence is that there is no regular progression in the values of the energies arising from successive operations because each is independent of its neighbours. There is therefore no way in which such sequences can be readily simplified before numerical work begins.

Detailed examples of the application of these ideas will be presented later but it is helpful at this stage to illustrate the way in which they arise in practice using a simple numerical example. Even the simplest operation can lead to complex systems and so in order to highlight the desired effects without the confusion of large amounts of detail we have constructed a simple hypothetical case based on the water pumping example used earlier.

Let us concentrate on the importance to the overall system energy of the energy required to produce water pumps. To do this we have assumed that a pump will move 1000 litres of water for a total electrical energy consumption of 2 MJ and that a pump needs replacement after it has moved 10^9 litres of water, Figure 3.9 illustrates that part of the system considered.

In the first order will be the main water pumping operation with an input and an output of 1000 litres of water. In addition, there will be an input of energy of 2 MJ due to the provision of electricity from a production facility lying in the second order. There will also be an input of materials due to the provision of water pumps, again from an operation lying in the second order. As mentioned earlier, this input of pumps is an intermittent flow of 1 pump every 10^9 litres of water and so it can be regarded as a continuous flow equivalent to 10^{-6} pumps per thousand litres of water pumped.

The input of water pumps to this first order operation is from the pump manufacturer whose operation lies in the second order, as shown. The pump manufacturer however simply assembles the pumps from raw materials fed into his factory and we have assumed that he uses 10 MJ of energy to assemble one pump. We have also assumed that the pump consists of essentially three components only, 50 kg of steel, 2 kg of copper wire and 0.1 kg of some polymer for use as insulation. For each of these three commodities we can construct a horizontal sequence of operations as shown, tracing the materials flows back to the ground. Note that for clarity these sequences have been very much simplified; compare the copper wire sequence, for example, with that shown in Figure 3.2. It has also been assumed that no materials losses occur within the processing routes, again to simplify the description.

In Figure 3.9, eleven sub-systems can be identified in the second order and the materials flows needed to produce one pump are shown. The energy consumed by each of these sub-systems can be calculated by multiplying the energy per unit output by the mass of the output from the sub-system. Hence in the case of iron production, the energy per unit output is shown as 25 MJ/kg and because the actual output needed to produce 1 pump is 50 kg, the total energy

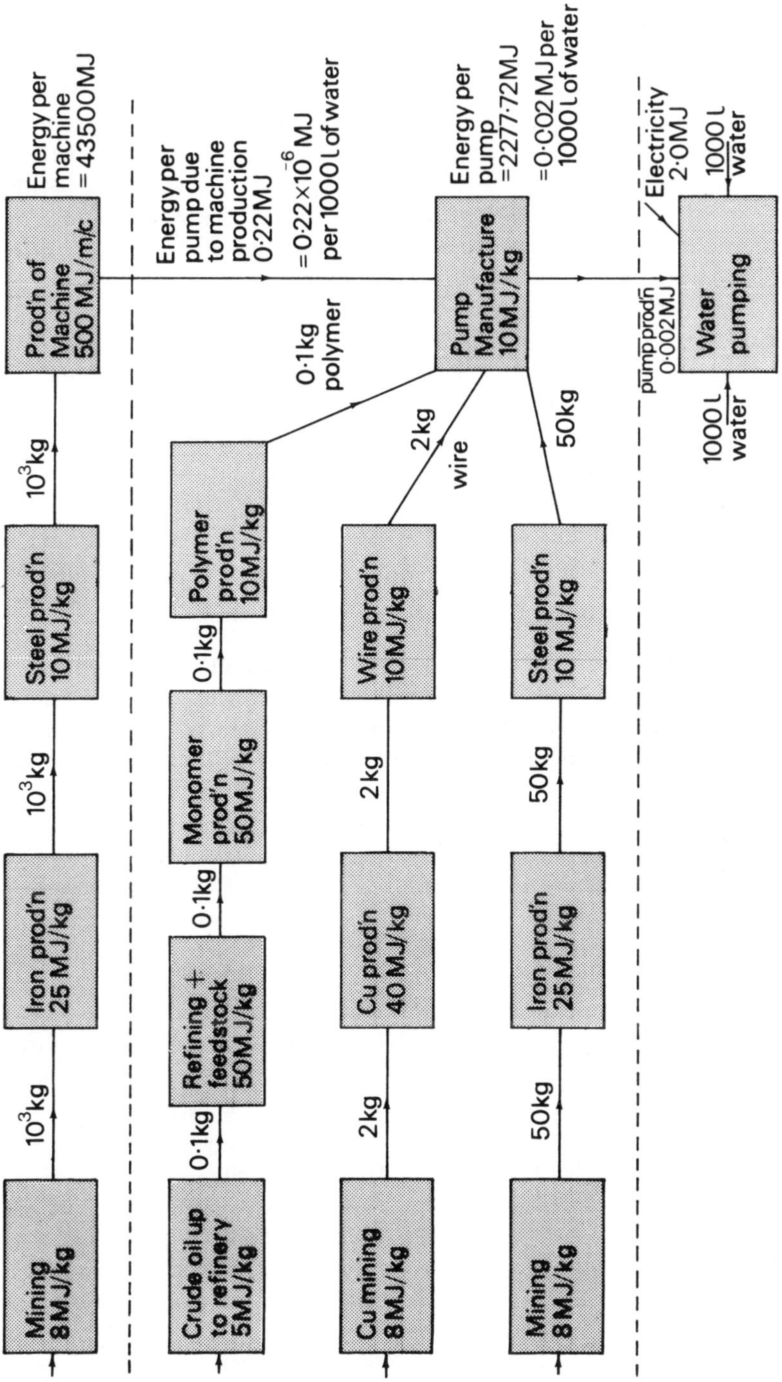

Figure 3.9 Simple numerical example of a portion of the extended system needed to pump water.

associated with this sub-system will be 25 × 50 = 1250 MJ. By calculating the energies associated with all eleven sub-systems of the second order and combining them, we arrive at an energy of 2277.50 MJ to produce 1 pump. Since the flow of pumps down to the first order is only 10^{-6} pumps per 1000 litres of water pumped, this production energy means that the energy contribution to the total system energy due to the production of pumps is only 0.002 MJ per 1000 litres of water, to give a total system energy of 2.002 MJ per 1000 litres of water pumped. Hence the second order contribution from pump manufacture is less than 0.1% of the total system energy.

It now becomes clear that it was not necessary to analyse the second order in as much detail as we did. In fact, with hindsight, we can now see that about 94% of this second order energy arises from the production of steel and so it would have been sufficiently accurate to ignore the production of copper wire and plastic insulation, because by neglecting them the error introduced into the second order energy is 5.6% and the error introduced into the overall system energy would have been 0.006%. In fact, knowing the amounts of plastics, copper and steel needed to produce one pump, this conclusion could have been deduced using only very approximate values for the energy requirements of the different sub-systems. Indeed the values shown in Figure 3.9 are only approximate, as a comparison with the values in Part 2 shows.

The effect of considering the third order inputs can also be demonstrated. One operation lying in the third order will be the production of the machines needed by the pump manufacturer to assemble the pumps. It is assumed that such machines weigh 1000 kg and are manufactured from steel. Hence the iron and steel sequence of the second order will also appear in the third order as shown. The calculations are performed in exactly the same way as previously and it can be shown that the energy to produce a single machine will be 43500 MJ. The feed of machines to the pump manufacturer is also an intermittent flow, and if it is assumed that a single machine will produce 200,000 pumps in its lifetime, this intermittent flow can be regarded as a continuous flow of 0.000005 machines per pump. This corresponds to an energy consumption per pump of 0.22 MJ. Inclusion of this energy raises the pump manufacturing energy to 2277.72 MJ per pump, but the change is so small that it exerts a negligible influence on the energy to pump 1000 litres of water. It is clear therefore that in this instance, this component of the third order analysis can be ignored.

Note that this could also have been deduced in advance, because the steel production sequence is identical with that appearing in the second order and so the numerical values would already be known from the second order analysis.

It is important to remember that the use of approximate values can often aid the simplification of the most complex systems. For this reason it is useful to have available a collection of production energy data such as that given in Part 2. Even when the values are of uncertain accuracy, they can be used to carry out approximate calculations and so determine whether there is any need to proceed with more detailed and accurate energy evaluations.

When a system or sub-system is complete in a vertical sense, the energy requirement associated with that system or sub-system is often referred to as the **process energy requirement**. If in addition the system or sub-system is complete in a horizontal sense, the energy requirement for that system is referred to as the **gross energy requirement** [69].

Chapter 4

Sources of numerical data

4.1 THE NEED FOR NUMERICAL INFORMATION

So far the discussion of industrial systems has been carried out in general terms and the numerical values have been used only by way of example. When we turn to examine real systems in practice, we need to know where suitable data can be obtained and we need to be able to assess the validity and accuracy of any data used to enable the accuracy of the system analysis to be estimated. We must be aware therefore of the bases from which the data were calculated and the purpose of this Chapter is to examine critically the information sources that the analyst may draw upon.

It is sometimes claimed that the type of data base available determines the methodology by which the analyst should approach a problem [29, 112]. It is certainly true that information derived from statistical sources will, in general, be less amenable to detailed analysis than information derived directly from a factory, but the essential difference is the limitation on the choice of systems that can be analysed rather than on the technique of analysis. If for example a source of information in discussing the iron and steel industry only implies those operations carried out from receipt of the raw materials at the site to the point at which finished products are despatched, this inevitably prescribes the definition of inputs and outputs to the system. In this example it would prevent such data being used to calculate say the energy associated with the blast furnace, which is only one of the many operations existing within this overall system.

There are four main areas in which information can be sought. These are process information obtained directly from manufacturers, published papers and books, general industrial statistics and input-output tables.

4.2 DATA FROM INDIVIDUAL OPERATORS

The prime source of data for analysis of the performance of industrial operations as they exist in practice is the individual plant operator. Analysts may therefore obtain data at first hand or may accept published numbers obtained by someone else from the same source. Unfortunately such data are usually published in summary form and the detail required for a complete analysis is thereby obscured. Whenever possible therefore data should be obtained personally by the analyst.

The simplest type of information usually held by individual companies is the total input of materials and fuels to their organisation together with detailed records of their saleable output. If such information can be obtained for a specific period, the average materials and fuels requirement per unit output can be calculated after allowances have been made for stockpiling inputs and outputs.

This information is valuable because it relates to the behaviour of a single company or production unit rather than to an average for the industry. This allows determination of the different energy requirements of factories producing the same commodity and often helps to indicate why one operation is more profitable than another. Furthermore, if a series of analyses are performed to cover a succession of consecutive time periods, it is possible to follow the changes in production efficiency of a company so that the effect of such influences as level of production can be examined. A further advantage is that a detailed breakdown of the total energy consumption of a plant by fuel type can often be obtained. Such details are useful in considering the effect of substituting one fuel for another as well as indicating the dependence of a particular producer on a specific fuel type.

If sufficient detail is available the performance of small areas of even single operations within a factory can be analysed. The relative importance of individual operations can then be assessed in energy terms and the data assembled to give an overall picture of the whole factory. Such detailed analysis is essential if the results are to be used for improving or modifying processing techniques. Furthermore, the capital energy requirement of buildings and machines can be estimated from site plans and manufacturers' specifications.

Most factories tend to produce more than one product and some means of partitioning the total energy amongst the various products is needed. Dividing the total energy input of a factory by the total output of all products is most unsatisfactory because it makes the, usually false, assumption that all products experience the same processing. Detailed process analysis of a factory allows calculation of the total energy input to each operation and hence the energy of specific products can be accurately calculated. The problem of partitioning energy amongst different products is considered in some detail in Chapter 10 but a simple hypothetical example highlights the problem. Suppose that a factory processes 3000 kg of raw materials in a period during which it also consumes an energy of 8000 MJ. Neglecting the supply of ancilliary materials, the average energy per unit output is 8000/3000 = 2.67 MJ/kg, assuming no materials losses during processing. Suppose however that a more detailed process analysis reveals that the factory makes two products employing four operations, as shown in Figure 4.1. The energies associated with each product are clearly different. A total energy of 5000 MJ from operations (2) and (3) is associated with product

A. In addition, two thirds of the throughput of operation (1) goes to product *A* so that two-thirds of the energy associated with operation (1) must also be attributed to product *A*, assuming that apportionment can be based on mass throughput (see Section 2.5).

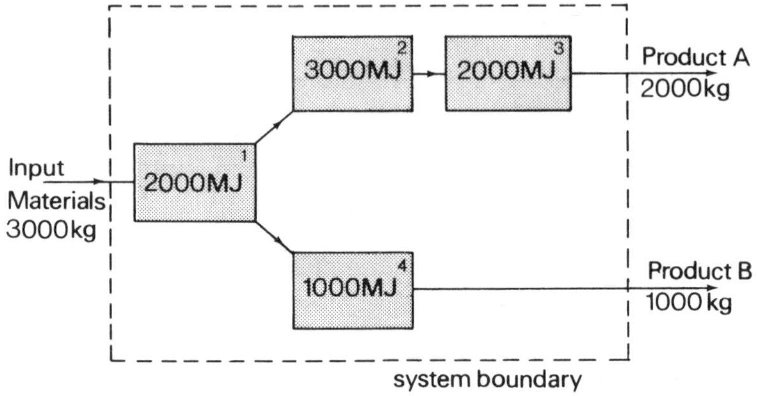

Figure 4.1 Simple hypothetical production system manufacturing two products *A* and *B*. The energies shown in each of the sub-systems represent those needed to maintain a throughput of 3000 kg of materials.

The energy to produce product *A* is therefore 6333/2000 = 3.17 MJ/kg, and by a similar argument that for product *B* is 1.67 MJ/kg. By using an average value of 2.67 MJ/kg for all products, the energy for product *A* would have been underestimated by 16% and that for product *B* would have been overestimated by 60%. Although this example may seem relatively trivial, it represents a recurring problem in energy analysis and highlights the ease with which very large errors can be introduced by making seemingly simple assumptions. Such assumptions can be eliminated when data is obtained directly from plant operators.

The need to partition the consumption of materials and fuels between different products gives rise to serious difficulties when inadequate metering of supplies is encountered. For example, it is not uncommon for electricity and gas supplies to be monitored only as they enter the site. Moreover, the consumption of central facilities such as steam and compressed air are seldom measured in terms of their ultimate destination. If physical measurements are not possible, the next best situation is to obtain estimates from the plant engineers of the requirements for fuels and services in different areas of the plant. In this way a more sensible apportionment between production areas can be achieved than is possible by simply averaging over the whole plant.

Finally, one major advantage of obtaining data directly from plant operators is that any suspect values may be checked directly with the source. Admittedly, as with all data, information provided must ultimately be accepted at face value, but at least with direct collection there is the possibility of questioning the originator, an opportunity seldom available to the user of published data.

The assembly of suitable information from production plant rests heavily on the amount of co-operation obtained from the plant operator. In general, time is the limiting factor. The analyst must also take great care to explain his needs exactly so that misleading information is not prepared unwittingly by the operator. Probably the best sources of information are the daily or weekly production records. When taken together with verbal comments on any factors such as power cuts or strikes which may have disrupted production, they usually give the most complete picture of the way in which a factory is operating. Such records form the basis for all management information within the company and although it would be naive to assume that they are not subject to 'adjustment' by production workers, they probably present the best estimate of production performance available. Moreover they are subject to some simple tests of accuracy; for example, a machine cannot process more materials than were delivered to the factory nor can a machine with a fixed operating cycle of 60 seconds produce more than 60 components in one hour.

One advantage of allowing the analyst to examine detailed production records is that they often reveal practices of which he was initially unaware. For example, a specific factory may have found a saleable use for one of its products which would be regarded as waste by similar factories elsewhere. Alternatively, it may have developed a simple method of re-using some of its own waste products internally. Such factors are seldom known to the analyst, who at the outset of the study may only be aware of general practice in the industry as a whole and who might not uncover such practices if the information were not volunteered.

One factor which inhibits co-operation by companies is confidentiality. To provide sufficient information for the analysis to be meaningful, an operator must reveal considerable detail about his process and this information is frequently potentially useful to a competitor. Not surprisingly therefore there is still considerable reluctance on the part of many operators to release detailed information for free publication, even though they may be prepared to release it for internal use. For this reason there is little reliable detailed process information for specific plants in the literature and published analyses are usually based on statistical information.

A common problem is that much company information is recorded in a form of little use to the energy analyst. In the commercial world, economic profitability is the yardstick of success and records are usually maintained in financial terms rather than in terms of physical quantities such as gallons of oil or kilogrammes of materials. Even though financial records are initially derived from physical quantities, the detailed sources of information are often discarded once the conversion has been made. Physical quantities can in principle be derived by calculating back from the financial records, but this can be a very hazardous exercise. Expressions such as 'electricity cost' do not necessarily mean the same to an accountant as to an energy analyst. To an accountant this may mean the cost of the units of electricity consumed plus the costs associated with maintenance and depreciation, whereas the analyst may want only the number of units of electricity consumed and records of replacement *materials* used. Attempts to separate the costs of electrical consumption and maintenance in retrospect represent an almost impossible task. A further difficulty is that accountants apportion costs to cost centres which seldom coincide with the system of interest to the energy analyst.

In such cases there is no really satisfactory way of extracting the desired information from the financial data. Even when cost centres do coincide with the energy analysis system, the accountant's method of cost apportionment is often only remotely linked to the consumption of materials and fuels because other factors, such as the costs of labour and capital, may dominate.

However, despite these limitations (and it must be remembered that they are also inherent in most other sources of information) the collection of process data directly from the operator is still the most satisfactory, if time consuming, means of obtaining information.

4.3 DATA FROM PUBLISHED PAPERS AND BOOKS

The confidential nature of much commercial practice limits the amount of detailed information published. Such published information as does exist often relates to novel plant, which is seldom representative of the industry as a whole, or it describes practices which are becoming obsolete and therefore commercially less sensitive. Such data is useful in establishing upper and lower bounds for the energy required to produce a specific commodity, but a changeover from one type of plant to another is usually determined of course by economic reasons in which factors such as labour and capital costs often dominate.

However, published papers often give valuable information on some of the basic operations common to many plants and for which secrecy is not important. Typical of such operations are the provision of services such as steam and compressed air as well as transport. A supply of very useful information concerning a large number of essential industrial operations that are needed in the more detailed type of analysis can be built up using such sources. Much of this type of information is contained in Part 2.

Review articles in journals aimed at the more specialist audience often give typical or average values for the energies of industrial operations for which detailed information is not available. Whilst such data must be treated with caution as far as accuracy is concerned, it is useful in establishing approximate values for the energies of operations.

In similar manner, specialist textbooks, particularly in the areas of chemical and production engineering (for example Ref. [121]) often quote typical requirements of materials and fuels for specific processes which are valuable in the absence of detailed process data from operators. The main problem with most published process data is that the system is frequently defined so that it is not clear which operations in a given sequence have been included.

4.4 DATA FROM GENERAL STATISTICS

Most trade associations and large companies provide regular statistics relating to performance, and Governments usually publish national statistics itemising production in the major industrial sectors. Such information can often provide a useful basis for calculating approximate energy values. Chapman [29], for example, quotes an energy of 256.7 therms/ton for steel production. This was calculated using the value of energy consumption by the steel industry of 6871×10^6 therms taken from the 1968 U.K. Digest of Statistics and a total steel production in the same year of 25.86×10^6 tons derived from the U.K. Iron and Steel Industry Annual Statistics. Such a value is of course a first order analysis of an average system for producing steel and hence is of only limited application. Moreover, as Chapman points out, the U.K. Iron and Steel Industry is also in the business of selling surplus gas and electricity and such energy credits may not be taken into account when the production energy is calculated.

While such problems may be overcome by careful analysis of the statistics, there are more general difficulties associated with the use of statistical information irrespective of source.

Statistical data are compiled from information supplied by the actual users and producers of commodities, usually in response to a questionnaire. The response to questionnaires is frequently less than 100% but if the statistics are to describe the whole of a particular industry, the actual response must be scaled up to cover the whole of the industry. A variety of techniques are available but that used in a particular situation may well be unsuitable for energy analysis. For example, the Census of Production produced by the U.K. Government scales up its values using the number of employees of the industry. Whilst such a technique is as reasonable as any other, the number of employees seldom bears any direct relationship to the energy consumption of a particular industry. Indeed the reverse may be the case because a highly automated plant may have few employees but his may be offset by the consumption of more energy than a less automated plant in order to produce the same output.

Because of the limited amount of time and money available to compilers of statistics they are able to check only a very small sample of any data collected. Consequently incorrect returns remain undetected. Moreover some information included in the tables may be simply guesswork. For example, the instructions sent to companies supplying information for the U.K. Census ask establishments to give reasonable estimates when exact figures are unavailable. As a result the uncertainties in much statistical information may be significant.

The energy analyst faces a further problem in the use of statistical information arising from the need to present statistics in a manageable form. The 1972 U.K. Census of Production, for example, recognised 155 sectors of industry ranging from wall coverings and lace manufacture to inorganic chemicals and electricity supply. Although the number of sectors seems large, considerable aggregation of information is needed to restrict the entries even to this level. Such aggregation raises three serious problems. First, even when a sector of industry is well defined, aggregation of the data within that sector eliminates variation in production efficiency from plant to plant. Secondly, many sectors are ill-defined; organic chemicals as a group encompasses an enormous range of products. Hence the information is only of limited value in energy analysis. Thirdly, at some stage an industry must be assigned to one particular sector even when it produces a range of products. If the principal product is chosen it excludes the others from more appropriate classification and if a subsidiary product is chosen it causes a serious imbalance in the statistics even though the numerical values may be subsequently refined by recognised mathematical techniques.

Finally, much statistical information is collected and presented in a form suitable for economic analysis. Whilst some of the information may be adapted for use by energy analysis, the conversion from monetary value to energy requirement is very difficult, a factor discussed in more detail in the next Section.

Despite these limitations, statistical information has been and continues to be widely used in energy analysis. Whilst it is easy to criticise results based on such data, it is important to recognise that in the recent past it has been responsible for providing very useful approximate values for energies required to manufacture certain products or groups of products. However, the contined use of statistics does tend to sustain the practice of referring energy requirements to products rather than systems.

4.5 INPUT-OUTPUT ANALYSIS

An input-output table is a square matrix of numbers showing the transactions in commodities taking place between different sectors of industry in a specified time period. Such tables are a compact form of presenting a very large amount of data and the underlying principle can be simply illustrated by reference to Table 4.1. This shows a matrix containing only two industries, steel and electricity. The left hand column shows the producers of goods and the top row shows the consumers. This Table indicates that the production of 1 unit of steel requires the consumption of 0.1 units of steel and 0.2 units of electricity. Similarly, the production of 1 unit of electricity requires the consumption of 0.3 units of steel and 0.4 units of electricity.

Table 4.1
Simple input-output table showing steel and electricity. The numbers have no real significance.

Consumer Supplier	Steel	Electricity
Steel	0.1	0.3
Electricity	0.2	0.4

Using such a table it is possible to carry out a detailed analysis of the resource requirements for the production of any commodity and to reach any desired order of analysis. Thus since the primary resource requirements for unit steel production are 0.1 units of steel and 0.2 units of electricity, the second order requirements can be calculated by using the Table a second time. Thus the supply of 0.1 units of steel requires a second order consumption of:

$$0.1 \times 0.1 = 0.01 \text{ units of steel}$$
$$\text{and } 0.1 \times 0.2 = 0.02 \text{ units of electricity.}$$

Similarly the supply of 0.2 units of electricity requires a second order consumption of:

$$0.2 \times 0.3 = 0.06 \text{ units of steel}$$
$$\text{and } 0.2 \times 0.4 = 0.08 \text{ units of electricity.}$$

Hence the total second order requirements are:

0.07 units of steel,
0.10 units of electricity.

Third order analysis may now be carried out using the input-output table again to calculate the requirements for the second order inputs. In this case, the third order requirements are:

0.037 units of steel,
0.054 units of electricity.

This process may be repeated until the requirements of further orders becomes negligible. The total resource requirements for any operation can then be calculated by summing the requirements for each order of analysis.† Hence for the three orders of analysis outlined above, the total requirements for the production of 1 unit of steel are:

0.207 units of steel,
0.354 units of electricity.

In reality input-output tables contain many more elements than the 2 X 2 matrix of Table 4.1. The 1968 tables for the U.K. [65] contain 90 industrial sectors and the 1967 tables for the U.S.A. contain 357 sectors. Amongst these are the fuel producing industries so that the tables can be used to analyse completely the materials and fuels requirements of any industrial sector listed in the tables. Such tables therefore appear to be a valuable source of data and a number of analyses based on such tables have been carried out [23, 141].

There are, however, problems associated with the use of these tables and the References quoted above show how some of them can be overcome. Without examining in detail the methods used to overcome these difficulties, it is well to be aware of the limitations of this type of analysis because they influence the validity of the energy values so derived.

In common with many statistical tables, input-output tables are usually published in monetary rather than physical units and so only yield the cash cost of the energy supplies to a process. In principal such costs can be converted to give the energy units required by the process if the unit costs of the various fuels are known. However, fuel costs usually vary during a 12 month period and large consumers often negotiate concessionary fuel rates. It is therefore extremely difficult to arrive at a satisfactory average fuel cost, which in any case may vary from one industrial sector to another, depending upon the size of the average production unit within each sector.

Once an average cost per unit of fuel is assigned, the results are still in the form of say megajoules per dollar's worth of product. This is frequently referred to as the **energy intensity** of the product. Conversion of energy intensities to energy per unit product requires a value for the cost of the product. The major problem here however is that most industrial sectors identified in the tables do not produce homogeneous products.

For example, a sector such as boat building would cover everything from fibre glass dinghies to cabin crusers as well as many of the ancilliaries used in sailing. It is therefore extremely difficult to assign an average value to the outputs from such sectors. For the 1963 U.S. input-output tables, Wright [141] has identified only 29 industrial sectors, out of a total of 363 sectors, which result in products sufficiently homogeneous to allow conversion to energy per unit product in physical terms.

†In general, if the input-output matrix is represented by A, then the second order requirements will be represented by A^2, the third order requirements by A^3, and so on. The total requirement will theretore be

$$A + A^2 + A^3 + A^4 + \ldots\ldots$$

Input-output tables cannot be derived from the economic statistics normally available to Governments and require a special census. Moreover, the compilation of the tables from the collected data takes a long time; the 1968 tables in the U.K. were not published until 1973. This time delay can be serious particularly in developing industries, although Bullard and Herendeen [23] have considered techniques for up-dating the information which should be available for most sectors.

On the credit side, the immediate benefit of the input-output tables is the availability of a large body of data. However, although the results are exhaustive within the sectors of industry listed in the tables, they are in general applicable only to large scale analyses and are inapplicable, except in a very few cases, to specific small scale systems. For these detailed data from other sources must be obtained.

In addition to the difficulties in interpretation of results mentioned above, the results from input-output analysis can never be better than the quality of the information contained within the tables. Such tables will contain errors arising from incomplete or inaccurate data, as discussed in the previous Section.

The need to consider secondary products raises further difficulties. Input-output sectors are identified primarily by company rather than by production process and few companies produce single products. Allowance for secondary products must be made because they would appear in a different sector of the table if they were the primary product of the company. Bullard and Herendeen [23] quote the example of a company producing primary aluminium as its principal product but which also produces aluminium castings as a secondary product. The secondary product, that is the aluminium castings, is transferred to the secondary aluminium products sector as a sale, but the corresponding input is not transferred. This convention is satisfactory from an economic viewpoint but from a resource viewpoint it means that the cost of the output has been entered twice into the tables while the energy used has been entered only once. Such products must therefore be identified and a correction made if satisfactory results are to be obtained.

Two very important omissions from input-output tables must be mentioned. First, capital goods such as buildings and machinery used by manufacturers do not appear as inter-industry transactions within the tables and so input-output analysis cannot be used directly to calculate the energy attributable to such inputs to manufacturing processes. Secondly, it is not possible to distinguish between fuels that are used as a source of energy and those that are used as sources of materials, that is feedstocks (see Section 7.3). Therefore, although input-output analysis will include feedstock energy, it cannot separate this feedstock energy from the fuel energy.

Finally, it is important to bear in mind that the data in input-output tables represents the average of all available technology during the period to which the tables relate. In effect this implies that the system corresponding to each industrial sector is very crudely defined by a product or product group only. Steel production, for example, would probably include production from pig-iron or scrap by all available routes. Moreover, since most countries import commodities, the tables also include contributions from overseas technology.

In conclusion we should note that, while use of input-output data is restricted in its application to detailed energy analysis of some industries, the tables do provide a wealth of information and the data form a useful basis for approximate calculations when no other sources of information are available.

4.6 LIMITATIONS IMPOSED ON ANALYSIS BY THE DATA SOURCE

It should be clear from the discussions in the previous Sections of this Chapter that the type of industrial system that can be analysed is determined by the source of information available. At one extreme, data collected directly by the analyst from an operator can, in principle, be as detailed as desired. Here the only limitations imposed are those of the time that the analyst is prepared to devote to extracting the necessary detail from company records and the amount of detail that the company is prepared to release.

At the other extreme, the use of published information limits the choice of system to that which matches the detail contained within the publication. Moreover, although all data irrespective of source must be treated at face value, there is a provision for questioning data collected directly which is absent from published information.

In general therefore, detailed analysis of specific processes can only be carried out satisfactorily by process analysis using directly collected information. For approximate values of production energies averaged over a number of plants, statistical information and input-output analysis can quickly provide a large amount of useful information, although in many cases the precise system is extremely ill-defined.

Chapter 5

The energy content of materials

5.1 COMBUSTION

In spite of the advent of nuclear technology and the harnessing of water power in hydro-electric schemes, by far the greatest source of energy for industrial processing is derived from burning naturally occurring materials. This pre-eminence of combustion as a source of energy is of such importance that it is worth looking at the combustion behaviour of all materials because it raises a number of important and surprising features that we shall need to examine more closely later.

We live on a planet where about one fifth of the atmosphere is composed of oxygen molecules. If therefore any reaction carried out in contact with the atmosphere involves reactants or products which can chemically interact with oxygen, then they will do so because there is essentially an unlimited supply of oxygen available to participate in such reactions. Many materials, both natural and manufactured, will react with oxygen in this way. Some, such as phosphorus, will ignite spontaneously at room temperature, whereas others have to be heated in a finely divided state to a high temperature before the reaction will proceed rapidly. Iron is among these. Materials such as wood and coal must be heated before the reaction will start, but once started sufficient heat is generated to maintain the reaction. In general, **when a material reacts with oxygen and in doing so liberates heat and possibly light, the reaction is said to be a combustion reaction.**

In all cases where combustion occurs, the initial material is converted into products which contain oxygen, combined chemically with the original material. For example, in the case of phosphorus the reaction is:

$$2P + \frac{5}{2}O_2 = P_2O_5 + \text{heat}, \tag{5.1}$$

and in the case of more complex compounds such as butane, the reaction will be:

$$C_4H_{10} + \frac{13}{2} O_2 = 4CO_2 + 5H_2O + \text{heat.} \tag{5.2}$$

Because of the importance of such oxidation reactions as (5.1) and (5.2) in everyday industrial processing, it is useful to have some means of estimating the heat generated in reactions. In some instances it is harnessed as a source of energy, as when oil and coal are burnt in a furnace for example. In other instances it is unwanted and some means of cooling may have to be provided to control temperature as in the addition of scrap to a steel-making furnace.

5.2 CALCULATING ENERGY CONTENT

Whenever a chemical reaction occurs there is a change in the free energy of the reactants and products. In such instances as the reduction of iron oxide to metallic iron, the change requires a supply of energy to enable the reaction to occur. Under these conditions, the products are in a higher energy state than the reactants and the reaction is said to be **endothermic**. Conversely, for reactions where the products are in a lower energy state than the reactants, the energy difference must have been given out during the reaction, usually in the form of heat. Such reactions are said to be **exothermic** and all combustion reactions are of this type.

The only unique value for the energy change between reactants and products is the change in the free energy of the materials. However, the definition of free energy change requires that the reaction be carried out in such a way that it is reversible. This means that for a reaction of the type:

$$A + B = C + D, \tag{5.3}$$

the reverse reaction,

$$C + D = A + B, \tag{5.4}$$

should be just as likely. In the case of practical combustion reactions such as those occurring in a furnace, these conditions do not exist because the products are dispersed almost as soon as they are formed. Thermodynamic calculations can be performed to evaluate the energy changes under such non-reversible conditions, but the calculations are extremely involved. However, it is possible to calculate another parameter which in most cases differs from the free energy by only a small amount. This is the **enthalpy** or **heat content**. The precise definition of enthalpy and the difference between enthalpy and free energy requires a more detailed knowledge of thermodynamics than is developed here and so the reader is referred to one of the standard textbooks on the subject (for example References [46] and [97]). For a brief discussion of the implications of measuring enthalpy rather than free energy see Reference [69].

Provided that the chemical composition of a material is known, it is possible to calculate the heat energy released during a combustion reaction using published thermodynamic data. It is interesting to note that although thermodynamic values are usually calculated from combustion reactions, the data are used for a variety of theoretical calculations of which the reaction of a compound with oxygen is only one. In an attempt to provide data in a more generally usable form, numerical values are usually tabulated as **enthalpies of formation**. For example, the enthalpy of formation of methane will be the energy needed to produce methane from the constituent atoms in the free state, that is:

$$C + 4H = CH_4 + \text{(enthalpy of formation)}. \qquad (5.5)$$

It is assumed that such a reaction will be carried out under standard conditions, which by international agreement have been defined as a temperature of 298.15K (25°C) and a pressure of 101.325 kPa (1 atmosphere). Enthalpies of formation calculated under such conditions are referred to as **standard enthalpies of formation** and are denoted symbolically by $\Delta H_f^\ominus$ where the plimsoll symbol ($\ominus$) means standard conditions. It should be noted that a number of different standard states are still to be found in use in published data (for example Reference [69]) and these give rise to small variations in the values in different sets of tables.

Calculations are usually based on written chemical equations such as Equation (5.5). Because the atomic mass of carbon is 12 and that of hydrogen is 1, Equation (5.5) tells us that 12g of carbon will react with 4g of hydrogen to produce 16g of methane. To simplify the calculations, it is usual to define a new unit of mass for use in such calculations. This is the **mole** and is **the molecular mass of the element or compound expressed in grammes**. Thus 12g of carbon is equal to 1 mole of carbon and 4g of hydrogen is equal to 4 moles of hydrogen. Using this new mass system, Equation (5.5) tells us that 1 mole of carbon with react with 4 moles of hydrogen to produce 1 mole of methane. The corresponding enthalpy change is then expressed as energy per mole and for methane this is 0.0784 MJ/mol.

Table 5.1 gives the standard enthalpies of formation of a number of inorganic oxides and Table 5.2 gives the standard enthalpies of formation of a number of organic compounds. More comprehensive lists can be found in Reference [128]. Note that the values in both Tables are prefixed by either a positive or a negative sign. This sign is important because it determines which way heat will flow when the formation reaction is carried out. A positive sign means that heat energy must be supplied to effect the reaction (that is the reaction is endothermic) and a negative sign means that energy is released by the reaction (that is exothermic). The enthalpies of formation of molecules such as O_2, H_2 are defined as zero.

Table 5.1

Standard enthalpies of formation of some inorganic oxides

Substance	State	$\Delta H_f^{\ominus}$ MJ/mole	Molecular mass	$\Delta H_f^{\ominus}$ MJ/kg
Ag_2O	solid	−0.031	231.7	−0.134
Al_2O_3	,,	−1.674	102.0	−16.412
Au_2O	,,	+0.081	441.9	+0.183
B_2O_3	,,	−1.273	69.6	−18.290
BaO	,,	−0.554	153.3	−3.614
BeO	,,	−0.610	25.0	−24.400
CO	gas	−0.111	28.0	−3.964
CO_2	,,	−0.394	44.0	−8.954
CaO	solid	−0.635	56.1	−11.319
Cr_2O_3	,,	−1.140	152.0	−7.500
Cs_2O	,,	−0.318	281.8	−1.128
Cu_2O	,,	−0.170	143.1	−1.188
CuO	,,	−0.156	79.6	−1.960
FeO	,,	−0.272	71.9	−3.783
Fe_2O_3	,,	−0.821	159.7	−5.141
Fe_3O_4	,,	−1.118	231.6	−4.827
H_2O	liquid	−0.286	18.0	−15.889
HgO	solid	−0.091	216.6	−0.420
K_2O	,,	−0.363	94.2	−3.854
K_2O_2	,,	−0.496	110.2	−4.501
KO_2	,,	−0.283	71.1	−3.980
Li_2O	,,	−0.599	29.9	−20.033
Li_2O_2	,,	−0.633	45.9	−13.791
MgO	,,	−0.602	40.3	−14.938
N_2O	gas	+0.082	44.0	−1.864
NO	,,	+0.090	30.0	+3.000
N_2O_3	,,	+0.083	76.0	+1.092
NO_2	,,	+0.033	46.0	+0.717
N_2O_4	,,	+0.009	92.0	+0.098
Na_2O	solid	−0.418	62.0	−6.742
Na_2O_2	,,	−0.513	78.0	−6.577
NiO	,,	−0.241	74.7	−3.226
P_4O_{10}	,,	−2.984	283.9	−10.511
PbO	,,	−0.219	223.2	−0.981
Pb_3O_4	,,	−0.734	685.6	−1.071
PbO_2	,,	−0.270	239.2	−1.129

Table 5.1 – *continued*

Substance	State	$\Delta H_f^\ominus$ MJ/mole	Molecular mass	$\Delta H_f^\ominus$ MJ/kg
Rb_2O	,,	−0.330	186.9	−1.766
SO_2	gas	−0.297	64.1	−4.633
SO_3	,,	−0.396	80.1	−4.944
SiO_2(quartz)	solid	−0.911	60.1	−15.158
SnO	,,	−0.286	134.7	−2.123
SnO_2	,,	−0.581	150.7	−3.855
SrO	,,	−0.592	103.6	−5.714
ZnO	,,	−0.348	81.4	−4.275

Table 5.2

Standard enthalpies of formation of some carbon compounds

Name	Empirical formula	State	$\Delta H_f^\ominus$ MJ/mole	Molecular mass	$\Delta H_f^\ominus$ MJ/kg
acetaldehyde	C_2H_4O	g	−0.166	44.1	−3.764
acetic acid	$C_2H_4O_2$	1	−0.484	60.1	−8.053
acetone	C_3H_6O	1	−0.248	58.1	−4.270
acetylene	C_2H_2	g	+0.052	26.0	+2.000
benzene	C_6H_6	1	+0.049	78.1	+0.627
butane	C_4H_{10}	g	−0.126	58.1	−2.167
carbon tetrachloride	CCl_4	1	−0.133	153.8	−0.865
cyclohexane	C_6H_{12}	1	−0.156	84.2	−1.853
1,2 dichloroethane	$C_2H_4Cl_2$	g	−0.130	96.0	−1.354
diethylether	$C_4H_{10}O$	g	−0.252	74.1	−3.401
dimethylamine	C_2H_7N	g	−0.019	45.1	−0.421
dimethylether	C_2H_6O	g	−0.184	46.1	−3.991
ethane	C_2H_6	g	−0.085	30.1	−2.824
ethanol	C_2H_6O	1	−0.277	46.1	−6.009
ethyl acetate	$C_4H_8O_2$	1	−0.479	88.1	−5.437
ethylene	C_2H_4	g	+0.052	28.1	+1.851
formaldehyde	CH_2O	g	−0.116	30.0	−3.867
methane	CH_4	g	−0.078	16.1	−4.845
methanol	CH_4O	1	−0.239	32.1	−7.445
propane	C_3H_8	g	−0.104	44.1	−2.358
propylene	C_3H_6	g	+0.020	42.1	+0.475
toluene	C_7H_8	g	+0.050	92.2	+0.542
vinyl chloride	C_2H_3Cl	g	+0.035	62.5	+0.560
vinylidene chloride	$C_2H_2Cl_2$	g	+0.001	96.9	+0.010

Because there is usually only a small difference between enthalpy and free energy, the values of $\Delta H_f^\ominus$ given in Table 5.2 in MJ/kg also represent the approximate minimum thermodynamic energy requirement for the production of these compounds from their elements. However, production of these compounds is usually carried out starting with naturally occurring raw materials possessing at least a part of the desired final structure so that the thermodynamic energy requirement in practice will be less than that indicated by Table 5.2.

Since the products of combustion of the elements will be oxides, energies of combustion can be calculated directly from Table 5.1. For example, when iron is burnt in air the reaction will be:

$$2\text{Fe} + \frac{3}{2}\text{O}_2 = \text{Fe}_2\text{O}_3 - 0.8210 \text{ MJ/mol.} \tag{5.6}$$

The atomic mass of iron is 55.85 and Equation (5.6) shows that 2 moles of iron (that is 2 × 55.85g) react with oxygen to give 1 mol of iron oxide. The energy liberated is 0.8210 MJ. Hence when 1 kg of iron is burnt in air, the energy release will be:

$$\frac{0.8210 \times 1000}{2 \times 55.85} = 7.35 \text{ MJ},$$

and this is the origin of the value that has been used in the examples so far.

In a similar way, the energies associated with the combustion of organic compounds can be calculated using Table 5.2, although the procedure is somewhat more lengthy. The technique relies on a discovery by the nineteenth century chemist, Hess. He found that if a reaction can be carried out in a number of steps, then the enthalpy change of the overall reaction is the algebraic sum of the enthalpy changes of the individual steps. Suppose that we wish to calculate the enthalpy change of a reaction occurring between two compounds A and B which follows the overall equation:

$$2A + B = A_2B + \Delta H_1^\ominus. \tag{5.7}$$

Suppose also that this overall reaction can be produced by the two steps:

$$A + A = A_2 + \Delta H_2^\ominus \tag{5.8}$$

$$A_2 + B = A_2B + \Delta H_3^\ominus. \tag{5.9}$$

By adding together Equations (5.8) and (5.9), we would end up with the overall required reaction shown by Equation (5.7) but we should now have:

$$\Delta H_1^{\ominus} = \Delta H_2^{\ominus} + \Delta H_3^{\ominus}. \tag{5.10}$$

Hence if $\Delta H_2^{\ominus}$ and $\Delta H_3^{\ominus}$ are known, then $\Delta H_1^{\ominus}$ can be calculated.

Applying these ideas to the combustion of methane, for example, the overall combustion reaction is shown by Equation (5.11):

$$CH_4 + 2O_2 = CO_2 + 2H_2O + \Delta H^{\ominus}. \tag{5.11}$$

From Tables 5.1 and 5.2, we can write down the following formation equations where the enthalpies of formation are in MJ/mol.

$$C + 4H = CH_4 \quad -0.075, \tag{5.12}$$

$$O + O = O_2 \quad +0.000, \tag{5.13}$$

$$C + 2O = CO_2 \quad -0.394, \tag{5.14}$$

$$2H + O = H_2O \quad -0.286. \tag{5.15}$$

Inserting the appropriate multipliers and rearranging Equations (5.12) to (5.15) gives Equations (5.16) to (5.19), which when added together give Equation (5.20).

$$CH_4 = C + 4H \quad +0.075, \tag{5.16}$$

$$2O_2 = 2O + 2O \quad +0.000, \tag{5.17}$$

$$C + 2O = CO_2 \quad -0.394, \tag{5.18}$$

$$4H + 2O = 2H_2O \quad -0.572, \tag{5.19}$$

$$CH_4 + 2O_2 = CO_2 + 2H_2O \quad -0.891. \tag{5.20}$$

Equation (5.20) therefore tells us that for every mole (that is 16.1g) of methane that is burnt in oxygen, there will be an energy release of 0.891 MJ. Hence when 1 kg of methane is burnt, the energy release will be:

$$\frac{0.891 \times 1000}{16.1} = 55.34 \text{ MJ}.$$

The calculated enthalpy changes for the combustion in oxygen of 1 kg of a selected number of the compounds listed in Tables 5.1 and 5.2 are shown in

Table 5.3. Under these conditions the values in Table 5.3 represent **the maximum energy that can be derived from the combustion reaction** and they are known as the **gross calorific values** of the compounds. It may be surprising to discover from Table 5.3 that a large number of metals (for example silicon and magnesium) appear suitable for use as energy sources because they release significant quantities of heat during combustion.

Table 5.3

Calculated standard enthalpy changes during combustion in oxygen of selected materials – based on the data of Tables 5.1 and 5.2

Material	Gross calorific value MJ/kg	Material	Gross calorific value MJ/kg
acetaldehyde	27.07	ethylene	50.25
acetic acid	14.58	formaldehyde	18.80
acetone	30.84	hydrogen	143.00
acetylene	43.30	iron	7.35
aluminium	31.00	lead	1.06
barium	4.03	magnesium	24.77
benzene	41.88	mercury	0.45
boron	58.94	methane	55.34
butane	49.56	methanol	22.65
calcium	15.83	nickel	4.10
carbon (graphite)	32.83	nitrogen (to NO_2)	2.36
carbon tetrachloride	2.56	phosphorus	24.09
chromium	10.96	propane	50.38
copper	2.45	propylene	48.93
cyclohexane	46.60	silicon	32.43
1,2-dichloroethane	12.81	silver	0.14
diethylether	37.16	sodium (to Na_2O)	9.09
dimethylamine	61.44	sulphur	9.26
dimethylether	31.71	tin (to SnO)	2.41
ethane	51.86	toluene	42.86
ethanol	29.70	vinyl chloride	20.03
ethyl acetate	25.44	vinylidene chloride	11.09

It is however important to remember that few metals occur in the elemental state. Instead they are found in combination with other elements such as oxygen and sulphur. Considerable energy has therefore to be expended in converting these compounds to the metallic element so that far more energy is used in producing the metal from the ore than is ever realisable by combustion of the finished metal. It would not therefore be economic to produce metals with the express intention of using them later for combustion. However, during melting operations reaction between hot metal and oxygen may well occur. The energy released during such reactions contributes to the overall heat balance of the melting vessel, as in the oxidation of iron in steelmaking mentioned earlier.

Despite the ease with which the combustion energy values can be calculated for methane and other materials of known composition, there remains a significant number of products which are either complex mixtures of known compounds, as for example fuel oil, or are materials of ill-defined composition such as paper. In these cases, the heat of combustion can only be sensibly determined by direct experiment.

5.3 MEASURING ENERGY CONTENT EXPERIMENTALLY

The values presented in Tables 5.1, 5.2 and 5.3 are based on experimental measurements of the energy available from the combustion of a particular material expressed in terms of a standard parameter known as the **calorific value**. Calorific value is defined as **the quantity of heat released in a special vessel, known as a calorimeter, by the combustion of a unit quantity of material under given conditions** [20]. In practice the reactants start at, and the products are maintained at, a standard reference temperature of 298.15K and the reaction is carried out at a pressure of 101.325 kPa. The precise experimental details of the procedure can be found in the appropriate British Standard [20] and adoption of these standard conditions allows comparable absolute values to be established for the energy content of most materials likely to be used. In general, for solid and liquid fuels, calorific values are determined using oxygen as the burning medium. For gaseous fuels, the burning medium is air.

Before stating the calorific values of any materials we must consider one important practical point which applies when one of the products of combustion is water, that is when the starting material contains hydrogen. At a temperature of 298.15K and a pressure of 101.325 kPa, water is a liquid. So in cooling the product gases to the standard conditions, the water must change from the gaseous state (steam) to the liquid state (water). In doing so it will give up its heat of vaporisation. Calorific values determined under such conditions, that is when the water is a liquid, are referred to as **gross calorific values** and represent **the maximum energy that can be obtained from materials containing hydrogen**.

However, in practice equipment burning hydrogen-containing material as a fuel will not be operated at a temperature at which the water will condense to liquid. Although greater heat energy could then be extracted from the fuel, the incidence of corrosion within the equipment would be increased. Usually the temperature is controlled so that the water remains in the gaseous state and is carried away with the waste flue gases, thus minimising potential corrosion problems. However, by adopting such a procedure the waste gas stream must be kept at a significantly higher temperature than that specified for the standard conditions defined above. Heat energy of the order of 2.6 MJ per kg of water will be carried away with the flue gases. To take account of this, values are often quoted for the **net calorific value** for a material representing **the heat energy obtained from a unit mass of material when the residual water is kept gaseous.**

Throughout this book we shall use the **gross** calorific value because this represents the maximum energy that can be obtained from any material. For comparative purposes, Table 5.4 gives the measured gross calorific values for a number of materials commonly used as fuels.

Table 5.4
Typical measured gross calorific values for a number of fuels and other materials of complex composition.

Material	Gross calorific value MJ/kg	Burning medium
Coal	29.5	oxygen
Coke	25.5	oxygen
Heavy fuel oil	42.6	oxygen
Diesel oil	44.8	oxygen
Light petroleum gas (LPG)	50.0	air
Lignite	17.0	oxygen
Rubber	27.5	oxygen
Garbage	6.0	oxygen

5.4 MATERIALS AS FUELS

Quite apart from the problems associated with liquid and gaseous water and any inefficient use of the energy released during a combustion reaction, other factors can reduce the level of useful energy available from combustion. Of these the two most common arise from the use of air instead of oxygen for promoting combustion and the frequently insufficient supply of air or oxygen to permit complete combustion.

Of these factors, the use of air in combustion is less efficient than the use of pure oxygen because the oxygen in air is accompanied by about four times its volume of nitrogen plus small traces of other gases. Nitrogen and these other gases do not participate in the combustion reaction to any great extent, but because of their presence some of the energy released during combustion is transferred to them in order to raise their temperature to the reaction temperature. As a result, for a given fuel input the product gases reach a lower temperature than they would if oxygen alone had been supplied, or alternatively more fuel must be burnt to achieve the same reaction temperature. In most applications, the higher the temperature of the combustion gases, the greater will be the energy that can be extracted from them. Hence the use of air in combustion is less efficient than the use of pure oxygen.

The second factor reducing the energy available from carbon-containing materials arises from incomplete combustion. This is well illustrated by considering the combustion of methane. Equation (5.21) shows the reaction for complete burning.

$$CH_4 + 2O_2 = CO_2 + 2H_2O \quad \Delta H^{\ominus} = 55.34 \text{ MJ/kg } CH_4. \qquad (5.21)$$

However, as the supply of oxygen is reduced, the burning reaction is modified to that shown in Equation (5.22), resulting in the formation of carbon monoxide rather than carbon dioxide.

$$CH_4 + \frac{3}{2} O_2 = CO + 2H_2O \quad \Delta H^{\ominus} = 37.76 \text{ MJ/kg CH}_4. \qquad (5.22)$$

If the oxygen supply is further reduced, the carbon content of the material may not be oxidised at all but will instead form soot or coke on the walls of the burning chamber. Equation (5.23) shows this reaction.

$$CH_4 + O_2 = C + 2H_2O \quad \Delta H^{\ominus} = 30.87 \text{ MJ/kg CH}_4. \qquad (5.23)$$

The enthalpy changes in Equations (5.21), (5.22) and (5.23) have been calculated from the data in Tables 5.1 and 5.2. It can be seen that the energy available from the combustion reaction decreases markedly as the oxygen supply is restricted.

Chapter 6

Fuels and fuel production

6.1 PRIMARY AND SECONDARY FUELS

At present, energy is derived from materials occurring in or on the surface of the earth, although it is arguable that at some time in the future we may tap non-materials sources such as sunlight on a large scale. Materials from which energy is derived are usually referred to as **fuels**, a word derived from the old French word 'fouaille', meaning material for burning. Despite this original meaning, its use nowadays tends to include not only materials that are burnt to release heat energy but also materials such as uranium, which produce heat by nuclear reaction, and other forms of energy such as electricity.

There have been various attempts to redefine the meaning of the word fuel to take account of this diversity and one such recent definition is 'a technologically useful source of energy' [31]. It is however important to recognise that materials like iron, which produces heat when oxidised in a steel converter, would tend to fall outside the scope of such a definition even though in such specific instances they are acting in the same way as fuels. Moreover, some materials which have potential as fuels are not so used, either for economic reasons or because of their limited availability. Wood is a good example. Although it is used as a fuel in certain areas of the world where it is in plentiful supply, its use is by no means widespread in the developed world. There must therefore be some doubt as to whether it can really be regarded as a 'technologically useful source of energy' for the purposes of this book.

Rather than attempt to provide some rigid definition of the word 'fuel', it is helpful to use it in as broad a sense as possible so that it can provide a collective name for all materials and phenomena from which energy may be drawn. If we do this then we need some means of distinguishing between the different types of fuels that are available. This need to distinguish is fairly obvious when we compare coal with iron, oil with paper or natural gas with electricity.

Two quite distinct groups of fuels can be recognised and these are usually referred to as **primary** and **secondary** fuels. To show how this distinction arises, examine Figure 6.1. This is a very much simplified diagram showing the production of different fuel types. The left hand side of the diagram shows the three major materials inputs extracted from the earth that are capable of being burnt in air to produce useful heat energy. These are the primary fuels. We can therefore define a **primary fuel** as a **naturally occurring raw material which can be used as a technologically useful source of energy without modifying its chemical structure prior to the reaction which releases the energy.**

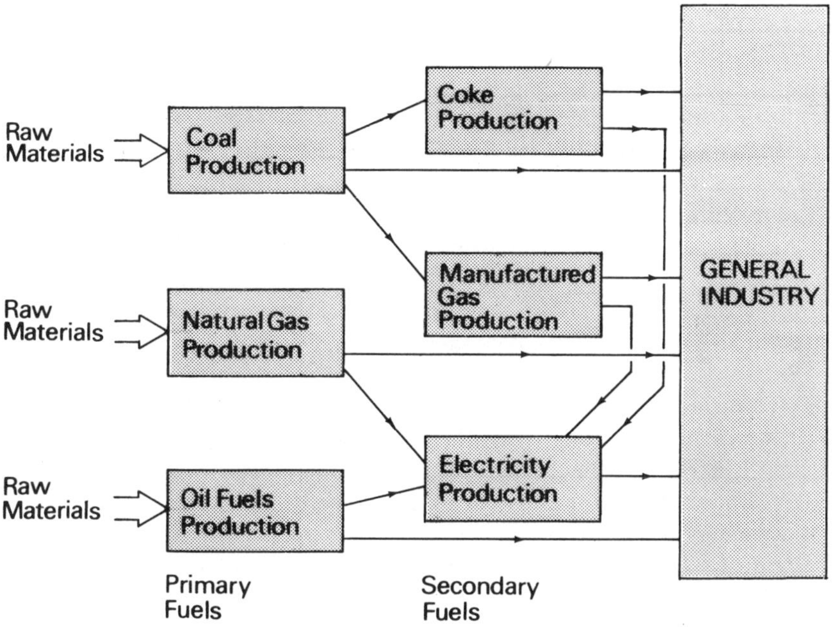

Figure 6.1 Simplified flow diagram for the fuel flows in the fuel producing industries; not all flows are shown.

The most obvious example of this type of fuel is coal. Apart from washing, removing stones and possibly crushing and grading by size, coal is used in exactly the same chemical form as it is mined. Natural gas likewise is subjected to only minor treatment to remove corrosive impurities before it can be used. In contrast, crude oil is usually subjected to extensive treatment before being used, yet it can still be regarded as a primary fuel because the treatment given in the oil refinery is essentially a physical separation of components of different molecular mass. There is little or no change in the chemical structure of these components and this is the reason for including the words 'without modifying its chemical structure' in the definition. However, because some processing and blending is carried out, some workers do tend to refer to oil fuels as secondary. This distinction is of no significance in energy analysis.

This definition of a primary fuel would also allow us to include other natural materials in the group, such as wood, camel dung and straw, which are used to provide energy in some parts of the world. It is also important to recognise that in all the examples of primary fuels mentioned so far, energy is extracted from them by burning in air. That is, the energy release originates from the change which takes place in the chemical structure of the compounds making up the material during burning. However, primary fuels do exist in which the energy release results from a change in some property other than the chemical structure. For example, uranium liberates heat energy by nuclear fission and here it is a change in the structure of the atomic nucleus that is the source of energy. Similarly, the water in a mountain reservoir can also be regarded as a primary fuel. In this case it is a physical property, the mechanical potential energy, which is the source of energy. This energy could be converted to heat directly simply by allowing the water to fall through the required distance, but the heat produced by the resulting agitation of the water molecules would be distributed throughout the mass of the water and so would be of no practical use as an energy source. Instead, it is more usual to use the kinetic energy gained when the water falls through this height to turn the rotor arm of a generator so that the energy is converted into the much more concentrated energy form, electricity.

Returning now to Figure 6.1, the three primary fuels shown take one of two routes after their extraction and treatment. The first and most direct route is the one in which they are sent straight to the block labelled 'General Industry', on the right hand side of the diagram where they are burned to release their energy content. The second route is to one of the three operations collectively referred to as secondary fuel production systems. In these operations, the energy content of the primary fuels is converted to some other form.

In the case of town gas production, the input coal is chemically modified to give a gaseous product which, for some applications, is a more convenient source of energy. In the case of electricity, the input chemical energy is transformed into another energy form which is not only easier to transport, but is also capable of performing functions such as electrolysis, that the primary chemical fuel cannot perform. Coke production is essentially the removal of most of the hydrogenous and aromatic content of the coal to produce a chemical fuel with much more closely defined burning characteristics and with greater mechanical strength for metallurgical applications.

We can therefore define a **secondary fuel** as a **source of energy which has been derived from a primary fuel**. Again, as shown in Figure 6.1, the output from the secondary fuel industries are also fed to the 'General Industry' block. One very important feature of the fuel producing industries not shown in Figure 6.1 is the interchange of fuels between the different fuel producers. Secondary fuels are used in the production of primary fuels. Coal mines, for example, use electric lights and electrically driven machinery. Primary fuels consume other primary fuels in their production; lorries and other oil consuming machines are used in the the production of coal. Moreover, any one fuel producing industry may well consume part of its own fuel output, electric lights are used in power stations.

Thus the primary and secondary fuel producing industries present an interlinked network.

When a consumer receives a primary or secondary fuel he is concerned with the maximum energy that he may potentially derive from it. He is therefore interested in the gross calorific value and, with the exception of electricity which is monitored in terms of energy delivered (for example in kWh), other fuels have calorific values as shown in Table 5.4. It must be remembered that commercial fuels are not chemically pure compounds and so the exact calorific value of a fuel type will vary from sample to sample. Where standards are laid down they usually specify a lower limit for calorific value, but the manufacturer usually produces a fuel with a calorific value considerably higher than this lower limit rather than risk rejection of his product. However, as a result of this variation in calorific values for fuels of the same type, there will always be small variations from time to time in the calculated values for the associated energy. Over a sufficiently long time period however, these variations will be smoothed out so that the average values for calorific values stated in Table 5.4 are reasonable representations of the total energy available per unit mass of fuel.

When an industrial operator uses a primary or secondary fuel, the gross calorific value represents the energy that is potentially available to him. The fact that he may not use the whole of this energy efficiently is an inherent problem of the operation, but by taking in the fuel he has essentially deprived any other user of its use. All of the fuel intake must therefore be attributed to the operation and not just that fraction which survives conversion to yield useful work. The energy associated with the consumption of fuel by a process is known as the **direct energy consumption** of the operation and is equal to the energy content of the fuel. From the above observations, it should be clear that the direct energy consumption of an operation can be decreased by increasing the efficiency with which the available energy in the fuel is converted to useful work. This idea is usually summarised in an **overall thermal efficiency** of the plant, defined as the ratio:

$$\frac{\text{Work output}}{\text{Fuel energy supplied}}$$

As a simple example, consider a steam boiler which has an overall thermal efficiency of 60%. Hence for every 100 MJ of energy directly fed in as fuel, only 60 MJ will be absorbed by the water in the boiler; the remaining 40 MJ will pass out of the boiler flue as waste heat. In this instance, the direct energy consumption will be 100 MJ and not 60 MJ. If the efficiency of the boiler were increased to 70%, then 70 MJ of every 100 MJ fed in would be absorbed by the water in the boiler but the direct energy consumption will still be 100 MJ. However, the difference between the performance of the two boilers would soon become apparent if we were to calculate the energy per unit mass of steam output from the boiler, because the more efficient boiler would produce the greater mass of steam for the same energy input.

6.2 ENERGY TO PRODUCE A FUEL

The energy content of a fuel is the energy present in the fuel that is actually available to the ultimate consumer. However, in order to produce fuels in a usable form and deliver them to those consumers, additional energy must be expended. Figure 6.2 shows the sequence of processes through which an oil fuel must pass before it reaches the final user. As discussed in Chapter 3, there will be an energy and materials input to each of these operations. Therefore the energy associated with the use of any fuel will include not only the fuel content but also the energy arising from the provision of the fuels and materials used throughout the processing sequence from extraction of the fuel from the ground through to delivery to the final consumer.

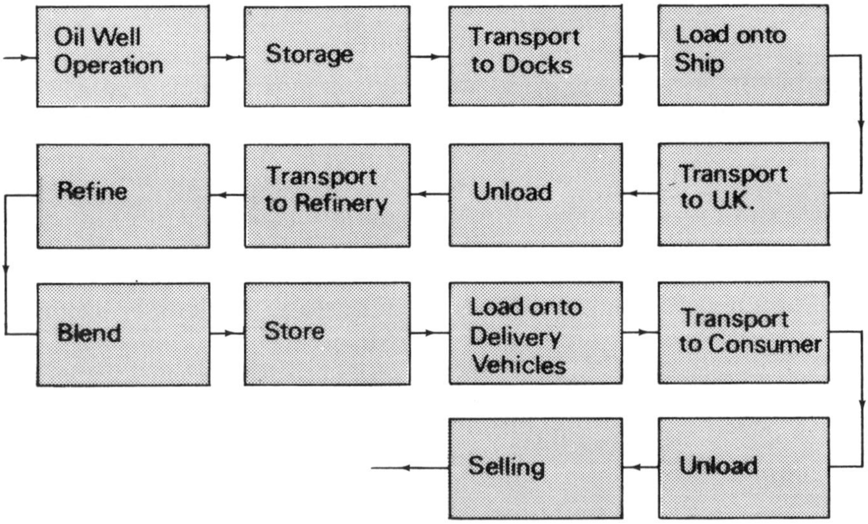

Figure 6.2 Flow diagram showing a typical sequence of operations for the production of an oil fuel from oil well to retail sale.

The energy associated with fuel processing is the first example we shall encounter of analysing industrial operations. The products of the fuel producing industries are inputs to all materials processing sequences, so their analysis is clearly of fundamental importance. We can however simplify the treatment by recognising that although some fuels are also used as materials, interest is centered on their energy content rather than their mass or volume. The use of materials other than fuel-bearing materials in the system is then confined to that necessary to provide the processing plant and machinery and usually contributes only a very small amount to the overall production energy.

So to a first approximation, a fuel producing industry can be represented by a simple system of the type shown in Figure 6.3, where all of the inputs and outputs are measured in terms of energy. The fuel input is E_i and the energy required to process it is E_p. The output energy received by the consumer is E_o and, as a consequence of the second law of thermodynamics, there will also be a waste energy output E_w. The advantage of this simple representation is that it allows the definition of an overall energy efficiency, η, for the production of a fuel as:

$$\eta = \frac{\text{Total delivered energy available to the consumer}}{\text{Total input energy}}. \quad (6.1)$$

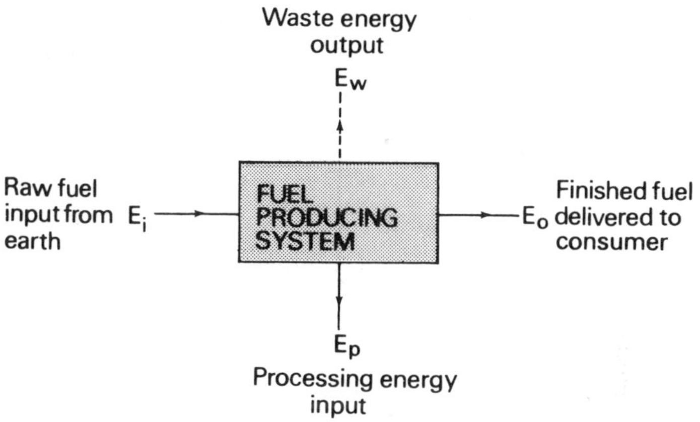

Figure 6.3 Inputs and outputs in a fuel producing system.

Using the parameters defined in Figure 6.3, this gives:

$$\eta = \frac{E_o}{E_i + E_p}. \quad (6.2)$$

The definition of an efficiency in this manner eliminates the need to consider further the parameter E_w, the waste energy, which is seldom known except by subtraction of the measureable parameters E_i, E_p and E_o. The use of an energy efficiency of production is therefore a convenient way of expressing production energies for fuels.

For a secondary fuel, the overall production system will be a two stage process. The first will be the production of the primary fuel with an energy efficiency η_p and the second will be the conversion of this primary fuel into the secondary fuel with an energy efficiency for the conversion of η_s, as shown in Figure 6.4. If the overall system has an energy efficiency η, then:

$$\eta = \eta_p . \eta_s . \tag{6.3}$$

Since both η_p and η_s will be less than unity, it follows that the overall production efficiency will always be less than either η_p or η_s alone. Hence the energy efficiency of producing a secondary fuel will *always* be less than the energy efficiency of producing the primary fuel from which it made.

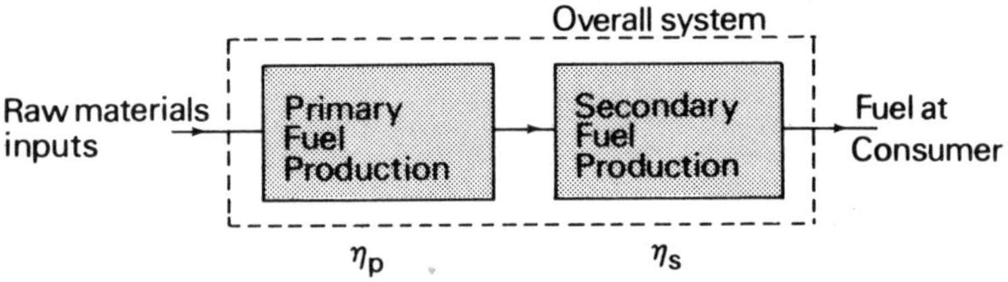

Figure 6.4 Overall system for producing a secondary fuel showing the breakdown into primary fuel production and conversion to secondary fuel with efficiencies of production of η_p and η_s as shown.

The calculation of η is central to energy analysis because all subsequent computations require some numerical value for this parameter. However its evaluation is not a simple task and a large number of values exist in the published literature. Since there is still considerable argument in some sectors of industry over which values should be adopted, it is important to understand the basis of such arguments so that appropriate choices can be made.

The main cause of difference between results arises inevitably from the choice of system describing the fuel producing process. Some are more complete than others and a few simple examples will illustrate this. For the electricity supply industry, the systems boundary may be defined as passing through the output terminals of the power station or through the point at which the consumer extracts electricity from the supply. The latter system includes the distribution network whereas the former does not. Consequently the energy efficiencies of the two systems are different, with the system which excludes the distribution network appearing the more efficient because power losses in the distribution cables are excluded. Such differences are essentially differences of definition since the two systems are strictly non-comparable. The lack of precision in defining systems makes comparison of published values difficult and the situation is exacerbated when data from different countries are compared.

A second source of difference arises from the inclusion or exclusion of the capital energy needed to set up the production facility in the first place. For example, the construction of a power station or an oil refinery requires the expenditure of considerable energy in the form of building materials before any fuel can be produced. In the same way as an accountant would subsequently spread the cost of such a plant over the estimated revenue from the output of the plant, so the energy analyst must spread the initial capital energy requirement over the expected outputs from the production facility. In some calculations this contribution is not included. Capital energy is not specifically shown in Figure 6.3 but it is one component of E_p. It has already been mentioned that it is usually only a small contributor, of the order of a few percent of the production energy, but for more complete accuracy it must be included.

A third difference between approaches is the inclusion within the system of the sales and administrative functions of the undertaking. If offices and local shops are regarded as essential adjuncts to a particular undertaking by the operators themselves, then the energy requirement of such internal functions must be set against the output of the undertaking. This again is effectively the same as including their energy requirement in the parameter E_p in Figure 6.3.

In this book we have adopted the now generally recognised criterion that the fuel supply industries should encompass all of the operations needed to sustain the industry as it exists where all inputs are raw materials in the ground and the output is fuel delivered to the consumers premises. Because of the use within these industries of materials needed in the construction, maintenance and running of the plant, such a definition means that sectors of the industries which manufacture basic inputs such as iron and steel and cement must be included within the overall fuel producing system. The complete system will therefore be similar to the extended system shown in Figure 3.6 where the box labelled 'Main Processing Operation' represents the fuel producing operation and the input will be fuel-bearing material derived from the earth. Such an energy when added to the energy content of the fuel is called the **gross energy requirement of the fuel.**

We must therefore consider how to calculate a value for the energy efficiency of production for each type of fuel. This task is somewhat simplified by the very nature of the fuel producing industries themselves. Note that this statement suggests that an energy must be calculated for a **product** rather than an energy to operate a production **system**. This apparently conflicts with the statement in Chapter 2 that the only unit for which a production energy can be calculated is a system. There are however two reasons to justify our use of a product in the case of fuels. First, from the earlier discussion we have automatically defined the system as the complete system with the fuel derived from raw materials in the earth. Moreover, the technology used within the fuel producing industries is more of less standardised; where any significant variations occur, the product fuel is usually identified in some way. For example, electricity is at present produced by thermal generation from fossil fuels, by generation from nuclear fuels and by hydro-electric generation. In each case, the electricity is labelled thermal-, nuclear- or hydro-electricity to identify its source.

The second reason why we are able to define the fuel producing system by specifying only the output is that, with the exception of small coal mines supplying the immediate locality and of on-site electricity generation facilities, the production of fuels is in the hands of large organisations which aggregate their output. A consumer of electricity cannot identify the specific power station from which his supply has come, nor can the user of an oil fuel identify the source of the crude oil or the specific fractionating column used to refine it.

Because of this aggregation of output, it is therefore sensible to calculate an average energy production efficiency for each type of fuel. Such a value will represent a description of some average production technology, which will of course vary from country to country and will also be time dependent. In the case of electricity, most large networks involving a number of power stations will be gradually updating their generation and distribution facilities and the more modern plants tend to be more energy efficient than older equipment. The average efficiency is therefore expected to improve with time because of this up-dating. At the same time, the average efficiency of electrical generation will also depend upon the mix of production techniques. The mix may include thermal-, nuclear- and hydro-electricity generation and the relative proportions from each of these sources will vary with time thus affecting the overall average efficiency. Moreover, because of the different relative proportions in different countries, the average efficiency is expected to change from country to country. Norway and Switzerland, for example, satisfy a significant proportion of their electricity requirement with hydro-electricity and so the average efficiency in these countries is expected to be considerably higher than in the United Kingdom, where most electricity is produced by thermal generation.

From the point of view of calculating a production efficiency, the physical aggregation of fuel outputs immediately allows use of industry-based and Government-based statistics, and almost all information concerning the fuel producing industries are derived from these sources.

In later Sections of this Chapter each fuel type is considered in turn, but it is useful here to consider the two most widely used methods of calculating production efficiencies. Despite the limitations mentioned in Section 4.5, **input-output analysis** provides one of the most comprehensive techniques for evaluating energy production efficiencies. This is because it takes into account the use of materials as well as fuels in the production process and, within the limitations of the tables themselves, allows an evaluation of the complete production system of the type shown in Figure 3.6. Moreover, by the very nature of the tables, it automatically includes the interchange of fuels between the different fuel producing industries, such as the use of electricity to drive coal cutting machinery which is in turn producing coal for the electricity power stations. Detailed analyses for the U.K. and the U.S.A. have been reported [23, 33].

An alternative source of data, which is usually more readily available on an annual basis, is the total inputs of materials and fuels to any fuel-producing industry together with a record of the total output of finished fuel from that industry. Using such data it is possible to calculate the average requirements of materials and fuels per unit quantity of fuel produced. In essence this is a form of **gross process analysis** and the onus is upon the analyst to ensure that all inputs are included within the calculation. An example of this type of calculation is given later for coal production.

One persistent problem encountered in the calculation of an energy production efficiency for a fuel is the manufacture by the aggregated industry of more than one product. The town-gas manufacturing industry also produces coke, and in many instances chemicals too, which both form an essential though small mass of physical output. The other industries similarly manufacture co-products. Such additional products are frequently ignored on the assumption that they represent so small a fraction of the total throughput of materials that the omission of an allowance for their production will not significantly affect the overall result. Such may well be the case for the more conventional fossil fuels, but in the nuclear power industry the ancilliary chemical processing of spent fuel rods is a significant factor in determining the viability of the process. Although input-output tables take such co-products into account, the high monetary value of some co-products, particularly chemicals, often makes them appear more significant than their actual mass warrants. This factor, coupled with the wide range of such co-products, can make their isolation from the input-output matrix difficult.

A second problem area in these calculations is the energy needed to provide the capital equipment employed and this is one of the most difficult parameters to estimate for a utility that may be operating for many decades. There are however three possible ways of evaluating this capital energy.

For a single production unit such as a coal mine, it should be possible to use historical data to calculate the energy required to produce the materials used to construct and maintain the equipment throughout the working life of the mine. If the total output from the mine during its lifetime is also known, then an energy per unit output average over the lifetime of the mine can be calculated with a fair degree of accuracy. This simple idea is appealing but is not really very satisfactory. Because the lifetime of many production facilities is measured in decades, the technological changes during the lifetime of say a single coal-mine are such that the averages become meaningless when applied to present day technology since they contain a contribution from practices which are now obsolete. Such an average would therefore underestimate current energy use because the trend has been towards more energy-intensive practices and away from the more labour-intensive practices of former years.

An alternative approach is to assume that the consumption of materials and energy do not vary significantly from year to year so that representative current values can be obtained by averaging the consumption of materials and fuels in a recent year over the saleable output in that same period. An example of this is given in Table 6.1.

Table 6.1

Estimate of the energy required to produce 1 kg of coal in the U.K. in 1975 based on data supplied by the National Coal Board

Cause of energy consumption	Energy in MJ per kg saleable coal	
	Used by collieries	Used by other activities
Coal†	0.29	0.05
Electricity††	0.54	0.03
Steel	0.24	negligible
Other materials	0.08	negligible
Other fuels	0.05	0.08
Lubricants	0.03	negligible
Totals	1.23	0.16

†Used for on-site generation of electricity.
††Assuming an energy conversion efficiency of 25%.

Because of the difficulties encountered with the estimation of capital energy, many workers adopt a third strategy and omit this element completely from their calculations. Capital plant depreciation, in energy terms, for the U.S. coal-mining industry is about 5% of the total production energy or about 0.25% of the total fuel energy (that is fuel energy content plus production energy) [88]. The omission of capital energy in this instance would therefore introduce only a very small error into the overall energy associated with this fuel and similar results are found for the production of other fuels. However it is also clear that if even a very crude estimate of capital energy is included, this small error will be significantly reduced.

Even though the system needed to describe a fuel production system completely is complex, a relatively simple representation of the calculated values may be obtained. If the total energy of a fuel is E_f, it may be regarded as the sum of two components, E_c representing the energy content of the finished fuel and E_p representing the energy needed to produce this fuel from the fuel-bearing mineral in the ground. Hence:

$$E_f = E_c + E_p. \tag{6.4}$$

As far as the consumer is concerned, E_c is the maximum energy that can be derived from the fuel. So for every unit of fuel consumed in an operation, the parameter E_c represents the **direct energy** consumed in this operation (see Section 6.1). In order that energy E_c be produced and delivered to the consumer's premises, the fuel producing industry must already have expended an energy E_p. Hence as a consequence of consuming an energy E_c, the consumer has indirectly been responsible for the consumption of an energy E_p. The energy E_p is therefore the **production energy** of the fuel.

Using the above parameters, the energy efficiency of production of a fuel, η in Equation (6.2), can be written as:

$$\eta = \frac{E_c}{E_c + E_p}, \qquad (6.5)$$

and this representation is used in the following Sections.

6.3 COAL PRODUCTION

Coal, of all the fuels used by industry, is subjected to the least processing because it is used in essentially the same form as extracted from the ground. Any treatment is aimed at reducing the amount of associated dirt, ash and sulphur, comminuting and grading according to size.

The energy required to produce coal varies significantly from one mine to another because of the different mining techniques used and the different physical characteristics of the deposits worked. Moreover, since the outputs from a number of mines are often mixed together, it is unlikely that a detailed study of a single mine would yield results generally applicable to the industry as a whole. It is therefore usual to calculate a value for the production efficiency of coal averaged over the whole industry. Data are usually obtained from trade or Government statistics.

As an example of the type of data available, Table 6.1 gives an estimate of the energy required to produce 1 kg of coal in the U.K. in 1975. The basic data have been collected by dividing the total consumption of all commodities by the total saleable output of coal during the same period. Energy requirements have been calculated using data such as those in Part 2 of this book. From Table 6.1, it is clear that the energy required to produce 1 kg of coal in 1975 was 1.39 MJ.

Because this production energy relates to 1 kg of coal and because the energy content of 1 kg of coal will vary with the type of coal as shown in Table 6.2, the energy efficiency of production as defined by Equation (6.5) will obviously vary from one type of coal to another. To show this variation Table 6.2 calculates the energy efficiency using the production data from Table 6.1. It can be seen that the value varies from 94.5% to 96.0%. The energy required in the mining and processing of 1 kg of coal with a high calorific value such as anthracite is likely to be very little different from that required for coal with a much lower calorific value such as general industrial coal. Any differences in production energies are more likely to be due to the nature of the mine than the energy content of the fuel. Therefore more energy is brought to the surface as fuel in anthracite per unit of energy expenditure than in a low grade coal.

Table 6.2

Total energy of coal fuels produced in the U.K. in 1975

Type of coal	Energy content of fuel MJ/kg	Production energy of fuel MJ/kg	Total energy of the fuel MJ/kg	Energy efficiency of production %
Untreated power station smalls	23.94	1.39	25.33	94.5
Washed power station smalls	26.73	1.39	28.12	95.1
Power station average	23.86	1.39	25.25	94.5
Coking coal	29.06	1.39	30.45	95.4
Coal for gas generation	29.16	1.39	30.55	95.5
General industrial washed smalls	27.43	1.39	28.82	95.2
General industrial graded	28.59	1.39	29.98	95.4
Domestic average	29.52	1.39	30.91	95.5
Anthracite	33.24	1.39	34.63	96.0
Dry steam coal	32.54	1.39	33.93	95.9

The values of energy content quoted in Table 6.2 are derived from data supplied by the British National Coal Board as typical values. However individual grades of coal fuel may vary substantially from time to time. For example Waller [139] quotes the average calorific value for coking coal as received by the British Steel Corporation during the period 1967–71, as 31.54 MJ/kg. Compare this with the value of 29.06 MJ/kg in Table 6.2.

It is important to note that the data given in Table 6.1 contain only a small element of the energy used in the final delivery of the coal to the consumer. Within the U.K. the National Coal Board does not perform a complete delivery service to all consumers. Apart from large consumers such as power stations, retail distribution is carried out by a network of small delivery merchants. The delivery energy of coal will obviously depend upon delivery distance. If an average delivery distance of 100 miles is assumed, this would act to decrease the production and delivery efficiency by about 1% so that the provision of coal for an average consumer would be expected to exhibit an overall energy efficiency lying between 93.5% and 95.0%.

Table 6.3
Published values for the energy efficiency of coal production

Country	Production efficiency (%)	Notes
U.S.A.	99.75	(1)
U.S.A.	99.40	(2)
U.S.A.	99.30	(3)
U.S.A.	99.10	(4)
U.K.	99.0	(5)
U.S.A.	98.6	(6), (7)
U.S.A.	98.02	(8)
Belgium	97.7	(9)
U.K.	97.0	(10)
U.K.	96.82	(11)
U.K.	96.3	(12)
U.K.	96.2	(13)
U.K.	96.0	(14)
U.K.	95.99	(15), (16)
U.K.	95.5	(6), (17)
U.K.	95.5	(15), (18)
U.K.	95.49	(15), (17)
U.K.	95.15	(19)
U.K.	95.0	(20)
U.S.A.	93.8	(21)
U.K.	83.3	(22)

Notes on Table 6.3

(1) Reported by Hannon, B. M. in *Environment,* 14, No. 2, p 11 (1975). Estimated value for power station coal covering all processes up to delivery at the power station. The exact system is not defined.

(2) Reported by Makhijani, A. B. and Lichtenberg, A. J. in *Environment,* 14, No. 5, p 10 (1972). Derived from analysis of national statistics for general industrial coal. Allowance has been made for the energy of capital plant estimated at 5% of the running energy. No allowance for transport to consumer. If delivery energy is included using the values for Note 8 below, the efficiency decreases to 98.14%.

(3) Reported by Bullard, C. W. and Herendeen, R. A. in *Energy Policy,* p 268, December 1975 for values based on the 1967 U.S. input-output tables.

(4) Reported by the U.S. Environmental Protection Agency (1974) report number EPA/530/SW-19C entitled, 'Resource and environmental process analysis of nine beverage container alternatives'. The calculation is based on the direct fuel requirements to mine 1 ton of coal of 0.001 ton coal, 0.287 gallons of oil fuels, 3.7 cu. ft. of natural gas and 10 kWh of electricity.

(5) Reported by Gartner, E. M. and Smith, M. A. in *Energy Policy,* p. 144, June 1976.

(6) Reported in 'Energy Analysis' published by the International Federation of Institutes of Advanced Study Workshop No. 6, Stockholm (1974).

(7) 1963 value.

(C'td. next page)

An alternative technique for calculating the energy efficiency of coal production using input-output tables has been applied by Chapman [27] for the 1968 production in the U.K. This represents a complete system analysis averaged over all coal types and yielded a value for energy efficiency of 95.99%. Again this excludes the final retail delivery to all but the largest consumers and it is interesting to note that it is in good agreement with the values based on industry-generated statistics.

It is useful to compare these energy efficiencies with other published values and a selection of published efficiencies are given in Table 6.3. It can be seen that the U.K. values all lie close to the values derived in Table 6.2 and such variations as do occur can probably be accounted for by variations in the different systems chosen for analysis. It is also interesting to note that the values 95.49%, 95.99% and 95.50% quoted for the U.K. industry for the years 1963, 1968 and 1971/72 show no significant movement either up or down. The other feature indicated by Table 6.3 is the generally higher production efficiency for U.S. practice. Such values do not necessarily indicate better mining performances because there is some uncertainty in the definitions of the systems analysed.

Notes on Table 6.3 *(C'td.)*

(8) Reported by Samples, D. K. in a seminar to the Institute of Sciences and Technology at the University of Michigan, 23 August 1974. The system includes extraction and all mineral processing of coking coal. Transport has been assumed to be half rail and half road over a distance of 200 miles. The energy associated with capital equipment and ancilliary functions of the industry are excluded.

(9) Report by Decker, A. to the 9th T.N.O. Conference, Rotterdam (1976).

(10) Reported by the National Economic Development Office, London, in 'Energy conservation in the United Kingdom', H.M.S.O. (1974).

(11) Reported by Smith, H. in *Transations of the World Energy Conference,* Section. E, **18**, (1969). The total production and delivery system energy is 220 kcal/kg (that is 0.92 MJ/kg); average energy content of coal taken as 28.01 MJ/kg for general industrial coal. It is not known whether this report allows for capital energy.

(12) Reported by Leach, G. and Slesser, M. in *Energy equivalent of network inputs in food production* issued by the University of Strathclyde (1973). This value applies to the complete production and distribution system and makes an allowance for capital energy.

(13) Reported by Leach, G. in *Energy and food production* published by the International Institute for Environment and Development (1975). Value assumed to be applicable for the period 1968–1972.

(14) Reported by Chapman, P. F. and Hemming, D. F. at the 9th T.N.O. Conference, Rotterdam (1976).

(15) Reported by Chapman, P. F., Leach, G. and Slesser, M. in *Energy Policy,* September 1974, p 231. Calculations are for the whole system and are based on Government statistics with some inputs from the fuel producing industries themselves.

(16) Value for 1968.

(17) Value for 1963.

(18) Value for 1971/72.

(19) Value calculated from Table 6.1 for 1975 production in the U.K.

(20) Reported in The Transport and Road Research Laboratory Supplementary Report No. 251, (1976) U.K.

(21) Reported by Maddox, K. P. in the *Mineral Industries Bulletin of the Colorado School of Mines,* **18**, July 1975.

(22) Reported by McKillop, A. in *Ecologist,* **2**, No. 12, p 1 (1972).

None-the-less such differences may indicate differing technologies in the two countries.

6.4 OIL FUELS

Oil fuels are derived predominantly from crude oil which is a complex mixture of hydrocarbons. Crude oil itself is unsuitable for use as extracted from the ground because of the very wide variations in composition that occur from one deposit to another. Instead it is usually broken down into a series of fractions or cuts, each of which boils within a defined temperature range. Although these fractions possess a defined boiling property, they are complex mixtures often containing varying proportions of many hundreds of hydrocarbons. Commercially available oil fuels are usually blends of fractions obtained from different sources and in many cases include additives such as the anti-knock lead compounds in some gasoline products.

Although the technology of oil fuels is well known [19], the calculation of a production energy for any oil fuel is complicated by a number of factors. Few countries are self-supporting in crude oil and so many must import it. Supplies of oil are seldom imported from a single source for political and economic reasons. The different compositions of crude oils from different sources are also important because composition affects the potential end-uses to some extent. Moreover, the proportions of crude oil derived from different suppliers are usually a commercial secret. It is therefore extremely difficult to calculate an average *delivery* energy because of the uncertainties in origin of the crude oil.

The oil industry in the Western industrialised nations is operated by large multi-national companies and it is by no means certain that the oil fuels imported into any country are in fact produced there from crude oil. In Europe for example, the large refining complex at Rotterdam supplies finished oil products to a number of countries. It is therefore extremely difficult to trace the precise route by which any particular oil fuel was derived from crude oil. Consequently calculations of oil production efficiencies are frequently made using assumed sources of crude oil [129] or national statistics [33].

It is however important to realise that because oil processing is an international trade, national statistics may be misleading. For example, a significant proportion of the oil fuels used in Western Europe are derived from Middle Eastern crude oil which is transported in tankers registered outside Europe. A considerable amount of the total energy required to convert crude oil to processed oil products is consumed in the shipment of the crude oil from the well-head to the refinery, yet this fuel may well be picked up by the tankers in any part of the world and so would not appear in European statistics. Similarly, as already mentioned, finished fuels are imported into the U.K. from refineries in Continental Europe and in such instances the refining energy will not appear in the U.K. statistics.

After processing, the distribution of the finished oil products is no less complex. Within the U.K. distribution is by pipeline, sea, canal, road and rail, and the distribution network for white oils such as gasoline may well be different to that for black oils such as fuel oil. The distribution energy will therefore depend not only on the type of product but also on the distance of the consumer from the refinery and the mode of transport used in the distribution. Superimposed on this are the reciprocal arrangements that oil companies may make with their 'competitors' to distribute to the more inaccessible customers.

In addition to the above complications, most oil companies maintain extensive sales and retail networks of which petrol stations are but one example. As noted earlier, the construction and maintenance of this network represents a further charge to the energy of the fuels' production and delivery.

There is a bewildering array of published data concerning the production of oil fuels and bearing in mind the variety of possible production and delivery methods, all of these data are incomplete to some extent. It is therefore extremely difficult to derive a satisfactory value for the production and delivery of oil fuels that is applicable in all circumstances.

In general, there have been two approaches to this problem; one based on a very approximate form of process analysis and one based on input-output or statistics analysis. As an example of the process analysis approach, Table 6.4 gives a typical type of calculation using information from a number of sources. This is not proposed as a definitive calculation but typifies the use of this technique. Note that the calculation is based on the *mass* throughput of oil and not upon the calorific value of the fuel. As with coal, when an energy efficiency of production and delivery is calculated according to Equation (6.5), this will vary with the calorific value of the specific fuel. Table 6.5 shows the values for the energy efficiency of production and distribution of the various oil fuels using the data in Table 6.4.

Input-output and statistical analysis of oil fuels production have been reported for the U.S. [23] and for the U.K. [33]. Quite apart from the problems generally associated with this type of analysis, there is also some uncertainty in the various oil industry operations that are included in the tables as well as difficulties in adding those production and delivery elements known to be excluded.

It is not surprising that a wide variety of values for production and delivery efficiencies of oil fuels are reported in the literature and Table 6.6 shows a selection of these values. It is worth noting that because of the physical complexity of this industry and the limited nature of much of the available data, an average energy efficiency is often quoted for all oil fuels. In view of the absence of a definitive value and the wide spread of reported values (Table 6.6), the use of an average would appear to be a sensible compromise. In this book we shall use a value of 83% for such an overall average, although alternative values can be found in the literature.

Table 6.4
Typical calculation of the production energy for oil fuels in the U.K. from Middle Eastern crude oil

Process	Energy requirement MJ/kg
Pumping from oil well†	negligible
Transport from Persian Gulf to Western Europe†	3.28
Refinery processing apportioned on a mass basis†	4.96
Average delivery energy in the U.K.††	0.23
Total energy excluding capital energy	8.47
Capital energy calculated as 5% of processing energy	0.42
Total production energy	8.89

†Sundstrom (1973)
††Boustead and Hancock (1975)

Table 6.5
Total energy of oil fuels

Oil fuel	Energy content (Typical gross calorific value) MJ/kg	Production energy MJ/kg	Total fuel energy MJ/kg	Energy efficiency of production %
Heavy fuel oil	42.60	8.89	51.49	82.7
Medium fuel oil	42.85	8.89	51.74	82.8
Light fuel oil	43.20	8.89	52.09	82.9
Gas oil	45.21	8.89	54.10	83.6
Kerosine	46.53	8.89	55.42	84.0
Diesel	44.84	8.89	53.73	83.5
LPG (propane)	50.00	8.89	58.89	84.9
LPG (butane)	49.30	8.89	58.19	84.7

It is also important to note that a number of different calorific values are often quoted for the same oil fuel. The values given in Table 6.5 are typical values based on oil company data, obtained by the authors. However, in general oil companies quote a specification for an oil fuel. This **specified gross calorific value** represents **the minimum calorific value guaranteed to the buyer of the fuel.** Invariably the product will possess a higher calorific value than the specification to ensure compliance with the specified minimum. Experimentally determined calorific values of actual finished fuels are usually referred to as **typical gross calorific values** and will be higher than the specified value. When using industrial data it is therefore essential to ascertain the nature of any quoted calorific value.

Table 6.6

Published values for the energy efficiency of production of oil fuels

Country	Production efficiency (%)	Notes
U.S.A.	97.5	(1)
U.K.	92.4	(2)
U.K.	91.6	(3)
U.K.	89.6	(4), (5)
U.S.A.	89.3	(6)
U.K.	89.2	(7)
U.K.	88.21	(4), (8)
U.K.	88.2	(9)
U.K.	88.2	(10)
U.K.	87.5	(11)
Holland	87.0	(12)
U.S.A.	84.0	(13), (14)
U.S.A.	82.8	(15)
U.K.	82.5	(16)
U.K.	82.0	(17)
U.K.	81.3	(13), (14)
U.S.A.	81.2	(18)
U.K.	80.82	(4), (14)

Notes for Table 6.6

(1) Reported by Makhijani, A. B. and Lichtenberg, A. J. in *Environment,* 14, No. 5, p 10 (1972). Average value for the production energy of fuel oil of 44.5 kWh/barrel. Efficiency calculated on a nominal energy content of 40 MJ/kg for illustration purposes only. If values in Table 6.5 for the energy contents of the different fuels are used, energy efficiency varies from 97.3% for petrol to 97.6% for heavy fuel oil. The system is from crude oil at the refinery to finished product also at the refinery and so the refining energy of 1.01 MJ/litre (that is 1.06 MJ/kg at an oil density of 950 kg/m^3) should be compared with the value of 4.96 in Table 6.4. Note that if Table 6.5 were to be recalculated using this revised energy, then the energy efficiencies would lie in the range 89.9% to 91.2%.

(2) Reported by Gartner, E. M. and Smith, M. A. in *Energy Policy,* June 1976, p 144.

(3) Reported in 'Energy conservation in the United Kingdom' by the National Economic Development Office, H.M.S.O., London (1974).

(4) Reported by Chapman, P. F., Leach, G. and Slesser, M. in *Energy Policy,* September 1974, p 231. Values are based on statistical analysis of Government data and therefore excludes delivery of crude oil to the U.K.

(5) Value for 1971/72.

(6) Reported by Maddox, K. P. in the *Mineral Industries Bulletin of the Colorado School of Mines,* 18, July 1975.

(C'td. next page)

6.5 NATURAL GAS PRODUCTION

Natural gas is the term applied to naturally occurring hydrocarbon gases suitable for recovery as fuels. Such gas deposits occur throughout the world and usually contain a preponderance of methane together with smaller quantities of gaseous hydrocarbons of higher molecular mass [19]. Many natural gas sources are suitable for use immediately upon extraction, but any processing carried out is usually of two types. First, because of variations in the calorific values of gases from different sources, in some cases the gas may be enriched or diluted by mixing with other gases to obtain a more uniform calorific value as well as reproducible burning characteristics. The second type of treatment occasionally required is the removal of sulphur-containing impurities such as hydrogen sulphide which cause excessive corrosion in the burning equipment as well as being a source of pollution.

Unambiguous data for natural gas production and distribution is somewhat scarce since its large scale distribution in many countries is in association with manufactured gas (see Section 6.7). In general it is expected that the energy involved in the production and distribution of natural gas will be considerably lower than that for manufactured gas because the latter requires the high temperature treatment of coal. Hence data which combine these two types of gas will tend to give lower average efficiencies than data for natural gas alone.

Notes on Table 6.6 *(C'td.)*

(7) Reported by Smith, H. in *Transactions of the World Energy Conference*, Section E, 18 (1969). The efficiency is based on a value of 5.18 MJ/kg to produce and deliver oil fuels. This value should be compared with the sum (4.96 + 0.23) = 5.19 MJ/kg in Table 6.4; there is close agreement after allowance for crude oil delivery and capital energy.

(8) 1968 value based on input-output analysis.

(9) Reported by Leach, G. in *Energy and food production* published by the International Institute for Environment and Development (1975). The value relates to the period 1971/72.

(10) Reported by Chapman, P. F and Hemming, D. F. at the 9th International T.N.O. Conference, Rotterdam (1976).

(11) Reported by Smith, H. in *Transactions of the World Energy Conference*, Section E, 18 (1969) as an average value for all oil fuels. See also Note (7) above.

(12) Reported by Tuininga, E. J. at the 9th International T.N.O. Conference in Rotterdam (1976).

(13) Reported in *Energy Analysis* published by the International Federation of Institutes of Advanced Study, Stockholm (1974).

(14) Value for 1963.

(15) Reported by Bullard, C. W. and Herendeen, R. A. in *Energy Policy*, p 268, December 1975. The value is based on the 1976 input-output tables for the U.S.

(16) Reported by Leach, G. and Slesser, M. in *Energy equivalent of network inputs in food production*, University of Strathclyde (1973). This includes an allowance for capital energy and for fuel loss in distribution. If capital is excluded the efficiency rises to 82.7%.

(17) Reported by McKillop, A. in *Ecologist*, 2, No. 12, p 1 (1972). It is stated that there is a 10% loss in production drilling and transport and up to 8% loss at the refinery.

(18) Reported by Samples, D. K. in a report of a seminar at the Institute of Sciences and Technology at the University of Michigan, 23 August 1974. The value is for gasoline based on the total energy requirements for U.S. production. The system is from crude oil entering the refinery up to and including retail sale. No allowance is made for capital energy.

It is possible to calculate a notional value for an overall efficiency by combining data from different sources. Leach and Slesser [84] quote an energy efficiency for the production of natural gas at the well-head of 98.0%. Assuming that there is no treatment of this gas, Makhijani and Lichtenberg [88] have considered distribution and have deduced from statistical evidence that 3% of natural gas is lost from pipelines during distribution, a further 4% is lost in vented and waste gases and the energy consumed in powering the pipelines is equivalent to a further 3.5% of natural gas production. Combining all of these factors, an energy efficiency for the production and delivery of natural gas of 87.5% as indicated.

As already mentioned, because of the use of manufactured and natural gases alongside one another, care must be taken when using published data because many references give average values for systems employing both types of fuels. Table 6.7 gives a selection of published values relating to natural gas production and delivery and it can readily be seen how the co-use problem affects the values for an overall efficiency.

Table 6.7

Published values for the energy efficiency of production of natural gas

Country	Production efficiency (%)	Notes
U.K.	98.0	(1), (2)
U.K.	95.2	(3)
U.K.	94.4	(4)
U.S.A.	93.0	(5)
U.K.	92.2	(6)
U.S.A.	90.9	(18)
U.S.A.	89.5	(7)
Average	87.5	(8)
U.S.A.	86.2	(9)
U.S.A.	85.5	(10)
U.K.	81.1	(11), (12), (13)
U.K.	76.5	(1), (14)
U.K.	71.92	(11), (13), (15)
U.K.	68.9	(14), (16)
U.K.	64.74	(11), (13), (17)

Notes on Table 6.7

(1) Reported by Leach, G. and Slesser, M. in *Energy equivalents of network inputs in food production,* University of Strathclyde (1973).

(2) Values for natural gas at the well-head.

(C'td. next page)

6.6 COKE PRODUCTION

Coke is a secondary fuel produced by the pyrolysis of coal in the absence of air. A small quantity of specialised cokes are produced using oil as the basic feedstock but this is usually employed in applications where high purity is required and the amounts are sufficiently small to be of no great significance in the calculation of an overall production efficiency. When coal or oil are heated in the absence of air, the more volatile and aromatic components are driven off or converted into an almost pure carbonaceous material with well defined burning characteristics which can be used either as a smokeless fuel or as a reducing agent in metallurgical reactions. Because of its secondary nature, the production of coke can be divided into two quite separate and distinct operations, the production of coal and the conversion of this coal to coke. The overall production efficiency will therefore be the product of the production efficiencies of these two processes as shown in Equation (6.3).

The precise value for the production energy efficiency for coke will vary with the size, age and type of coking plant and depend upon the use to which the co-products are put but values are usually available from existing coking plants, especially those within the iron and steel industry. As an example of a

---

Notes on Table 6.7 *(C'td.)*

(3) Reported by Leach, G. in *Energy and food production* published by the International Institute for Environment and Development (1975). Values relate to the period 1968/72.

(4) Reported in 'Energy conservation in the United Kingdom' by the National Economic Development Office, London, H.M.S.O. (1974).

(5) Reported by Maadah, A. G. and Maddox, R. N. in *Energy Communications,* 2, 237 (1976).

(6) Reported by Gartner, E. M. and Smith, M. A. in *Energy Policy,* p 144, June 1976.

(7) Reported by Makhijani, A. B. and Lichtenberg, A. J. in *Environment,* 14, No. 5, p 10 (1972). This reference considers the losses as 3% in pipelines, 4% as vented and waste gases and an energy equivalent of 3.5% for power to drive the pipeline system. Values are based on statistical evidence.

(8) Values derived in the text from a combination of published values.

(9) Reported in *Energy Analysis* published by the International Federation of Institutes of Advanced Study, Stockholm (1974).

(10) Reported by Maddox, K. P. in the *Mineral Industries Bulletin of the Colorado School of Mines,* 18, July 1975.

(11) Reported by Chapman, P. F., Leach, G. and Slesser, M. in *Energy Policy,* September 1974, p 231. Values are based on U.K. national statistics for the overall gas supply system, that is natural gas plus manufactured gas.

(12) Value for 1971/72.

(13) It is interesting to compare the three values of 64.74%, 71.92% and 81.10% for the years 1963, 1968 and 1971/72 since these cover the period when the U.K. gas system was turning over from manufactured gas to natural gas.

(14) Composite value for manufactured and natural gas.

(15) Value for 1968.

(16) Reported by McKillop in *Ecologist,* 2, No. 12, p 1 (1972).

(17) Value for 1963.

(18) Reported by Bullard C. W. and Herendeen, R. A. in *Energy Policy,* p 268, December 1975. Values based on the 1967 input-output tables for the U.S.

typical situation, Figure 6.5 shows the flows in a coking plant using the data of Thornton and Williams [131]. The unit produces four major outputs, all of which are usable or saleable products. These are coke, gas, breeze and by-product chemicals. A portion of the gas output is used within the coking ovens to provide the necessary heating and the consumption of energy for this purpose will largely depend upon the design of the coke ovens and the operating procedures. The input energy of the coal is the energy content of this fuel and does not contain any allowance for the production of the coal or its delivery to the coking plant.

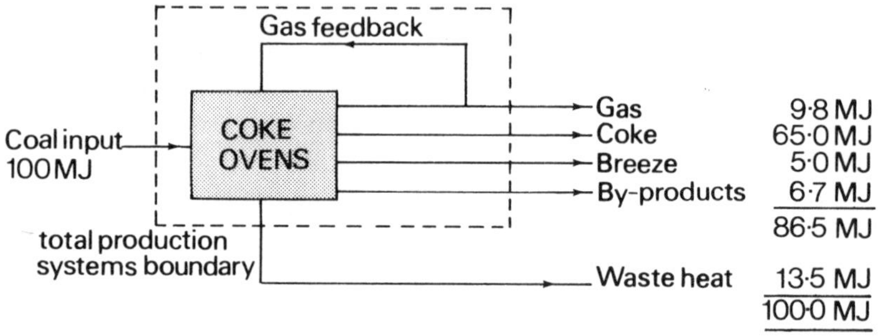

Figure 6.5 Energy flows in a typical coking plant based on the data of Thornton and Williams (1975).

Because of the incidence of four saleable products, the total input energy of the coal is not associated directly with the coke alone. The portion of the input energy attributable to coke production can be calculated on the basis of the energy content of the output coke plus a proportion of the waste energy. In Figure 6.5, if the waste energy of 13.5 MJ is apportioned on the basis of the relative magnitudes of the energy contents of the ouput products, then the waste energy associated with coke production will be 10.14 MJ. So for coke, an input energy of 75.14 MJ produces a coke energy of 65.0 MJ giving an overall energy efficiency of $65.0/75.14 = 0.8651$ or 86.5%. This efficiency contains no contribution due to the construction and maintenance of the plant nor does it include the energy required to transport the coal from the mine to the coking plant. If an allowance of 2% is made for these components then a typical coke production efficiency of 84.5% is expected.

An overall production efficiency for coke production may now be calculated by combining this value with that for coal production from Section 6.3, of 95.4%. The overall efficiency will therefore be $0.954 \times 0.845 = 0.806$ or 80.6%. It is important to recognise that this value is based on the data for a specific coking plant. Other plants may well produce different values. Barnes [6] for example quotes values which yield an overall production efficiency of 78.0% when calculated in the same way as that outlined above.

Using statistical data, Chapman, Leach and Slesser [33] have deduced a value of 75.54% for the energy efficiency of coke production in 1963 in the U.K. and have shown that this efficiency increased to a value of 88.0% in 1971/72. During this same period the iron and steel industry in the U.K. was undergoing extensive modernisation and the efficiency of coke production would be expected to rise as more efficient plant came on stream. At the same time, the supply of coke as a by-product of gas production was gradually diminishing. Since such gas plants tended to be small, old and relatively inefficient, their disappearance is reflected also in the improvement of the overall energy efficiency. The values calculated above should be compared with the selection of published values from the literature given in Table 6.8.

Table 6.8
Published values for the energy efficiency of production of coke

Country	Production efficiency (%)	Notes
U.K.	88.0	(1), (2), (3)
U.S.A.	86.48	(4)
U.K.	84.71	(1), (3), (5)
U.K.	84.7	(6)
U.K.	81.3	(7)
U.K.	81.1	(8)
U.K.	78.0	(9)
U.K.	75.54	(1), (3), (10)
U.K.	73.3	(11)
U.K.	70.7	(12)

Notes on Table 6.8
(1) Reported by Chapman, P. F., Leach, G. and Slesser, M. in *Energy Policy* September 1974, p 231. Values based on national statistics.
(2) Value for the year 1971/72 but is expected to be less accurate than the years 1963 and 1968 because of the limited information available in the statistics.
(3) Note the gradually increasing efficiency of production over the years thought to be due to the employment of more up-to-date plant.
(4) Report by Samples, D. K. in the report of a seminar at the Institute of Sciences and Technology at the University of Michigan, 23 August 1974. The system considers all processing and transport steps from coal in the ground through to the production of finished coke at the coke ovens. Coal transport has been included as half road and half rail over an average distance of 200 miles. No allowance has been made for the energy of capital plant nor for any ancilliary functions of the industry. The efficiency has been calculated from the data in the paper using the value of 29.06 MJ/kg as the energy content of coking coal.
(5) Value for 1968 based on input-output tables.
(6) Reported by Leach, G. in *Energy and Food Production* published by the International Institute for Environment and Development (1975). The value quoted is thought to be an average for the period 1960–1972.

(C'td. next page)

6.7 MANUFACTURED GAS PRODUCTION

Manufactured gas, also referred to as town-gas or city-gas, is a secondary fuel produced by heating coal in the absence of air in retorts. The process is identical with that used for the production of coke but plants designed primarily to generate gas, particularly in urban areas, tend to be smaller than those designed to provide a continuous supply of coke where gas is merely a by-product. As a consequence, these smaller plants tended to be less efficient than the more modern coke ovens. The use of manufactured gas has been largely replaced by natural gas particularly for domestic consumption, but for a considerable part of this century manufactured gas was the dominant gas fuel in many countries and this situation could return as natural gas becomes increasingly scarce.

The production energy efficiency for manufactured gas is expected to be similar to that for coke and so a typical modern value is expected to be 80.6% for the conversion from coal to gas (see Section 6.6). However, the distribution of the gaseous product is subject to a variety of losses (see Section 6.5) and these are expected to amount to 10.5%, giving an overall production and delivery energy efficiency from coal in the ground of 70.1%.

As for coke, the actual production efficiency from any specific plant will obviously vary with the detail of the plant and the method of operation. However, when manufactured gas is used on a large scale, the output from a number of production units is usually mixed in large gas-holders. This effectively means that the gas delivered to the consumer is more appropriately described by an average for the producing units involved. Values derived from company or national statistics are therefore more appropriate than those calculated for specific plants.

Table 6.9 gives published values for the energy efficiency of production of manufactured gas. These have been derived from a variety of sources, as indicated. As with natural gas data, it is often difficult to ascertain from some of the references whether the efficiency quoted truly represents an efficiency for manufactured gas alone or whether it is a value incorporating some use of natural gas.

Notes on Table 6.8 *(C'td.)*

(7) Reported by Chapman, P. F. and Hemming, D. F. at the 9th International T.N.O. Conference, Rotterdam (1976). The value is based on a coal production efficiency of 96% and a coke oven efficiency of 84.7%.

(8) Values derived in the text for the complete production system but with no allowance for capital energy.

(9) Reported by Barnes, R. S. in *Ironmaking and Steelmaking,* No. 4, p 271 (1975). The calculation is similar to that in the text but uses different data for the operation of coke ovens.

(10) Value for 1963.

(11) Reported by Waller, R. F. in 'Primary energy requirements for the production of iron and steel in the U.K.' published by the Energy Research Group of the Open University, England. Report number 11 (1976).

(12) Reported by Gartner, E. M. and Smith, M. A. in *Energy Policy,* p 144, June 1976.

Table 6.9
Published values for the energy efficiency of production of manufactured gas

Country	Production efficiency (%)	Notes
U.K.	81.1	(1), (2)
U.K.	76.5	(3)
U.K.	76.3	(4)
U.K.	73.0	(5)
U.K.	71.92	(1), (6)
U.K.	71.9	(7)
U.K.	64.74	(1), (8)

Notes on Table 6.9
(1) Reported by Chapman, P. F., Leach, G. and Slesser, M. in *Energy Policy*, September 1974, p 231.
(2) Value for 1971/72.
(3) Reported by Leach, G. and Slesser, M. in *Energy equivalent of network inputs in food production*, University of Strathclyde (1973). The value includes an allowance for capital energy, distribution losses and distribution energy. The system relates to gas production from coal in the ground through to the consumer.
(4) Reported by Leach, G. in *Energy and food production* published by the International Institute for Environment and Development (1975).
(5) Reported by Gartner, E. M. and Smith, M. A. in *Energy Policy*, p 144, June 1976.
(6) Value for 1968.
(7) Value calculated in the text for the production efficiency from coal in the ground to gas at the consumer. No allowance has been made for capital energy in the gas-making areas of production.
(8) Value for 1963.

6.8 ELECTRICITY PRODUCTION

Of all the more commonly used fuels, electricity is the most difficult to which to assign a satisfactory production efficiency. This is particularly unfortunate because electricity possesses the lowest average production efficiency of all fuels and any uncertainty in the value of production efficiency introduces a significant error into the fuel production energy requirement associated with the use of this fuel. The origin of this uncertainty in production efficiency for electricity arises from the variety of generating techniques available within any one country and the variation in practices from country to country.

We can discover the origin of these difficulties by considering the production sequence for electricity as shown in Figure 6.6. It is essentially a four stage process and if the energy production efficiency of each stage is as shown, the overall efficiency, η, will be given by:

$$\eta = \eta_1 . \eta_2 . \eta_3 . \eta_4 . \qquad (6.6)$$

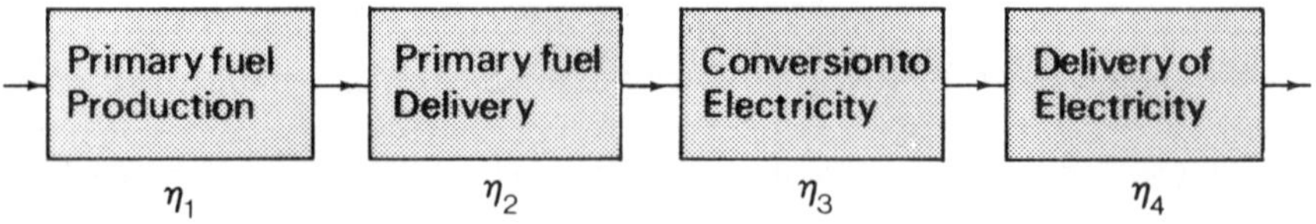

Figure 6.6 Production sequence of four operations in the production of electricity. Each sub-system possesses an energy conversion efficiency as shown.

By far the greatest proportion of all electricity used in industrial countries is derived from thermal generating techniques using fossil fuels such as coal, oil and natural gas. The values of the various parameters in Equation (6.6) can therefore be evaluated for these primary fuel inputs. The product $\eta_1\eta_2$ for coal can be derived from Table 6.2 as 0.945. Similarly, from Table 6.5, a typical value for heavy fuel oil will be 0.827 and again, from Table 6.7, a suitable value for natural gas will be 0.875.

For thermal generating sets the value of η_3 will vary with the size and age of the plant, but will typically lie in the range 30% to 40% for modern plant, though older plant probably gives values below 30%. In addition there will be an extra charge due to the supply and maintenance of machines and buildings used in the generation. Transport of the fuel to the power station amounts to about 2.5% of the the input energy [33]. The values quoted above already contain an allowance of 1% for the delivery of coal (see Section 6.3) and so an allowance of 1.5% must be attributed to the capiral energy of the plant. A similar argument can be advanced for oil and gas fuels. If therefore a capital allowance of 1.5% is included for all generating techniques, then η_3 for thermal generating plant will take a value lying in the range 28.5% to 38.5%.

The distribution of electricity from the power station to the final consumer is a two component parameter. First there are the losses from the transmission lines. These have been variously estimated as 8.5% [84], 7.5% [33] and 10% [134] for the U.K. The exact value will obviously depend upon the size and type of distribution network, but a reasonable average is thought to be 8%. The second component of η_4 is electricity used in the sales function of the utility. In the U.K., where the distribution is in the hands of a series of area boards each maintaining offices, workshops and showrooms, it has been estimated that these consume about 7.6% of the total generated electricity [33]. Although it is arguable that some proportion of this energy consumption should be attributed to the sale of electrical consumer goods, such as function can only represent a small fraction of the total consumption by the industry. If therefore we assume that 2% of the 7.6% is due to sales functions of consumer goods, then the overall loss due to transmission losses, office support and so on will be 13.6%, so that η_4 will take the value of 0.864.

The overall production and delivery efficiency for thermally generated electricity can therefore be calculated using the data described above and the values are summarised in Table 6.10. Here it can be seen that the efficiency of production and delivery of electricity might lie in the range 20.4% to 31.4%.

Table 6.10

Typical values for the production and distribution efficiency of thermally generated electricity based on U.K. data using Equation (6.6)

Fuel	$\eta_1\eta_2$	η_3		η_4	η	
		max	min		max	min
Coal	0.945	0.385	0.285	0.864	0.314	0.234
Oil	0.827	0.385	0.285	0.864	0.275	0.204
Natural gas	0.875	0.385	0.285	0.864	0.291	0.215

The average efficiency of thermal generation and distribution for any country will therefore depend upon the relative proportions of the different primary fuels used. In the U.K. in 1968 for example, about 86% of all thermally generated electricity from fossil fuels was derived from coal with the remainder derived largely from oil [33]. Hence if the values given in Table 6.10 apply to the technology for that year, the average efficiency for thermal production lies in the range 23.0% to 30.9%. Similarly in the U.S. in 1973, the proportions of primary fuels used for the thermal generation of electricity were coal 56%, gas 29% and oil 15% [66]. Again using the data of Table 6.10, the average efficiency of thermal production would then lie in the range 22.4% to 30.1%. Although such calculations make the assumption that the technologies in the two countries at two different times are similar, it is clear that the average energy efficiency of production and distribution can be changed simply by altering the mixes of primary fuels used.

It should also be noted that the calculations in Table 6.10 are based on the assumption that modern plant is employed so that thermal generation efficiency lies in the range 30% to 40% for the condensing set alone. In many countries with long established electricity supply systems this is not necessarily the case and the lower overall production efficiencies given in Table 6.10 would represent overestimates. However, as more modern and more efficient plant is installed, the average production efficiency is expected to rise.

It is also important to recognise that many large processing complexes operate their own internal generating sets when the demand is sufficiently large and stable to merit the necessary capital expenditure. In these cases, the losses due to transmission and sales functions are eliminated and higher overall production and delivery efficiencies are therefore expected.

A small but increasing proportion of thermally generated electricity is produced from steam generated by nuclear reactions. The conversion of the steam energy to electricity employs conventional generating sets so that the parameters η_3 and η_4 in Equation (6.6) will take the values already assigned. However, the production of nuclear fuels presents something of a problem because it is difficult to calculate a sensible value for the production energy of a fuel and it is also difficult to assign a sensible value to the 'calorific value' for a nuclear fuel.

The source of these difficulties lies in the very different mechanisms whereby energy is extracted from nuclear fuels compared with the extraction of energy from conventional fossil fuels. To understand their origin, we need to know something of the way that nuclear power stations operate. It is not our intention here to provide a detailed description of nuclear power, but a simple description of the types of reactor is helpful. Essentially **energy is released when the atomic nucleus divides.** This is called **fission**. Some nuclei can be readily made to undergo fission reactions. For example, if a nucleus of uranium-235 absorbs a neutron it will break up into two parts of roughly equal mass. At the same time it releases energy and further neutrons which are then available to induce further fission reactions.

The basic fuel for such a reaction, uranium-235, occurs naturally but only in a widely dispersed form. Moreover, its occurrence is not as the native metal but as a chemical compound. Hence unlike conventional fossil fuels it requires extensive chemical and metallurgical treatment before it is suitable for use in a reactor. This incurs the expenditure of considerable energy and, in common with all metals, the leaner the ore available, the greater will be the processing energy requirements (see for example Figure 1.2).

In addition to the processing energy needed to produce the initial uranium, reactors are seldom operated until all of the fissile material has been used up. Usually the reaction is permitted to proceed until the reaction rate falls below a predetermined level, at which point the fuel rods are replaced. Spent rods are then reprocessed to reclaim any unspent fuel, which is made into further fuel rods. Consequently, after a sufficiently long time all of the uranium could in principle be used up but only after undergoing repeated reprocessing. With current practices, this seldom happens because only a proportion of the recovered uranium is mixed with virgin metal. Practices may however change as uranium becomes increasingly scarce.

Reactors using fission of uranium-235 as a source of energy represent the majority of commercial reactors currently operated. Because the fissile material in these reactors is gradually consumed, they are often referred to as **burner reactors**. For such reactors, the maximum energy that may be derived from the uranium charge may be readily calculated (see Table 1.1) and so it is possible to quote the equivalent of a calorific value for such fuels. Calorific values are however seldom quoted because problems arise in the case of **breeder reactors**. In these the initial uranium charge reacts to produce heat but at the same time is converted into further isotopes which are subsequently capable of being reacted to produce still more heat. In such a breeder reactor, **a fuel is essentially being consumed to give a further fuel in addition to any energy released.** This concept is quite alien to the idea of a fossil fuel being burnt and thereby destroyed for ever. This cannot be taken into account in the concept of a calorific value.

The only satisfactory way in which the energy associated with the production of nuclear electricity can be calculated is to consider the total energy needed to produce and process the uranium required to run a power station for a sufficiently long time period. Mortimer [100] has calculated the energy required to produce electricity by burner reactors, including all processes from mining and concentrating the uranium, enriching the U-235 isotope, constructing and operating the reactor and finally reprocessing the spent fuel rods. He has shown that the energy required depends upon three main factors:

(a) the grade of the uranium available,
(b) the load factor of the generating plant, and
(c) the proportion of electricity used in the operations which is derived from fossil fuels as opposed to that derived from the output of the reactor.

The grade of ore is perhaps the most significant of these factors because this determines the scale of the mining operations needed and the extent of the mineral processing, both very energy intensive processes. The ores currently available in the U.S.A. contain 0.2% uranium oxide (U_3O_8), whereas those used in South Africa contain only 0.05% oxide.

The **load factor** is the proportion of the time that the reactor is working at full loading. For greatest efficiency, a constant high loading is essential but in the U.K. a typical load factor is 62%. This factor is of importance in fossil-fuelled electrical generation but assumes greater significance in nuclear generation because of the difficulties inherent in controlling nuclear reactions. They cannot satisfactorily be stopped immediately as can fossil-fuelled reactions.

When a country embarks upon a nuclear power programme, all of the electricity used in the initial stages of setting up will be derived from fossil-fuelled generating plant. As the nuclear power stations come on stream, an increasing proportion of the electricity used in the construction of further power stations will be derived from them. If the energy efficiency of nuclear generation is better than that for fossil-fuelled generation, then the net energy requirement of nuclear power will decrease as more stations become operational. However, if nuclear electricity is, or becomes, less efficient than electricity based on fossil fuels, by the same token this inefficiency is fed back so that succeeding stations become progressively more inefficient.

Mortimer [100] has shown that for a nuclear power station operating with a load factor L and deriving its uranium from an ore of grade G, the energy required to produce unit output of electricity is given by an empirical expression of the form:

$$\frac{1}{L}\left\{\frac{a}{G}+b\right\}+\left\{\frac{c}{G}+d\right\}, \qquad (6.6)$$

when all electricity used in the construction and so on is derived from fossil fuels; *a*, *b*, *c* and *d* are constants. For the same power station deriving its electricity totally from nuclear power, the expression is of the form:

$$\frac{(pL + q)G + (rL + s)}{(tL + u)G - (vL + w)}, \qquad (6.7)$$

where *p*, *q*, *r*, *s*, *t*, *u*, *v*, and *w* are constants. For the Magnox reactor with a load factor of 62% (that is $L = 0.62$), these expressions reduce to

$$\frac{0.0064}{G} + 0.107 \text{ (MJ per MJ of electricity)}, \qquad (6.8)$$

for electricity from fossil fuelled plant, and

$$\frac{27580G + 740}{349400G - 400} \text{ (MJ per MJ of electricity)}, \qquad (6.9)$$

for self-generated electricity. Hence with an ore grade of 0.2%, (that is $G = 0.2$) the energy required to produce 1 MJ of electricity will lie between 0.139 MJ [(6.8)] and 0.090 MJ [(6.9)], which correspond to conversion efficiencies of 87.8% and 91.7%. Compare these with the fossil-fuelled generation efficiency of 30% to 40%. However, as the uranium ore grade decreases as the richer deposits are worked out, the nuclear generation efficiency decreases so that at an ore grade of 0.01%, the efficiency will lie between 57% [(6.8)] and 75% [(6.9)] until at a grade of 0.001% nuclear generation consumes as much electricity as it produces. At lower grades, assuming no improvements in nuclear technology, nuclear power represents a net energy sink.

Clearly then nuclear electricity presents an energy efficiency for production which will vary with time, and because nuclear power is still in its infancy the efficiency of generation is likely to vary significantly in a relatively short time. This is an area which must be followed closely.

For present purposes we have assumed an electrical generation efficiency of 50% by nuclear power plants. This represents an underestimate at the time of writing but may well be an overestimate within 10 years. Nuclear electricity is usually fed into the same distribution network as electricity generated from the more conventional fossil-fuelled power stations and therefore the distribution losses are expected to be the same. From the earlier discussion, these amount to about 14% and so an everage energy efficiency for the production and delivery of nuclear electricity will be 36%, somewhat higher than the value for fossil fuel generated power.

A further source of electrical generation that is widely used in some countries is hydro-electricity. For example, Canada produced about 75% of its electrical requirement from this source in 1974 [136]. The conversion of mechanical energy to electrical energy is reported to be 80 to 90% efficient [129]. If therefore an allowance is made for the distribution losses and maintenance energies, an overall efficiency for production and distribution will be of the order of 70%, a value that is considerably higher than that for either thermally generated electricity or nuclear electricity. Moreover, since water supplies are a renewable resource, the future of electricity from this source seems to be assured.

Because of the different ways in which electricity may be generated, each with a different efficiency, it is clear that the average efficiency of electricity production and distribution in any country depends upon the amount generated by each of these techniques. Even if the technology in different countries is the same, the average will vary from this source. Table 6.11 shows the different

Table 6.11

Electrical generation in 1974 for selected countries based on United Nations data (1976). Values are percentages of total electricity generated.

Country	Thermal	Hydro	Nuclear	Average efficiency†
World	73.3	23.0	3.7	35.7
Canada	19.7	75.3	5.0	59.4
U.S.A.	78.7	15.6	5.7	32.6
Japan	79.9	17.9	2.2	33.3
Belgium	98.0	1.6	0.4	25.6
Denmark	99.9	0.1	nil	25.1
France	60.8	31.5	7.7	40.0
West Germany	90.4	5.7	3.9	28.0
Italy	69.6	28.1	2.3	37.9
U.K.	85.9	1.8	12.3	27.1
Norway	0.1	99.9	nil	69.9
U.S.S.R.	85.6	13.6	0.8	31.2
Australia	80.6	19.6	nil	33.8

†Based on a thermal efficiency of 25%, a hydro efficiency of 70% and a nuclear efficiency of 35%.

mixes of electrical generation in a number of countries and gives values for overall national efficiencies based on a thermal efficiency of 25%, a nuclear efficiency of 35% and a hydro-electricity efficiency of 70%. It is clear from Table 6.11 that as a move away from fossil-fuelled generation plant occurs, the average national efficiency will rise. The effect of this can be seen in Figure 6.7, which uses the data for the U.K. where nuclear electricity has increased from zero in the mid 1950's to 12.4% of all electricity generated in 1974. However, it is also important to remember that overlaid on these variations will be influences due to the age of plant and the extent of distribution networks and these may often exert a much more significant influence on the national average than the mix of generation methods.

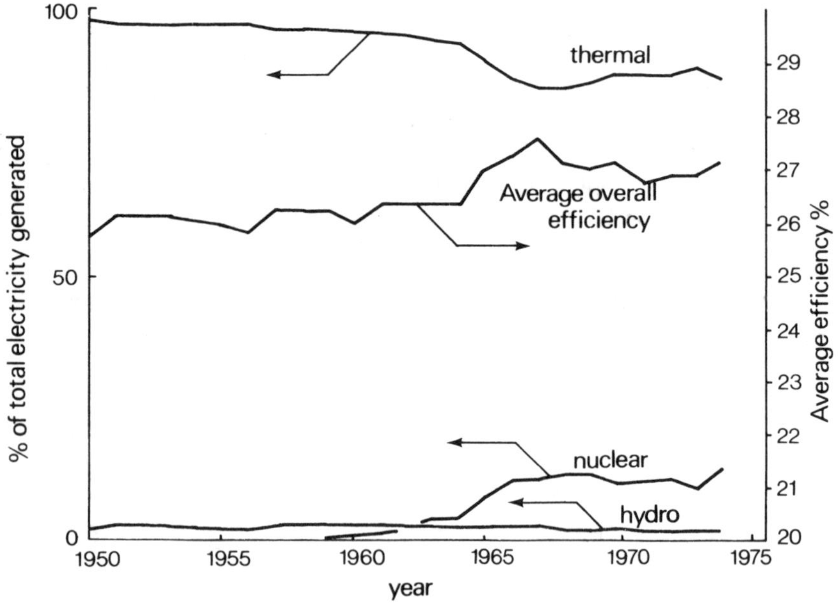

Figure 6.7 Proportions of electricity generated from thermal, nuclear and hydro-sources in the U.K. together with the resulting average production efficiency for the overall electricity supply.

A variety of energy efficiencies for the production and distribution of electricity have been reported in the literature and not only do they refer to averages calculated in different ways for different countries but they also represent values for differing systems of varying degrees of completeness. Care must therefore be taken in selecting a value for use in analysis. Table 6.12 gives a selection of published values.

Table 6.12
Published values for the energy efficiency of production of electricity

Country	Production efficiency (%)	Notes
Sweden	90	(1), (2)
U.K.	85	(2), (3)
Turkey	76.0	(2), (4)
Sweden	71.2	(1), (5)
Sweden	38.0	(6), (7)
U.S.A.	34.6	(8)
U.S.A.	33.0	(9)
Sweden	33.0	(1), (10)
U.S.A	32.5	(11), (12)
U.S.A.	32.0	(11), (13)
U.S.A.	31.5	(6)
U.K.	30.0	(14)
U.K.	30.0	(15)
U.S.A.	30	(16)
U.S.A.	30	(17)
U.S.A.	29.4	(18)
U.S.A.	29.3	(19)
U.K.	28.2	(8)
Turkey	28.0	(4), (20)
U.S.A.	28.0	(21)
U.K.	27.7	(22)
Turkey	27.0	(4), (23)
U.S.A.	26.6	(24)
U.K.	26.5	(25)
U.K.	26.5	(26)
U.K.	26.5	(27)
U.S.A.	26.3	(28)
U.K.	25.94	(29)
U.K.	25.7	(30)
U.K.	25.2	(3), (31)
Turkey	25.0	(4), (32)
U.K.	25.0	(33)
U.K.	23.9	(34)
U.K.	23.85	(35)
Turkey	16.0	(4), (36)

(C'td. next page)

Notes on Table 6.12

(1) Reported by Sundstrom, G. in 'Investigation of the energy requirements from raw materials to garbage treatment for four Swedish beer packaging alternatives', a report for Rigello Pak A.B. (1973).
(2) Value for hydro-electricity.
(3) Reported by Chapman, P. F., Leach, G. and Slesser, M. in *Energy Policy,* September 1974, p 231.
(4) Reported by Kavrak, I. in *Energy Communications,* **2**, 521 (1976).
(5) Average value for Sweden based on a mix of 67% hydro-electricity and 33% thermally generated electricity.
(6) Reported by Schipper, L. and Lichtenberg, A. J. in *Science,* **194**, 1001 (1976).
(7) Based on thermal generation for which, of the total heat energy input, 29% appears as electrical energy, 24% of the residual heat is utilised and the remaining 47% is wasted.
(8) Reported in *Energy Analysis* published by the International Federation of Institutes of Advanced Study, Stockholm (1974). Value relates to 1963 production.
(9) Reported by Atkins, P. R. in *Engineering and Mining* (1973). The system is not stated but probably refers to the generating set alone.
(10) Value for thermal generation in Sweden and probably refers to the generating set alone.
(11) Reported in *Minerals Yearbook 1971,* published by the U.S. Department of the Interior, Government Print Office, **1**, p 22 (1973).
(12) Values for fossil-fuelled steam plant 1970/71.
(13) Values for nuclear plant 1970/71.
(14) Reported by Smith, H. in *Transactions of the World Energy Conference,* Section E, **18**, (1969). This value is for the generating set alone. No allowance has been included for the capital energy.
(15) Reported by the Transport and Road Research Laboratory, Crowthorne, Berkshire, U.K. (1976) in Supplementary Report No. 251. The value is for production from coal in the ground to electricity at the consumer.
(16) Reported by McNabb, J. W. in a statement to the Illinois Pollution Control Board, Chicago, 20 February 1973.
(17) Reported by Makhijani, A. B. and Lichtenberg, A. J. in *Environment,* **14**, No. 5, p 10 (1972). The value is based on statistical information and assumes that all primary fuel is already at the power station. It includes an allowance for capital energy calculated by amortising plant over 20 years.
(18) Reported by Hannon, B. M. in *Environment,* **14**, No. 2, p 11 (1972). The system is not clearly defined but is assumed to be the whole system.
(19) Reported by Maddox, K. P. in the *Mineral Industries Bulletin of the Colorado School of Mines,* **18**, July 1975. The value is based on fossil-fuelled steam-electric plants with an average efficiency of 32% and with a further loss of 8.5% between generator and consumer.
(20) Generation from lignite.
(21) Reported by Maadah, A. G. and Maddox, R. N. in *Energy Communications,* **2**, 237 (1976).
(22) Reported by McKillop, A. in *Ecologist,* **2**, No. 12, p 1 (1972).
(23) Nuclear generation.
(24) Reported by Maddox, K. P., see Reference in Note 21.
(25) Reported by Smith, H. in *Transactions of the World Energy Conference,* Section E, **18** (1969). Value based on an energy loss of 8.397 MJ/kWh during transformation, 1.502 MJ/kWh for the provision of primary fuel, 0.0418 MJ/kWh for water and 0.0377 MJ/kWh for constructional materials for the plant.
(26) Reported by Gartner, E. M. and Smith, M. A. in *Energy Policy,* p 144, June 1976.
(27) Reported in 'Energy conservation in the United Kingdom' published by the National Economic Development Office, H.M.S.O., London (1974).
(28) Reported by Bullard, C. W. and Herendeen, R. A. in *Energy Policy,* p 268, December 1975. Based on the 1967 input-output tables for the U.S.
(29) Reported by Chapman, P. F. in *Energy Policy,* p 91, June 1974. The value refers to the total system based on statistical data.

(C'td. next page)

6.9 SUMMARY OF ENERGY PRODUCTION EFFICIENCIES

From the previous Sections, it is clear that there is a wide variation in the published values for the production and delivery efficiencies of any fuel type. Because of the very large nature of most fuel producing industries, coupled with their use of a variety of production methods each with a slightly different production efficiency, it is clear that no precise value is likely to be established for each type of fuel. Moreover, since efforts are constantly being made to improve production efficiency, either by the introduction of new techniques or by the installation of more modern machinery, values of production efficiency are expected to change with time.

In spite of these problems, suitable values for production efficiencies are needed in order to carry out any form of energy analysis and the inevitable problem is that of choosing from a range of available values. To some extent the choice is almost arbitrary since all published values have some basis in reality. If however the aim is to produce as complete an energy analysis as possible, values should be selected from those describing the most complete systems.

As a result of considering all the published data given earlier, we have chosen a set of values as indicated in Table 6.13. These are thought to represent a consistent set which, with minor variations, provides an adequate description of the gross energy requirement associated with the use of any fuel. It is these values that are used throughout the remainder of the book. As will be seen later, energy requirements calculated on the basis of Table 6.13 can be readily revised to accomodate alternative values of production efficiency.

Notes on Table 6.12 *(C'td.)*

(30) Reported by Leach, G. and Slesser, M. in *Energy equivalent of network inputs in food production*, University of Strathclyde (1973). This value is for the whole production and delivery system and is based on national statistics. An allowance has been made for capital energy and for losses in transmission.

(31) Average value for the whole of the U.K. production and distribution based on national statistics.

(32) Generation from petroleum.

(33) Reported by Leach, G. in *Energy and food production* published by the International Institute for Environment and Development, London (1975).

(34) Reported by Chapman, P. F. and Hemming, D. F. in the Proceedings of the 9th International T.N.O. Conference, Rotterdam, (1976).

(35) Reported by Chapman, P. F. in *Metals and Materials*, p 107, February 1974.

(36) Generation from coal.

Table 6.13
Typical values for the total energy associated with fuels

Fuel	Quantity	Production & delivery energy MJ	Energy content of fuel MJ	Total energy of fuel MJ	Energy production efficiency %
Coal	1 kg	1.39	28.01	29.40	95.0
Coke	1 kg	3.93	25.42	29.50	86.6
Electricity	1 kWh	11.40	3.6	15.00	24.0
Natural gas	1 therm	15.06	105.44	120.50	87.5
Manufactured gas	1 therm	41.21	105.44	146.65	71.9
Heavy fuel oil	1 kg	8.89	42.60	51.49	82.7
	1 imp gall	38.97	186.30	225.27	82.7
	1 litre	8.57	40.98	49.55	82.7
Medium fuel oil	1 kg	8.89	42.85	51.74	82.8
	1 imp gall	38.64	186.02	224.66	82.8
	1 litre	8.50	40.92	49.42	82.8
Light fuel oil	1 kg	8.89	43.20	52.09	82.9
	1 imp gall	37.68	182.66	220.34	82.9
	1 litre	8.29	40.18	48.47	82.9
Gas oil	1 kg	8.89	45.21	54.10	83.6
	1 imp gall	33.75	172.02	205.77	83.6
	1 litre	7.42	37.84	45.26	83.6
Kerosine	1 kg	8.89	46.53	55.42	84.0
	1 imp gall	31.63	166.07	197.70	84.0
	1 litre	6.96	36.53	43.49	84.0
Diesel	1 kg	8.89	44.84	53.73	83.5
	1 imp gall	33.88	171.44	205.32	83.5
	1 litre	7.45	37.71	45.16	83.5
Petrol (gasoline)	1 imp gall	31.15	163.53	194.68	84.0
	1 litre	6.85	35.97	42.82	84.0
LPG (propane)	1 kg	8.89	50.00	58.89	84.9
LPG (butane)	1 kg	8.89	49.30	58.19	84.7

Note: When oil fuels are not specified by type, an average value for the energy efficiency of production of 83% is reasonable.

Chapter 7

Representation of energy requirements

7.1 ENERGY ASSOCIATED WITH A PRODUCTION SYSTEM

The procedure for calculating the energy requirement of any industrial system is central to energy analysis. The method of analysis is quite general and can be applied to the description of *any* system. For example, we could define a large and complex system such as the U.S. iron and steel industry, or we could examine the operation of a single injection moulding machine within a factory. Both systems possess the common property of an input and an output of materials and energy but the essential difference as far as the energy analyst is concerned is one of complexity, with the former obviously being more complex than the latter. Moreover, for any chosen system, the degree of completeness may be defined in different ways ranging from simply measuring the energy directly involved in processing materials throughput, to a complete analysis which includes the energy associated with the construction and maintenance of plant and buildings (see Chapter 8). **The first essential step for the analyst is the definition of the system**; thereafter the procedure is the same for all systems.

Table 6.13 clearly shows that any system may have a variety of fuel inputs. The calculation of the system energy will therefore generate considerable detail describing the sub-division of the total energy requirement amongst the different types of fuels employed and also describing how it is split between energy actually used within the consuming plant and energy used within the fuel producing industries supplying the plant. The analyst must therefore decide whether it is of value to preserve all of the detail generated or whether it is possible to aggregate some of it into a simpler form for convenience of data handling.

7.2 REPRESENTATION OF ENERGY REQUIREMENTS

The problems associated with the detail of energy calculations are best illustrated by a simple numerical example. Figure 7.1 shows a hypothetical industrial system in which materials of mass 1 kg are processed without loss to yield products of mass 1 kg. At the same time, this system consumes the following fuels:

1 kg of heavy fuel oil,
1 kg of diesel oil,
1 kWh of electricity,
1 therm of natural gas.

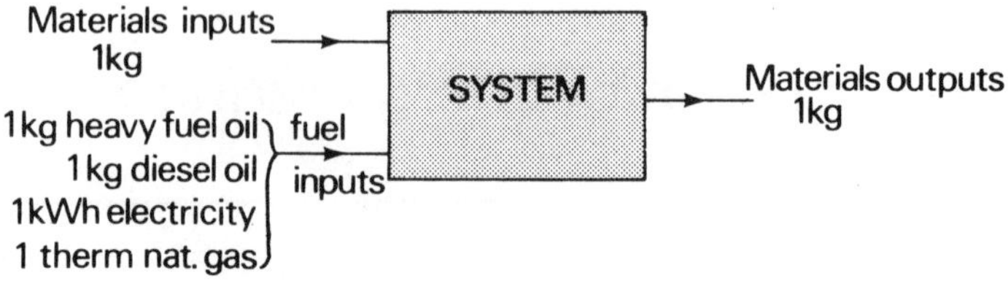

Figure 7.1 Hypothetical industrial system.

For an output of 1 kg, these fuel inputs to the system also represent the fuel requirements per unit product. They may be converted into energy requirements using Table 6.13 and this is shown in Table 7.1, which presents all of the available data concerning the energy requirements of this system. For simplicity in this example, contributions from the capital energy of the plant are assumed to be zero.

In this instance, the detailed energy requirements are described by the 15 entries in Table 7.1. However, in real systems there could be up to 14 conventional fuel inputs (see Table 6.13) resulting in a table containing 45 entries. To decide whether all of the numbers in such tables need to remain separately identified, we must consider the usefulness to the analyst of the various tabulated parameters.

Table 7.1
Energy required to produce 1 kg of output in the hypothetical system of Figure 7.1

Fuel type	Fuel production energy (MJ)	Energy content of fuel (MJ)	Total energy (MJ)
Heavy fuel oil	8.89	42.60	51.49
Diesel oil	8.89	44.84	53.73
Natural gas	15.06	105.44	120.50
Electricity	11.40	3.60	15.00
Totals/MJ	44.24	196.48	240.72
	Column 1	Column 2	Column 3

One of the most widely publicised uses of energy analysis, particularly of complete systems, is the description of the total resource requirements of particular processes. In Table 7.1, the total of Column 3 (that is 240.72 MJ/kg) is the parameter providing such a description. This represents the total energy required to operate the system and is a measure of the total energy that must be extracted from the ground to process 1 kg of materials. (Remember that capital energy has been excluded).

The total of Column 2 in Table 7.1 (that is 196.48 MJ/kg) is of immediate use to the plant operator because this represents the total fuel energy actually used within his processing plant. Irrespective of the mix of fuels, this is the energy directly consumed within the system and is therefore within his control. In the special case of a system defined to include only a specific processing operation, this is the energy requirement used to compare plant performance with similar machinery elsewhere in the factory or with that of a competitor. Moreover, if a minimum thermodynamic energy requirement for the process can be calculated, this represents the minimum value of the total of Column 2.

The entries in Column 1 are derived from those in Column 2 using the different fuel production efficiencies (Table 6.13). Because a variety of values exist for the production efficiency of any fuel, it is important to present the results of energy analysis in such a form that revised values may be readily calculated using different production efficiencies. This implies that Columns 1 and 2 in Table 7.1 be retained.

The detail in Table 7.1 is also of value when considering conservation strategies which involve fuel substitution. If the operator of the system shown in Figure 7.1 wishes to reduce the total energy requirement (240.72 MJ/kg) while continuing to operate the process essentially unchanged in terms of direct fuel energy requirement (196.48 MJ/kg), one possible way is the use of alternative fuels. From Table 7.1 it is clear that irrespective of the fuel mix employed, the energy consumed on the site is 196.48 MJ/kg of product. However, Column 3 shows that approximately half the total energy requirement of the system arises from the use of oil fuels. If therefore a more efficiently produced fuel which will perform the same function can be found, then the total energy requirement of the system can be reduced while keeping the direct energy requirement the same. Inspection of Table 6.13 shows that the energy efficiencies of production of heavy fuel oil and diesel oil are 82.7% and 83.5% respectively, whereas that for natural gas is 87.5%. Clearly then, if the oil fuels can be replaced by natural gas, the process will require a lower total energy requirement provided that the efficiency of use within the plant itself remains unchanged. Before making such a change in practice, the operator would wish to know the magnitude of the energy saving before assessing the economics of the changeover.

The energy saving arising from a change of fuel can be assessed as follows. To carry out the suggested substitution, the direct energy to be supplied by natural gas is (from Table 7.1):

$$42.60 + 44.84 + 105.44 = 192.88 \text{ MJ},$$

and from Table 6.13 this corresponds to a gas fuel consumption of:

$$\frac{192.88}{105.44} = 1.829 \text{ therms},$$

for which the fuel production energy is:

$$1.829 \times 15.06 = 27.54 \text{ MJ}.$$

The modified process will now have an energy requirement as shown in Table 7.2. The total *direct* energy remains unchanged at 196.48 MJ/kg so that the energy available to the plant remains unchanged. However the fuel production energy total as well as the total energy requirement has been reduced by 5.30 MJ/kg compared with Table 7.1. Whether such a saving merits the changeover of fuel types is now a matter of judgement for the plant manager, but the simple analysis and in particular the comparison of Tables 7.1 and 7.2 provides the information necessary for the decision to be taken on energy grounds. Whether or not this saving will be reflected in the economics of the process is another matter. Tables such as 7.1 and 7.2 are essential tools in the decision-making process. There is clearly some value in retaining a detailed breakdown of the total energy requirement by fuel type. However, situations arise where the tables become unmanageable because of the large number of fuels employed. Some method is therefore needed to reduce the number of entries whilst at the same time preserving the *essential* detail contained within the full table. This we can do by arranging the fuels into three groups which we shall refer to as **electricity**, **oil fuels** and **other fuels**.

Table 7.2

Energy required to produce 1 kg of output from the system shown in Figure 7.1 if all oil fuels are replaced by natural gas

Fuel type	Fuel production energy (MJ)	Energy content of fuels (MJ)	Total energy (MJ)
Natural gas	27.54	192.88	220.42
Electricity	11.40	3.60	15.00
Totals/MJ	38.94	196.48	235.42

Electricity needs separate identification because, with very few exceptions, it is consumed by all industrial processes to a greater or lesser extent. It is also the fuel with the lowest production efficiency, a fact which is reflected in its cost to industry and which in turn causes its use to be the subject of close scrutiny. Moreover, it is the one fuel most likely to experience a marked change in production efficiency in the foreseeable future. For these reasons, electricity is kept apart from other fuels by almost every energy analyst.

All oil fuels, that is fuel oils, gas oil, kerosine, diesel, petrol and L.P.G., are derived from crude oil by fractional distillation. Although each type is subjected to different treatments before delivery to the consumer, the over-riding factor in determining its cost and availability is the cost and availability of crude oil. Table 6.13 also shows that the energy production efficiencies of all oil fuels lie within a well defined band so that an average production efficiency of 83% may be assumed without introducing any significant error into the calculations. For these reasons, there is a strong case for grouping all oil fuels under the same heading.

Of the remaining fuels, natural gas, manufactured gas, coal and coke, the last three are closely linked since both coke and manufactured gas are predominantly produced from coal. Moreover, when used as fuels natural gas and manufactured gas are usually interchangeable. It is therefore reasonable to group these fuels together. Unlike oil fuels they have different production energy efficiencies and so, although the energies associated with their use may be aggregated in the final table, the calculation must be performed on the consumption of the individual fuel types before aggregation.

Adopting these three groups of fuel types, Table 7.1 for example can be written as shown in Table 7.3. By combining the oil fuels in this way we obtain an average production efficiency of 83.1%. If the values in Table 7.1 had been calculated using an overall average efficiency of 83% as suggested earlier, the total energy requirement would have been in error by only 0.05%, which is well below the level of any inaccuracies in the raw data likely to be supplied by any operator or derived from any statistical tables.

Table 7.3

Table 7.1 rewritten by grouping together fuels under three headings

Fuel type	Fuel production energy (MJ)	Energy content of fuels (MJ)	Total energy (MJ)
Electricity	11.40	3.60	15.20
Oil fuels	17.78	87.44	105.22
Other fuels	15.06	105.44	120.50
Totals/MJ	44.24	196.48	240.72

By adopting the above procedures and reducing the fuels into three groups, we have a means of producing energy tables in a relatively simple form. Remember however that this concise representation is derived from a more detailed breakdown and should the occasion arise when greater detail is required, the expanded form of table can be readily reconstructed. The representation is not however complete because it contains no provision for accounting for feedstock energy.

7.3 FEEDSTOCK ENERGIES

The materials inputs to any processing operation are often referred to as **feedstocks**, especially in those industries using organic materials. A proportion of the feedstock entering any system will eventually form part of the saleable product, the remainder will enter the waste products from the system. In many processes the input materials used as feedstocks can also be used as fuels, as for example oil or natural gas feedstocks for the petrochemical industry or wood in paper-making.

If therefore the calculated energy requirement of a system is to describe the total energy that must be taken from the ground in order to support the system, then it must include a contribution for any feedstock energy used. Unlike fuels which enter the system and are consumed, however, feedstock energy is not used but remains tied up in the mass of the final products. For this reason, feedstock energy appearing in the final product is often referred to as the **rolled-up feedstock energy**. In principle it is possible to reclaim that proportion of the feedstock energy present in the final product and so the feedstock energy requirement of a system is different from the fuel energy requirement. It is unsatisfactory to mix the energy which is actually consumed by a process in the form of fuels and therefore irretrievably lost, with the energy taken into the system as feedstock. This energy is potentially reclaimable, say by burning paper or plastics.

Feedstock energies can be accomodated in the energy table by adding a further column as shown in Table 7.4, where the letters A to V indicate the positions of the various energy terms. Note that the parameter K will always be zero because electricity cannot act as a feedstock.

There is clearly a connection between feedstock energy and the fuel content of the output materials from a system but a very important distinction must be made between them. The **feedstock energy is the energy content of the input materials** and may differ quite markedly from the energy content of the output materials for two reasons.

First, even with no materials losses within the system, the chemical structure of the output materials may differ from the chemical structure of the input materials. From Table 5.3, it is clear that the energy content of a material is very sensitive to its chemical structure. Therefore if the chemical structure is changed during processing, the energy that can be released from the products will be different from that which could be obtained from the input materials.

Table 7.4
Table showing the method of representing energy requirements which retains the detailed breakdown by fuel type

Fuel type	Fuel production energy (MJ)	Energy content of fuels (MJ)	Feedstock energy (MJ)	Total energy (MJ)
Electricity	A	E	K	R
Oil fuels	B	F	N	S
Other fuels	C	G	P	T
Totals/MJ	D	H	Q	V

As an example, consider the simple case of ethane being converted to ethylene and hydrogen in a cracking plant. The reaction is as shown in Equation (7.1).

$$C_2H_6 = C_2H_4 + H_2. \tag{7.1}$$

Using atomic masses for carbon and hydrogen of 12 and 1 respectively, this equation indicates that 30 kg of ethane produce 28 kg of ethylene and 2 kg of hydrogen. If this reaction is carried out without materials loss, then for every 30 kg of ethane fed into the system, the associated feedstock energy may be calculated from Table 5.3 as (30 × 51.86) = 1555.8 MJ. If both reaction products are collected as saleable products from the system, the energy content of the hydrogen is (2 × 143.0) = 286.0 MJ and the energy content of the ethylene is (28 × 50.25) = 1407.0 MJ. Therefore an input feedstock energy of 1555.8 MJ yields products with a total energy content of (1407.0 + 286.0) = 1693.0 MJ. Hence more energy may be recovered from the products than was put in as feedstock. The energy difference is that part of the processing energy fed in as fuels which is required to effect the chemical change. This energy is locked into the products by the structural change. Not all chemical processes however result in the energy content of the products being higher than those of the inputs. Indeed, in many cases it will be reduced.

The second reason why feedstock energy and the energy content of saleable products may differ arises from materials losses during processing. To illustrate this, suppose that in the ethane example above, 10% of all products is lost for one reason or another. As a result, the energy content of the recovered products will now be only 90% of 1693.0 MJ, that is 1523.7 MJ, which is of course less than the feedstock energy of 1555.9 MJ. Alternatively, the process may be operated so that only ethylene is recovered, in which case the energy content of the saleable products will be only 1407.0 MJ. This is again less than the feedstock energy.

From the above discussion, it is clear that in order to evaluate feedstock energy per unit output, we need to know the fuel content of the *input* materials and the materials conversion efficiency of the process as defined in Equation (2.9). It is misleading to write down feedstock energy based on the fuel content of the output materials.

Feedstock energies only arise when potential fuel materials are extracted from the ground for use as materials rather than fuels and the number of such materials is quite small. Those most commonly encountered are coal, oil, natural gas, wood and rubber. Since these are naturally occurring materials, their chemical composition and hence their calorific values will vary from sample to sample. Typical values are given in Table 7.5.

Table 7.5

Typical gross calorific values for some feedstocks. Because of the varying composition of these materials a range of values will be found in the literature

Feedstock	Typical gross calorific value MJ/kg
Coal	28.5
Oil	44.5
Natural gas	50.0
Wood	12.5
Rubber	27.5

A table such as Table 7.4 which has been constructed to show the *total* energy requirements of a particular system will describe the total energy that must be removed *from the earth* in the form of fuel-bearing materials in order to support a given system. The Table is assembled by considering the energy requirements of a whole series of linked sub-systems, one or more of which may represent the removal from the earth of materials which also have potential for use as fuels. Feedstock energy is only incurred when these potential fuels are removed *directly from the earth* and fed into the system as as materials inputs. Hence for systems defined so that input materials are derived from the earth, feedstock energy is entered in Table 7.4 by multiplying the appropriate calorific value in Table 7.5 by the input mass. The input mass will be the mass of feedstock required to yield unit mass of final product, that is it takes into account losses incurred in each processing sub-system.

Frequently however we need to write down an energy table for an operation which is an isolated sub-system of some larger and more complete system and for which the input materials are not derived directly from the earth even though they possess fuel energy content. In such cases, no feedstock energy should be shown in the energy table because the acquisition of the feedstock from the earth lies outside the boundary of the sub-system.

To illustrate the importance of this, consider a pair of operations A and B linked by a flow of materials as shown in Figure 7.2. Operation A draws its input materials directly from the earth and therefore possesses a feedstock energy E'_f. Operation B draws its input materials of energy content E''_f from the output materials of operation A. For simplicity assume that there are no materials losses during processing. For the overall system defined by the system boundary (1), the energy table describing the energy requirements of this system will contain a feedstock entry E'_f because the materials are derived from the earth. Similarly if we choose the system enclosed by boundary (2), the feedstock energy requirement will also be E'_f and the energy table describing this system will therefore contain a feedstock entry E'_f because these materials are derived from the earth as well.

However, the input materials to the system enclosed by boundary (3) are not derived from the earth although they possess an energy content E''_f. The energy table for system (3) does not therefore include a feedstock entry because the feedstock energy for system (1) should be calculable as the sum of the feedstock energies of systems (2) and (3). Because the feedstock energies of systems (1) and (2) are each E'_f, the feedstock energy for system (3) must be zero.

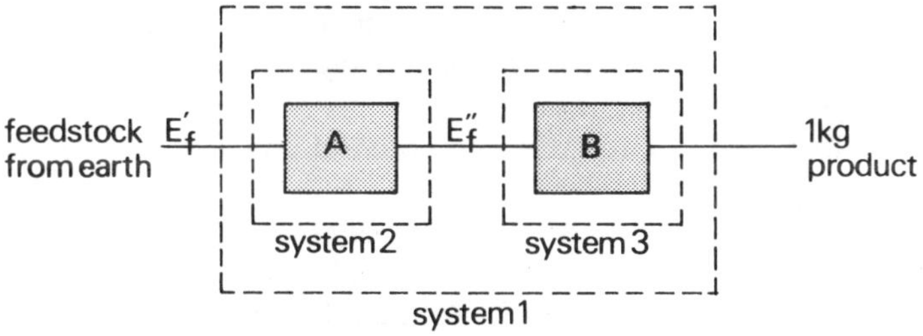

Figure 7.2 Feedstock energy requirements for a two-stage process. See the text for a detailed description of the working of this system.

It is therefore of crucial importance to recognise that feedstock energies are *only* included in the energy table when the materials enter the processing system directly from the natural state. In all subsequent stages of processing, the fuel-bearing feedstock is treated in exactly the same way as any other material. Rigid adherence to this rule prevents feedstock energies being counted more than once in complex processing systems. For this reason, the ethane example shown in Equation (7.1) and discussed in the text is incorrect because pure ethane does not occur naturally and so would not give rise to any feedstock energy in the table describing the energy requirement of the reaction. The methodology described above enables us to take account of the energy of a feedstock in a system. The feedstock energy which appears in the final summary table of the energy requirement of the system is *not* the energy which is potentially reclaimable from the products of the system because it reflects losses along the processing sequence. The energy which is available for reclamation can be obtained by multiplying the mass of product by the calorific value of the product.

For extended processing sequences starting with naturally occurring feedstock and ending with some product which possesses an energy content, a new parameter may be defined. The **net energy requirement** is the gross energy requirement of the system less the energy content of the finished product [69]. Thus net energy requirement represents **the energy that will be needed to support the operation of the system provided that the energy content of the product is eventually reclaimed.**

The need to introduce feedstock energy at the point at which the materials are extracted from the ground raises problems for those materials which in practice serve the double function of fuel and material. At the present time this is particularly important as far as oil, natural gas and coal are concerned.

Crude oil is extracted from the ground as a mixture of compounds of which one group, collectively known as naphtha, form the basic input for many petrochemical operations. Although naphtha is present in the crude oil at the well-head, it is not recognisable as such until the end of the refining stage. During all the processing from oil at the well-head up to this point, it experiences exactly the same treatment as the other components of crude oil. Because it can be identified only after refining, the petrochemical industry often specifies its materials requirements in terms of naphtha rather than in terms of crude oil. The problem that arises therefore is the conversion of a naphtha requirement into a crude oil requirement.

Naphtha is produced in an oil refinery simultaneously with other oil fuels. If therefore it is treated as a fuel during refining, its production energy efficiency will be 83% as for the other oil fuels. With a calorific value of 44.5 MJ/kg (see Table 7.5) then the production energy will be 9.11 MJ/kg. This energy will be consumed almost entirely as oil fuels which have also been produced with an efficiency of 83%. Thus of the total processing energy of 9.11 MJ/kg, 7.56 MJ/kg will be the direct consumption of fuel energy and 1.55 MJ/kg will be the fuel production energy requirement.

Using these values we can now construct Table 7.6, which represents the energy associated with the crude oil requirement needed to allow a consumption of 1 kg of naphtha by a petrochemical plant. For most purposes this Table is sufficiently accurate, but revised values can be calculated using the above procedure if more accurate data are available.

Table 7.6
Energy of crude oil requirements for a consumption of 1 kg of naphtha as feedstock

Fuel type	Fuel production energy (MJ)	Energy content of fuel (MJ)	Feedstock energy (MJ)	Total energy (MJ)
Electricity	nil	nil	nil	nil
Oil fuels	1.55	7.56	44.50	53.61
Other fuels	nil	nil	nil	nil
Totals/MJ	1.55	7.56	44.50	53.61

A similar though slightly more involved argument is needed to establish a similar table for natural gas. Although the treatment of natural gas is simpler than the treatment of crude oil to produce naphtha, distribution losses are greater. Using a feedstock energy of 50.0 MJ/kg from Table 7.5 and an energy efficiency for the production and delivery of 87.5% (from Table 6.13), the energy associated with the provision of 1 kg of natural gas feedstock will be 7.14 MJ/kg. This energy arises from the extraction energy, power for the pipelines and gas losses during distribution. Using the proportions indicated in Section 6.5, it can be shown that 1.14 MJ/kg arise during extraction (assumed to be oil fuel), 2.0 MJ/kg arise from powering the pipelines (assumed to be electricity) and 4.0 MJ/kg arise from gas losses. Hence using the appropriate production efficiencies from Table 6.13, the energy associated with the use of 1 kg of natural gas feedstock will be as shown in Table 7.7.

Table 7.7
Energy arising from the use of 1 kg of natural gas feedstock

Fuel type	Fuel production energy (MJ)	Energy content of fuel (MJ)	Feedstock energy (MJ)	Total energy (MJ)
Electricity	1.50	0.50	nil	2.00
Oil fuels	0.19	0.95	nil	1.14
Other fuels	0.50	3.50	50.00	54.00
Totals/MJ	2.19	4.95	50.00	57.14

The use of coal as a feedstock requires that we distinguish between an intake of virgin coal and an intake of coal tar. For virgin coal, the data from Section 6.3 can be used to give Table 7.8. For coal tar, which is essentially a by-product in the manufacture of coke or town-gas, the data in Sections 6.3 and 6.6 are needed to give the values shown in Table 7.9. In both cases the feedstock energy is derived from Table 7.5.

For materials such as wood and rubber there is no common processing route and so each must be calculated separately from the initial operation involved in their winning.

Table 7.8

Energy arising from the use of 1 kg of coal feedstock

Fuel type	Fuel production energy (MJ)	Energy content of fuel (MJ)	Feedstock energy (MJ)	Total energy (MJ)
Electricity	0.43	0.14	nil	0.57
Oil fuels	0.13	0.64	nil	0.77
Other fuels	0.02	0.32	28.50	28.84
Totals/MJ	0.58	1.10	28.50	30.18

Table 7.9

Energy arising from the use of 1 kg of coal-tar feedstock

Fuel type	Fuel production energy (MJ)	Energy content of fuel (MJ)	Feedstock energy (MJ)	Total energy (MJ)
Electricity	0.43	0.14	nil	0.57
Oil fuels	0.13	0.64	nil	0.77
Other fuels	0.22	3.97	28.50	32.69
Totals/MJ	0.78	4.75	28.50	34.03

7.4 CALCULATING ENERGY REQUIREMENTS

Tables such as Table 7.4 provide a convenient means of representing calculated energy requirements obtained from analysis of industrial systems of varying size and complexity. It is important to understand the way in which values are calculated for insertion in the Table as well as appreciating the significance of each of the entries. This is because most other representations of energy requirements may be regarded as simplifications of this Table.

The methodology can best be demonstrated by examining some simple examples. At the same time the way in which available data may limit the choice of system can also be highlighted.

7.4.1 Limestone quarrying

Limestone is widely used in blast furnace operations, agriculture, glass making and road building and so is an important raw material. It occurs naturally in a wide variety of different locations and is usually produced for commercial use by quarrying, crushing, screening and drying. Usually all operations are carried out on the same site and so it is relatively easy to define a system for producing limestone products from the naturally occurring rock by enclosing within the systems boundary all of the site operations, plus the fuel producing and delivery industries.

Approaches to a commercial limestone quarry are likely to yield not only information on the fuel consumption of the machines directly involved in handling the product but also on that used by such on-site facilities as the site office. Typically the following information might be supplied.

The quarry annually produces 300,000 tonne of crushed, graded and dried limestone and the fuels consumed by the whole site are 150,000 imperial gallons of medium fuel oil, 35,000 imperial gallons of gas oil, 4,000 imperial gallons of diesel oil and 2.5 million kWh of electricity. These values do not include the fuel used by the lorries delivering the limestone from the quarry to the user.

From this information we can immediately identify the system described:

System input materials

Limestone in the ground.

System output materials

Saleable limestone products.

Operations included in the system

(1) Energy used by all limestone handling operations on site.
(2) Energy used by all administrative and other non-producing functions of the site.
(3) Energy directly consumed in on-site maintenance of machinery, and so on.
(4) Energy associated with the production of fuels (using Table 6.13).

Operations excluded from the system

(1) Energy associated with the provision of miscellaneous materials such as lubricants.
(2) Energy needed to manufacture the machinery and buildings used on the site.
(3) Energy to deliver the products to the consumer.

Having defined the system we can now calculate the energy requirement. Working in tonnes rather than kilogrammes to avoid having to manipulate small numbers, the fuel requirements per tonne of limestone are:

medium fuel oil = 0.5000 imp. gall.,
gas oil = 0.117 imp. gall.,
diesel oil = 0.013 imp. gall.,
electricity = 8.333 kWh.

The energy content and the fuel production energies associated with each of these fuel consumptions can now be calculated using Table 6.13. Hence,

Fuel production and delivery energy

Medium fuel oil	= 0.500 X 38.64 = 19.32 MJ,
gas oil	= 0.117 X 33.75 = 3.95 MJ,
diesel oil	= 0.013 X 33.88 = 0.44 MJ,
electricity	= 8.333 X 11.40 = 95.00 MJ.

Energy content of fuels

Medium fuel oil	= 0.500 X 186.02 = 93.01 MJ,
gas oil	= 0.117 X 172.02 = 20.13 MJ,
diesel oil	= 0.013 X 171.44 = 2.23 MJ,
electricity	= 8.333 X 3.60 = 30.00 MJ.

The values for the three oil fuels may now be combined to give:
Oil fuels production and delivery energy = 23.71 MJ,
energy content of oil fuels consumed = 115.37 MJ.

These values, together with those for electricity, can now be used to give Table 7.10, which shows that the total energy requirement of 264.08 MJ per tonne of limestone produced is due to approximately equal contributions from electricity and oil fuels even though electricity supplies only about one fifth of the direct energy consumption of the process. This is a good example of the way in which the relatively low conversion and distribution efficiency of electricity leads to a high total energy requirement.

The energy requirement calculated in Table 7.10 represents an average system energy that can be associated with the production of all limestone products manufactured on the site. The original data do not allow any more detailed analysis and if these were the only data available, Table 7.10 represents the best results that can be obtained for this site. However, for many purposes this analysis may be too crude. Suppose for example that the limestone quarry produced four products used for road-making, iron production, glass-making and agricultural use. From the data provided so far, the energy required to manufacture each of these products would have to be assigned the same value even though the different products may have undergone different treatments.

Table 7.10
Energy required to produce 1 tonne of limestone. See text for details of the system involved

Fuel type	Fuel production energy (MJ)	Energy content of fuels (MJ)	Feedstock energy (MJ)	Total energy (MJ)
Electricity	95.00	30.00	nil	125.00
Oil fuels	23.71	115.37	nil	139.08
Other fuels	nil	nil	nil	nil
Totals/MJ	118.71	145.37	nil	264.08

More accurate energy requirements for these individual products would only be obtained by a more detailed analysis of the site operation. Suppose that the site operator were able to provide a more detailed breakdown of fuel consumption so that the flow diagram of Figure 7.3 could be constructed. This shows that all rock quarried is fed to a series of crushing and screening operations and the different products arise in different parts of the processing sequence. The limestone products supplied to both glass-making and agricultural use are dried before despatch. The system also includes the site office. The Table in Figure 7.3 shows the way in which the different fuels are consumed by the different operations. Using this more detailed breakdown of fuel consumption, the energies associated with each of the products can be readily calculated.

Consider first the production of limestone for road-making. All of the quarried rock is fed to a crusher and is then screened. The road-making limestone is only one third of the materials flow in this section of the production sequence and so only one third of the fuel consumption of these operations can be reasonably attributed to the road-making rock. The total consumption of fuels is 47,000 gallons of medium fuel oil and 4,000 gallons of diesel oil, so that the consumption attributable to limestone for road-making will be $47{,}000/3 = 15{,}667$ gallons of medium fuel oil and 1333 gallons of diesel oil. In addition a proportion of the fuel consumed by the site office should also be attributable to this product and the most sensible method of apportionment is on the basis of the proportion of the total output mass that road-making limestone constitutes. Hence there will be an electricity contribution of $0.005 \times 10^6/3 = 1{,}667$ kWh and a fuel oil contribution of 1,000 gallons. These fuel consumptions are responsible for producing 100,000 tonne of road-making limestone and so the consumption per tonne of this product will be:

0.158 gall. medium fuel oil,
0.013 gall. diesel oil,
0.016 kWh electricity.

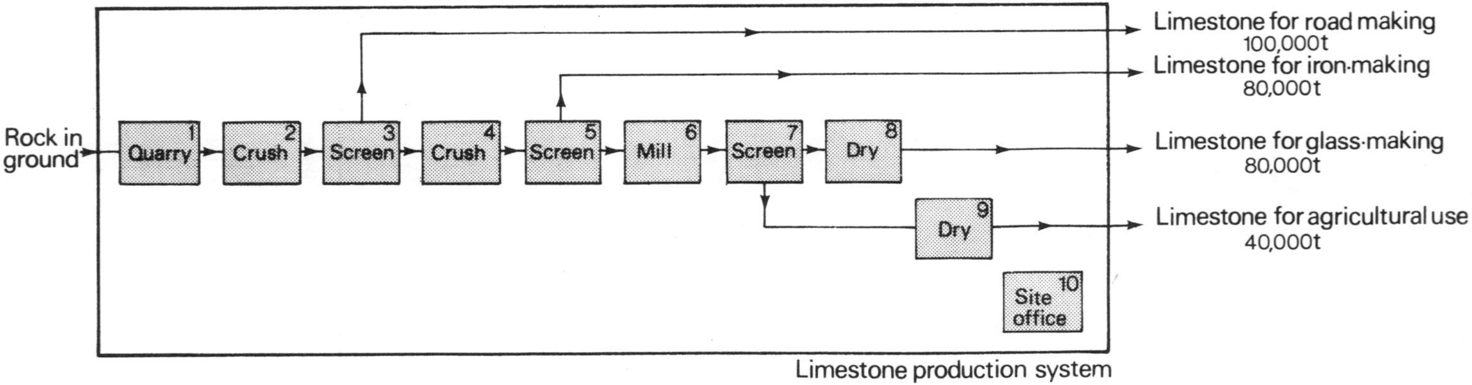

47,000	nil	nil	nil	nil	nil	nil	100,000	nil	3,000	Medium fuel oil gall
nil	nil	nil	nil	nil	nil	nil	nil	35,000	nil	Gas oil gall
2,000	1,800	200	nil	nil	nil	nil	nil	nil	nil	Diesel gall
nil	nil	nil	1·0	0·1	1·2	0·1	0·05	0·045	0·005	Electricity $\times 10^6$ kWh

Figure 7.3 Detailed flow diagram showing typical operations in the manufacture of limestone products together with the associated fuel requirements.

Using Table 6.13, these may be converted to units of energy and hence Table 7.11 may be constructed. It is now clear that the total energy requirement of 40.45 MJ/tonne is considerably lower than the 264.08 MJ/tonne predicted for the average of all limestone products by Table 7.10.

Table 7.11

Energy required to produce 1 tonne of road-making limestone using the system shown in Figure 7.3

Fuel type	Fuel production energy (MJ)	Energy content of fuels (MJ)	Feedstock energy (MJ)	Total energy (MJ)
Electricity	0.18	0.06	nil	0.24
Oil fuels	6.91	33.30	nil	40.21
Other fuels	nil	nil	nil	nil
Totals/MJ	7.09	33.36	nil	40.45

Now consider the production of limestone for iron-making. This passes through operations 1, 2 and 3 in Figure 7.3 in the same way as the road-making limestone and it will also attract a contribution from the operation of the site office. So Table 7.11 will represent the energy required to process 1 tonne of iron-making limestone through operations 1, 2, 3 and 10. In addition however, this limestone product also passes through processes 4 and 5 in Figure 7.3 in company with both glass-making and agricultural limestone. Hence a proportion of the energy used by these sub-systems must be attributed to the processing of the limestone used for iron-making. Of the total throughput of 200,000 tonne for these operations, iron-making limestone represents 80,000 tonne or 40%. The total fuel requirement for operations 4 and 5 is 1.1×10^6 kWh of electricity and so that attributable to all iron-making limestone will be $0.4 \times 1.1 \times 10^6 = 0.44 \times 10^6$ kWh or 5.5 kWh/tonne. Using Table 6.13 to convert this fuel consumption to units of energy and combining with Table 7.11, the energy required to produce iron-making limestone will be as shown in Table 7.12. Again the total energy required to manufacture this product is less than the average value given by the gross analysis of Table 7.10.

Table 7.12

Energy required to produce 1 tonne of iron-making limestone using the system shown in Figure 7.3

Fuel type	Fuel production energy (MJ)	Energy content of fuels (MJ)	Feedstock energy (MJ)	Total energy (MJ)
Electricity	62.88	19.86	nil	82.74
Oli fuels	6.91	33.30	nil	40.21
Other fuels	nil	nil	nil	nil
Totals/MJ	69.79	53.16	nil	122.95

This partitioning procedure may now be continued so that the fuel consumed by operations 6 and 7 will be divided between glass-making limestone and agricultural limestone. The fuel consumption of operation 8 will however be attributable only to glass-making limestone, and that from operation 9 to agricultural limestone. In the same manner as before, Tables 7.13 and 7.14 may be constructed. The results of this analysis shows that the energy requirements for both glass-making and agricultural limestones are considerably higher than the average value given by Table 7.10.

Table 7.13

Energy required to produce 1 tonne of limestone for glass-making using the system shown in Figure 7.3

Fuel type	Fuel production energy (MJ)	Energy content of fuels (MJ)	Feedstock energy (MJ)	Total energy (MJ)
Electricity	193.52	61.12	nil	254.64
Oil fuels	55.21	265.83	nil	321.04
Other fuels	nil	nil	nil	nil
Totals/MJ	248.73	326.95	nil	575.68

Table 7.14
Energy required to produce 1 tonne of limestone for agricultural use using the system in Figure 7.3

Fuel type	Fuel production energy (MJ)	Energy content of fuels (MJ)	Feedstock energy (MJ)	Total energy (MJ)
Electricity	199.22	62.92	nil	262.14
Oil fuels	36.44	183.82	nil	220.26
Other fuels	nil	nil	nil	nil
Totals/MJ	235.66	246.74	nil	482.40

It should now be clear why it is always preferable to perform the most detailed analysis allowed by the available data. In this instance the gross analysis leading to Table 7.10, whilst simple to perform, leads to a serious overestimate for two of the products and a significant underestimate for the remaining two products. In fact Table 7.10 can be derived as the average of Tables 7.11 and 7.14, weighted by the relative mass output of the different products. The shortage of sufficiently detailed information may prevent detailed analysis in all cases, but it is well to be aware of the errors likely to arise from oversimplified gross analyses. This is a common problem in much published information.

In this example, none of the materials involved in the production possess an energy content. Consequently there are no feedstock entries in Tables 7.10 to 7.14.

7.4.2 Naphtha cracking

Naphtha, one of the fractions from crude oil, is the starting material for many petrochemical products. As delivered from the oil refinery it is a complex mixture of hydrocarbons and therefore quite unsuitable for immediate use in the chemical industry. It must first be converted into a simpler mixture of well defined hydrocarbons which can be separated by fractional distillation. The process by which this is accomplished is known as **cracking**. In its simplest form, naphtha is mixed with steam, heated to a high temperature for a short time in the absence of air and then rapidly quenched. In this way the complex naphtha mixture is reduced typically to about half a dozen main products plus small amounts of heavier hydrocarbons. These products are easily separated by fractional distillation.

Cracking plants are frequently isolated units within petrochemical complexes and information concerning their performance is often available in an unambiguous form. Typically the sort of information that might be available to the energy analyst is:

For every tonne of naphtha processed, the cracking plant consumes 42 imperial gallons of medium fuel oil and 8000 kWh of electricity. From an input of 1 tonne of naphtha the product yield is 0.50 tonne of ethylene, 0.40 tonne of propylene and 0.10 tonne of mixed butenes.†

On the basis of this information we can describe the system as follows:

Materials inputs

Naphtha.

Materials outputs

Ethylene, propylene and mixed butenes.

Operations included within the system

(1) Energy used by all processing operations.
(2) Energy associated with the production of all fuels used in the processing operations (using Table 6.13).
(3) Energy used in lighting the plant.

Operations excluded from the system

(1) Energy associated with the production of water needed in steam raising.
(2) Energy associated with the provision of any ancilliary materials such as quench oil, lubricants, catalysts, and so on.
(3) Energy required to construct the plant.

It is not evident from the information supplied whether the fuel consumption includes an allowance for the maintenance of the plant. This is the sort of uncertainty that would have to be checked with the supplier of the information.

There are multiple products in this example, and because the output mass is equal to the input mass, all input naphtha may be assumed converted to saleable products. Suppose therefore that we wish to calculate the energy associated with the production of ethylene. Although 1 kg of naphtha only produces 0.50 kg of ethylene, the rest of the products are usable. Hence we can assume that of the input naphtha, only 0.50 kg goes to ethylene production. Apportioning fuel requirements on the basis of relative mass throughput, the fuels needed for the production of 1 kg of ethylene will be 0.042 imp. gall. of medium fuel oil and 0.8 kWh of electricity. Using Table 6.13, these fuel requirements can be converted to energy units giving Table 7.15 as the energy requirement for ethylene production. Following the arguments of Section 7.3, Table 7.15 does not include any entry in the feedstock column because naphtha is not a naturally occurring material. If however we redefine the system so that it now describes the production of ethylene from crude oil in the ground, then the energy requirement is obtained by adding together the corresponding elements of Tables 7.6 and 7.15 to give Table 7.16.

A comparison of Tables 7.15 and 7.16 should now underline the need to

†These are simplified data based on a specific plant. Energy requirements based on more detailed information are to be found in Part 2.

define a system not only in terms of the output materials but in terms of the input materials and the function of the systen as well. Both of the systems described above have an output of 1 kg of ethylene. However because the systems have different inputs and different functions they are not comparable operations and this is reflected in the very different energy requirements shown by Tables 7.15 and 7.16.

We may extend this example further to illustrate the effect of materials losses on system energies. If the information supplied by the plant operator included the statement:

There is a 5% loss of all materials inputs to the cracking plant,

then an input of 1/0.95 = 1.053 kg of naphtha would be needed to produce 1 kg of ethylene. Since the processing energy is related to input naphtha, this means that all of the elements in Tables 7.15 and 7.16 will be increased by a factor of 1.053.

Table 7.15

Energy required to produce 1 kg of ethylene from naphtha. See text for details of the system involved

Fuel type	Fuel production energy (MJ)	Energy content of fuels (MJ)	Feedstock energy (MJ)	Total energy (MJ)
Electricity	9.12	2.88	nil	12.00
Oil fuels	1.62	7.81	nil	9.43
Other fuels	nil	nil	nil	nil
Totals/MJ	10.74	10.69	nil	21.43

Table 7.16

Energy required to produce 1 kg of ethylene from crude oil

Fuel type	Fuel production energy (MJ)	Energy content of fuels (MJ)	Feedstock energy (MJ)	Total energy (MJ)
Electricity	9.12	2.88	nil	12.00
Oil fuels	3.17	14.74	44.50	62.41
Other fuels	nil	nil	nil	nil
Totals/MJ	12.29	17.62	44.50	74.41

Chapter 8

Describing real industrial systems

8.1 ENERGY ANALYSIS IN PRACTICE

The concepts and ideas outlined so far are sufficient to allow the definition of any system used to perform a specified function and then to allow energy analysis of that system. The job of the energy analyst can therefore be considered as consisting of three main steps:

(1) **definition of the system,**

(2) **energy analysis of the system, and**

(3) **presentation of the results.**

Each of these steps is, in principle, straightforward but difficulties are often encountered in applying the general ideas to real situations. It is therefore useful to examine the types of problems likely to arise in practice because much of the confusion arising in the interpretation of published energy analyses owes a great deal to the different ways in which different workers cope with these problems.

8.1.1 Definition of the system

If energy analysis is being carried out in response to a question, it is important to ensure that the question is sufficiently specific to allow construction of the appropriate system. Questions, and particularly those from industry, are usually expressed in terms of products but this alone is inadequate. It is not possible to calculate an energy requirement for steel unless the starting materials are specified and, in this example, unless the specific route is stated. Even when apparently adequate questions are formulated, some care is necessary. For example, the production of steel from scrap using an electric arc furnace is certainly a reasonable way of describing a process route but the analyst needs to know whether the scrap is at the furnace or at its point of origin. He needs to know whether the steel is liquid steel at the furnace or at the casting shop, or whether it is solid steel cast into some special form. Without this information the precise sub-systems to be included within the system boundary remain uncertain.

In the same way, he must also discover whether or not to consider the energy needed to produce and deliver other materials and services, to provide the capital plant and buildings needed to perform the operation and so on. It is therefore of great importance that the system eventually chosen for analysis matches the requirements of the questioner.

It has been proposed that the only systems for which satisfactory energy analysis can be performed are complete systems, in the sense discussed in Chapter 3 [69]. This proposal however ignores the different uses to which energy analysis may be put and so totally complete systems must be regarded as a special case of analysis. In general *any* system can be defined and analysed.

Once the requirements of the questioner are ascertained, it is the job of the analyst to choose the system boundary in such a way that it includes all the relevant sub-systems of interest. In general, these will be selected from existing physical operations.

The only limitation on the choice of sub-systems to be included is the availability of quantitative data to describe them. Whilst in principle numerical data should be available for all sub-systems, it soon becomes clear that data for many systems are not available, because they are either totally unknown or are confidential. In such cases the most sensible approach is to estimate the energy requirements of the sub-system approximately and if it appears that its contribution to the overall system energy requirement is negligible then the system should be defined to exclude that sub-system while making it clear that it has been excluded. This practice is commonly used in relation to capital energy requirement, as discussed later.

8.1.2 Calculating energy requirement

The **energy requirement** of any system is the energy content of the fuels that cross the system boundary plus the energy content of any feedstock materials that pass into the system **directly from the earth**. This should be regarded as an inviolable rule and much of the confusion that arises in the interpretation of the results of energy analysis is due to uncertainties in the definition of the system boundary and the rigidity, or otherwise, with which this rule is applied.

As an example of the way problems arise, let us consider the production of liquid steel at the furnace from scrap at the furnace using an electric arc furnace. Suppose that the system boundary has been defined as in Figure 8.1 such that it excludes the production of electricity. The fuel energy requirement of this system with the mass flows as shown is 0.45 kWh of electricity or 1.62 MJ. In this instance none of the input materials are fuel-bearing materials entering the system directly from the earth and so the electricity consumption is the only energy input to the system. Moreover, despite the arguments of Chapter 6 that the fuel processing industries, and particularly the electricity producing industries, are significant consumers of energy, it is strictly improper to construct Table 8.1 for the system of Figure 8.1 and state that the energy requirement is 6.75 MJ. This is because the description includes a sub-system which has been specifically excluded from Figure 8.1. In order that Table 8.1 be applicable as the system energy requirement, the system would need to be defined as in

Figure 8.2 so that the electricity producing industry is included within the system boundary.

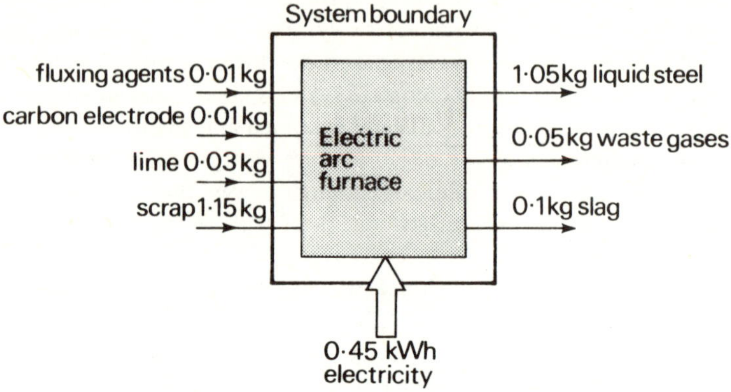

Figure 8.1 Simple system showing the production of liquid steel from steel scrap using the electric arc furnace. Note that the electricity production industry has been excluded from the system.

Table 8.1

Energy requirement for the system shown in Figure 8.2 to produce 1.05 kg of liquid steel from scrap using an electric arc furnace

Fuel type	Fuel production energy (MJ)	Energy content of fuels (MJ)	Feedstock energy (MJ)	Total energy (MJ)
Electricity	5.13	1.62	nil	6.75
Oil fuels	nil	nil	nil	nil
Other fuels	nil	nil	nil	nil
Totals/MJ	5.13	1.62	nil	6.75

However, because the fuel producing industries are significant energy consumers and because representations such as Table 8.1 readily allow identification of the energy contributions arising from them, it is usual to use such representations even when the system is defined as in Figure 8.1. Conversely, if the energy requirements are stated as in Table 8.1, there is no need to add blocks representing the fuel producing industries in system diagrams because they are automatically taken into account within the Table. This procedure simplifies the systems diagrams.

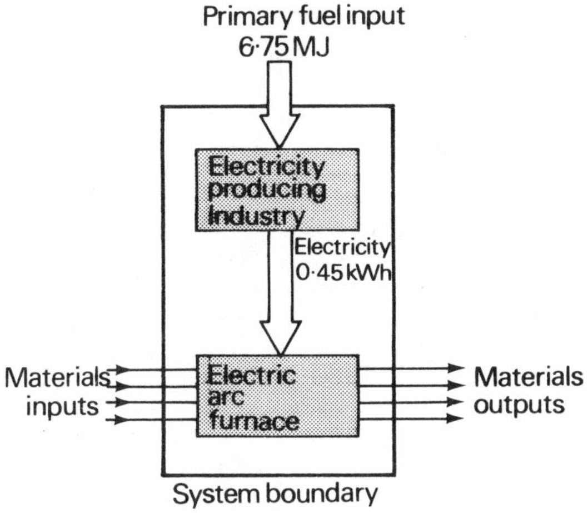

Figure 8.2 The simple system of Figure 8.1 extended to include the electricity producing sub-system.

It is important to recognise that energy requirements such as 1.62 MJ for Figure 8.1 and 6.75 MJ for Figure 8.2 are the energy requirements of the relevant *system*. They are *not* the energy requirements of the product. In both instances, it is clear that the same mass of saleable output arises from very different systems and it would be nonsense to express two quite different values of energy requirement for the same product manufactured by the same process.

It is also clear from Figures 8.1 and 8.2 that the system energy requirement will depend upon the throughput of materials. This dependence can be eliminated by normalising the system energy requirement with respect to some suitable flow parameter. The most obvious parameter to choose in this instance is the mass of steel produced, in which case the normalised system energy requirement for Figure 8.1 is 1.54 MJ/kg of steel and that for Figure 8.2 is as shown in Table 8.2. These are still *system* energy requirements and *not* product energy requirements because the choice of normalising parameter is arbitrary; it is simply an expedient for eliminating the dependence of energy on throughput.

In systems from which there is a single product, the system energy requirement when normalised with respect to the output of this product may be reasonably regarded as the energy required to manufacture that product using the defined system. Such is true of Table 8.2. This is as close as it is possible to come to answering questions asked in terms of products.

In practice, however, few systems of interest produce only a single product. More usually multiple products as well as waste form the outputs and then the aim is to sub-divide the identifiable sub-systems in such a way that a series of apparently independent systems is formed, each producing a single product.

Table 8.2

Energy requirement for the system shown in Figure 8.2 normalised with respect to the mass of the steel output

Fuel type	Fuel production energy (MJ)	Energy content of fuels (MJ)	Feedstock energy (MJ)	Total energy (MJ)
Electricity	4.89	1.54	nil	6.43
Oil fuels	nil	nil	nil	nil
Other fuels	nil	nil	nil	nil
Totals/MJ	4.89	1.54	nil	6.43

This was essentially the technique used in the limestone example of Figure 7.3, where four outputs exist. The derivation of Tables 7.11 to 7.14 required a sub-division of the physical system in Figure 7.3 into four separate systems *A*, *B*, *C* and *D* such as those shown in Figure 8.3. When a complex system is sub-divided in this way, each of the systems (*A*, *B*, *C* and *D* in Figure 8.3) may be analysed as separate systems and when the system energy requirements are normalised with respect to the system outputs, each can reasonably be stated as the energy required to manufacture the system product using the system as defined.

It is however important to remember that when complex systems are sub-divided in this way, the resulting single-product systems are only *apparently* independent. This is easily demonstrated in Figures 7.3 and 8.3. With the exception of sub-systems 8 and 9 in Figure 7.3, all other sub-systems contribute to more than one of the apparently independent systems *A*, *B*, *C* and *D* in Figure 8.3. Sub-system 1 for example contributes sub-systems *A*1, *B*1, *C*1 and *D*1 in Figure 8.3, and so on. Now the way that the fuel consumption of sub-system 1 is partitioned between sub-systems *A*1, *B*1, *C*1 and *D*1 in Figure 8.3 depends upon the relative magnitude of some suitably chosen flow parameter so that the energy requirement of sub-system *A*1 is influenced by flows in systems *B*, *C* and *D*. Hence system *A* cannot be regarded as totally independent of systems *B*, *C* and *D* and so the energy required to produce limestone for road-making depends, in this instance, upon the level of production of each of the other three limestone products. There is no way in which this interdependence can be eliminated and so it is important to be aware of it and to know how fuels have been apportioned between the different systems. The choice of parameter used in this apportionment is of crucial importance and is discussed more fully in Chapter 10.

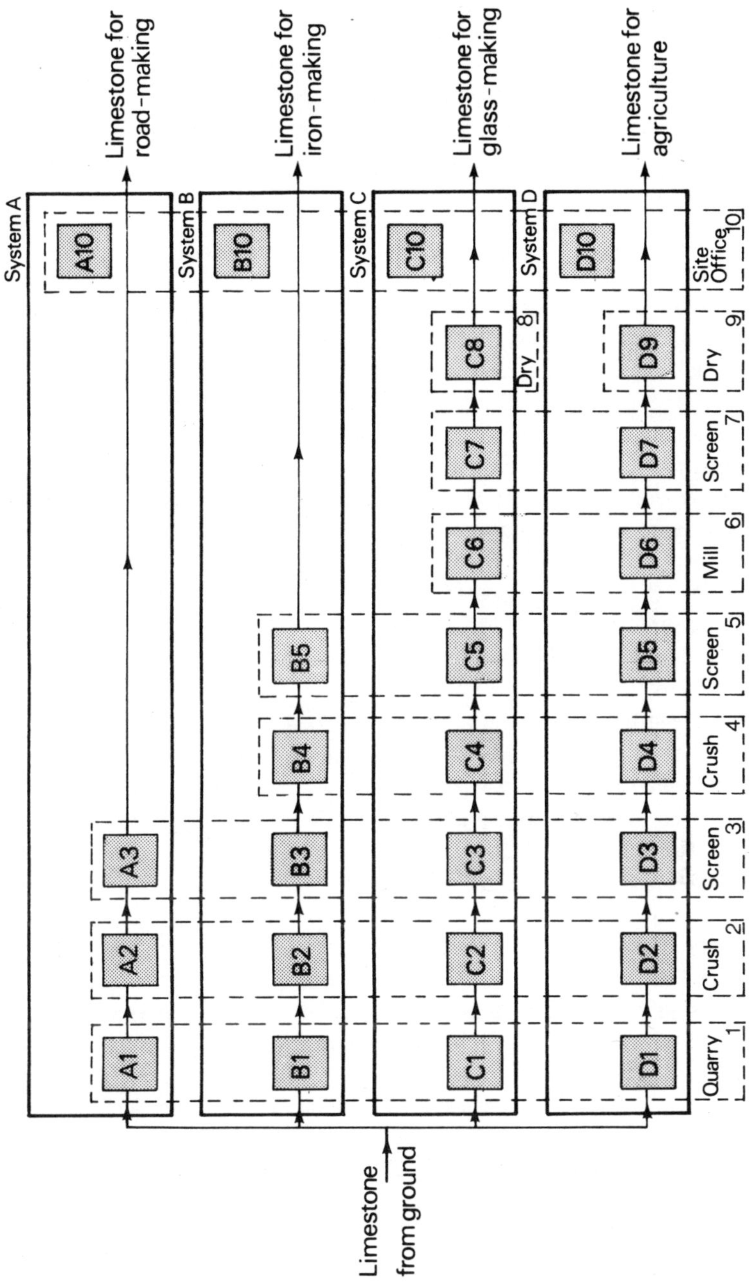

Figure 8.3 System diagram for the production of limestone showing how the physical operations of Figure 7.3 may be sub-divided and arranged to give four systems *A*, *B*, *C* and *D*, each of which produces only a single product.

8.1.3 Specifying the results of energy analysis

The results of energy analysis consist of two components; the numerical values of the normalised energy requirement and a statement of the system to which they relate. As described in Chapter 7, the presentation of results in the form of Table 7.4 conveys most of the essential information likely to be needed and this form is recommended. The use of a single number (that is parameter V in Table 7.4) is of little application, even though this is commonly found in published results.

A concise description of the system analysed is much more difficult to produce and the reason for this is easily demonstrated by a simple example. Suppose that the system shown in Figure 8.4 is defined for the production of liquid steel from scrap. Following the techniques outlined in Chapter 3, an energy requirement may be determined for each of the sub-systems shown and hence the overall energy requirement for the whole system may be calculated. If results are expressed in the form of Table 7.4, the fuel producing industries will be included even though they are not shown in Figure 8.4. The most adequate description of this system is of course the diagram shown by Figure 8.4, but it is clear that such representations can easily become unmanageable and a simpler representation is needed.

One widely used method is to collect the operations into groups. The **main operation**, here the electric arc furnace, together with the fuel supply industries feeding it, are usually the main contributors to the total system energy requirement and these are therefore formed into one group shown by sub-system 1 in Figure 8.4. This sub-system may usually be conveniently specified by the nature of the main input and output materials together with some identification of the operation, for example electric arc furnace, to distinguish it from possible alternative processes such as say a basic oxygen furnace.

The second group of sub-systems may be referred to as **the production and delivery of ancilliary materials and services**, as shown by sub-system 2 in Figure 8.4. The components of this group are responsible for the production of all materials inputs needed to carry out the main operation other than those already specified when defining the main process. In Figure 8.4 the main process input is specified only in terms of scrap and so lime, carbon electrodes, fluxing agents and water would be regarded as ancilliary materials and services. If the inputs to the main process are specified as scrap and lime, then the ancilliary materials and services would be carbon electrodes, fluxing agents and water.

The third group of sub-systems, referred to as **capital requirements**, represent those sub-systems needed to construct and maintain the furnace and any buildings involved. This group is shown by sub-system 3 in Figure 8.4.

By adopting this procedure, the extent of any system may be identified in a relatively straightforward manner. Thus the system of Figure 8.4 could be stated as the production of liquid steel at the electric arc furnace from scrap at

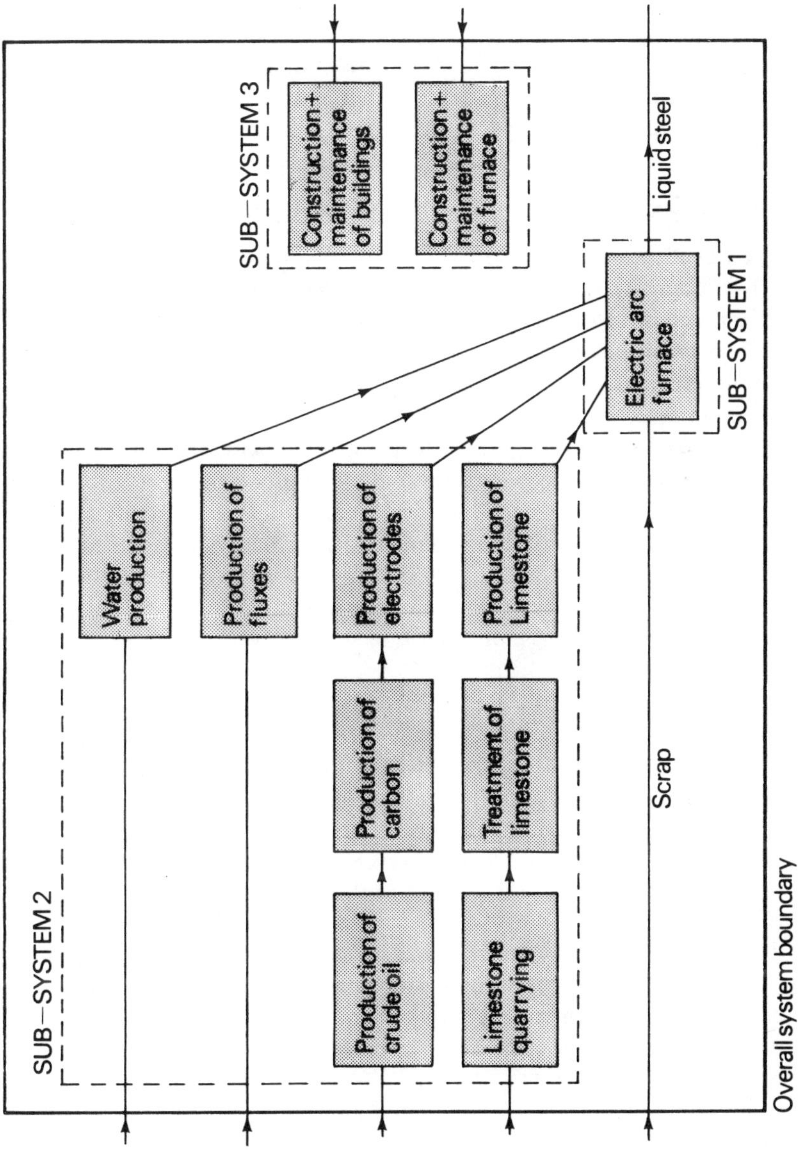

Figure 8.4 The system needed to produce liquid steel from steel scrap in the electric arc furnace showing the grouping of the component sub-systems into main operation (sub-system 1), the production of ancilliary materials and services (sub-system 2) and capital (sub-system 3).

the furnace with a system including the provision of ancilliary materials and services and an allowance for the construction and maintenance of the capital plant and buildings.

The energy requirements of ancilliary materials and capital are often omitted from energy analyses and it is useful to have some idea of the importance of these three groups of operations.

8.2 THE ENERGY OF ANCILLIARY MATERIALS

The classification of a materials input to a system as ancilliary is arbitrary. There is seldom any difficulty in identifying the *main* materials inputs because they usually constitute the greater part of the mass input. Hence the scrap input to the electric arc furnace system of Figure 8.1 is undoubtedly the main input. In general however, an input can be classified as ancilliary when it satisfies one of two conditions:

(a) the mass of the input is small in comparison with the total input mass,
(b) the material is not incorporated into the final product.

In the electric arc furnace, lime, carbon electrodes and fluxing agents may sensibly be regarded as ancilliary materials because their masses represent only a small fraction of the total input mass. Similarly the water input may be regarded as ancilliary because it is not incorporated into the finished steel but is used in cooling. In this case, the mass of water is likely to be very much greater than the mass of scrap input.

When systems descriptions use the method of grouping together ancilliary materials, often without specifying them in detail, it is important to have some standard procedure for defining the extent of the production sequences for these materials. In Figure 8.4, it can be seen that the production sequences have been extended back to production from raw materials in the ground and this is the procedure most commonly used. If however this procedure is not used, then it becomes necessary to specify the extent of the production sequences in detail to avoid misunderstanding.

The classification of materials as ancilliary suggests that in energy analysis they are in some way of less importance than the main inputs and contribute less energy to the system energy requirement. To see if this is so, consider the calculation of the energy required to produce the lime input in the system of Figure 8.4.

In choosing a value for the energy requirement of lime from data sources such as Part 2, we would seek one that took into account ancilliary materials feeds to all sub-systems in lime production so that the data included the same contributions to the energy requirement as were being counted for the main process operation. Similarly if capital energy is included in the energy requirement of the main process, it should also be accounted for in the energy requirement of ancilliary materials feeds. Data available to the analyst are however limited and the available energy requirements for the materials feeds may only include the fuel energy and fuel production energy. Remember however that the capital and materials contributions to the lime production sequence are an order of analysis above that for lime production, which is in turn an order above that describing the operation of the electric arc furnace. While their contributions to the energy requirements of a system describing *lime* production may well be significant, their omission is unlikely to introduce any serious error into the energy requirement for the electric arc furnace operation.

Table 8.3 shows the energy requirement for the production of 1 kg of lime from limestone in the ground. The electric arc furnace consumes 0.03 kg lime per 1.05 kg of liquid steel produced and so the energy requirement which takes into account the provision of lime is obtained by multiplying all entries in Table 8.3 by 0.03 and adding them to the entries in Table 8.1 to give Table 8.4.

Table 8.3
Energy required to produce 1 kg of lime from limestone in the ground. See Appendix 1

Fuel type	Fuel production energy (MJ)	Energy content of fuels (MJ)	Feedstock energy (MJ)	Total energy (MJ)
Electricity	0.64	0.20	nil	0.84
Oil fuels	0.13	0.64	nil	0.77
Other fuels	0.69	7.65	nil	8.34
Totals/MJ	1.46	8.49	nil	9.95

Table 8.4
Energy requirement for the system shown in Figure 8.2 to produce 1.05 kg of liquid steel from scrap using an electric arc furnace when the system boundary is extended to include the production of the ancilliary material, lime

Fuel type	Fuel production energy (MJ)	Energy content of fuels (MJ)	Feedstock energy (MJ)	Total energy (MJ)
Electricity	5.14	1.63	nil	6.77
Oil fuels	neg	0.02	nil	0.02
Other fuels	0.02	0.23	nil	0.25
Totals/MJ	5.16	1.88	nil	7.04

Accounting for lime production increases the energy requirement by some 4.3%. In practice all other materials inputs are treated in the same way so that the final table will include proportional amounts of all their energy requirements.

If the system encompassing the main operation is extended vertically in this way to take account of materials feeds from a higher order, the contribution to the energy requirement compared with the contribution of the energy consumed by the main operation varies considerably from system to system. Typically, ancilliary materials feeds will account for 10% or less of the total energy requirement, but there are examples where the reverse is true. Liquid steel for example is also produced from liquid pig iron by blowing oxygen through it in a basic oxygen furnace. In this instance. the external heat energy supply to the system is zero because all heat is generated within the furnace by oxidation of the carbon content of the pig iron. If the system is defined to take account of only the fuel supplied to the furnace, the energy requirement of the system is zero. Consequently if the system is extended to include the production of ancilliary materials, the system energy requirement arises totally from this source.

The provision of services such as water or compressed air are treated in exactly the same way as ancilliary materials. The energy associated with the provision of ancilliary materials and services is frequently called the **indirect energy** to distinguish it from that associated with the energy directly consumed by the main chain operations.

8.3 CAPITAL ENERGY

The energy analysis of a system can be completed by evaluating the energy associated with the construction of buildings and machinery. This contribution to a system energy requirement is often referred to as the **capital energy** of the system. It is also an indirect energy contribution to the system. One of the first reports of the significance of this contribution was the analysis of some sectors of the chemical industry [124]. The contribution varied from plant to plant so that in steam production, for example, the capital energy contribution was of the order of 2% whereas in the generation of electricity it was only about 0.4% of the total system energy requirement.

Despite the recognition of the need to include capital energy in complete energy analyses, published reports of detailed analyses of the capital energy of plant and buildings are scarce. Furthermore, such data as does exist is often derived from the monetary cost of constructing facilities assuming some relationship between monetary value and energy requirement so that no satisfactory fuel breakdowns are available.

In many respects this is understandable because a proper evaluation of capital energy demands a detailed analysis of the construction of every building and item of machinery involved in a specific production process. This exercise can only be regarded as of limited interest when the contribution of capital energy to overall system energy requirements is seldom greater than 5% and is more usually nearer 1%. This arises because, although the energy required to produce materials and erect plant is large because of the very large masses involved, lifetimes for buildings are frequently measured in decades and for machines in terms of many thousands of units of throughput. Two examples put this into perspective.

A plastics injection moulding machine of mass 5 tonne is made totally from steel for which a gross energy requirement is of the order of 50 MJ/kg. Hence a very approximate energy requirement for the production of this machine is 250,000 MJ. It is not unreasonable to expect such a machine to last for 10 years and to produce about one million components per annum. Therefore in its lifetime the average contribution of the manufacture of the machine to the total energy requirement of each component would be of the order of 0.025 MJ. For component of mass 0.05 kg made from polyethylene, the energy directly associated with moulding may typically be of the order of 2 to 3 MJ per component. Thus if the energy required to manufacture the machine were included, it would represent an increase of between 0.8% and 1.25% of the total energy requirement of the component. Even if the production energy for steel components were double the value quoted above to allow for machining and the use of specialist steels for some components, the contribution of the machine construction energy to the total energy requirement of each component produced by the machine would still not rise above 2%.

In the case of the electric arc furnace, no published data are available but estimates put the capital energy contribution as essentially zero [139]. Consequently for the system as defined in Figure 8.4, the total energy requirement will be unaffected by the inclusion or exclusion of the capital energy of the furnace itself.

The contribution from buildings to the capital energy is difficult to assess because of the paucity of data but once again approximate calculations are possible. The energy requirement for house building has been calculated as 7,029 MJ/m^2 [90] and for heavy industrial buildings incorporating massive steelwork and machinery such as overhead cranes, some 36,000 MJ/m^2 [26]. A building suitable for housing injection moulding machines will probably have an energy requirement somewhere between these two extremes and a value of 14,000 MJ/m^2 is likely to be reasonable. For a building of floor area 400 m^2 housing 10 machines such as those described above, the energy per machine will be $40 \times 14{,}000 = 5.6 \times 10^5$ MJ. If the building lasts for 30 years, then for a production rate of 10^6 components per machine per annum the energy requirement due to the building will be 0.019 MJ, which is less than 1% of the energy requirement of the finished product discussed earlier.

Although capital energy is usually a relatively small contributor to total system energy requirement, this is not always so. In situations where machines operate under particularly arduous and demanding conditions, the lifetime of the machines or parts of the machines may be very short indeed. A typical example is the use of drill bits in mining. Chapman [26] has calculated that for a direct electrical consumption of 8.04 MJ per ton of ore extracted (including fuel production energy) the production of replacement bits for the drill involves an additional indirect energy expenditure of 2.41 MJ/ton ore.

To summarise then, the energy requirement of an operation may be obtained in varying degrees of detail depending upon whether the system is defined to take account of the fuel energy only, the fuel energy plus the fuel production energy, or the latter plus services or plus services and capital. In this book and almost all other published data, the minimum system analysed takes account of the fuel energy and the fuel production energy.

8.4 THE ENERGY OF LABOUR

Almost all industrial processes involve the use of manpower which consumes energy in the form of food. Despite this, energy analysis of industrial operations usually ignores the energy associated with this food even though a proportion of its energy is used in industrial production activities.

One reason for ignoring the energy associated with the labour force is said to be conceptual. It is argued that if humans are included in an industrial system then not only are they producers of goods and services but they are also consumers. Thus the output of all industries are essentially consumed by the human 'machines' within the industrial system so that there is no net output. It is therefore conceptually unrealistic to identify the output of foodstuffs and their associated energies as 'energy debits' to production activities.

Such an argument is however more apparent than real because it presupposes that an industrial system is defined solely in terms of physical components. As discussed earlier, the nature of a system is defined in terms of identifiable functions and *not* in terms of physical components. There is no reason therefore why the production aspect of human behaviour cannot be separated from the consumption behaviour and included within the defined system. In the same way, the 'consumption' behaviour may also be sub-divided. Only if systems are defined solely in terms of physical components does it become necessary to include the whole of human behaviour, but at the same time the *total* behaviour of *all* machines must also be included.

If this were the only reason for excluding the energy associated with labour, then it is debatable whether its exclusion could be justified. There is however a more concrete and much less philosophical reason for ignoring it: this is the magnitude of the contribution. The average energy extracted from food per day by a human is of the order of 0.144 MJ per kilogramme of bodyweight [105]. For an adult of mass 70 kg this corresponds to a daily intake of about 10 MJ. By far the greater part of this consumption is required to maintain the man as a living organism rather than as an industrial component. At most, a contribution of the order of 5 MJ/day may be attributed to the industrial work actually performed.

If we compare this with the consumption of a machine operated by the man, we can put the problem into perspective. The injection moulding machine described earlier produced 10^6 components per annum, each with an energy consumption of 3 MJ. In one day therefore the machine will consume $3 \times 10^6/365 = 8{,}220$ MJ. Such a machine might be tended by a single man and his energy contribution would only amount to about 0.06% of the machine energy requirement. In fact, in most operations in highly industrialised systems the contribution from human labour is less than 0.5% and can safely be neglected. This may not however be the case in low energy consuming systems such as agricultural operations in developing countries, where human or animal labour may be the only input [83].

8.5 ALTERNATIVE REPRESENTATIONS IN ENERGY ANALYSIS

In the relatively short time that energy analysis has been carried out using extended systems, a surprising number of different conventions and representations have been used and many of these are still in use. Because fuel consumption of one form or another has been monitored in industry for a great number of years, it is hardly surprising that many engineering conventions have been extended into the wider form of analysis now practised. At the same time, this wider form of analysis is in many respects the energy analogue of cost accounting and economics and some of the terminology of these subjects has also crept in.

Since anyone wishing to pursue energy analysis beyond the scope of this book must inevitably delve into published literature, it is well to be aware of the different terminologies and representations in use. Some are straightforward and need little or no explanation; others are far from clear on first reading. In this Section therefore we have collected some of the more frequently used terminologies so that they can be explained in terms of the arguments already presented.

One of the earliest papers discussing the total energy requirements of processing in the chemical industry is that of Smith [124] and this is still widely used as a model, particularly within the chemical industries. Smith was primarily interested in evaluating the total energy requirement of complete systems starting with raw materials in the earth through to finished saleable chemical compounds. In the terminology of Chapter 3, he was interested in systems that are complete in both a vertical and a horizontal sense. To describe these total energy requirements he introduced the expression **cumulative energy**, although in the sense in which it is used in his paper it does not include any contribution for feedstock energies.

For any chemical compound the cumulative energy was considered as comprising two components; direct energy and indirect energy. Direct energy was defined as the energy content of the fuels used directly in the processing of the chemical intermediates. The energy associated with the fuels consumed by all other operations needed to support these chemical processing steps was referred to as indirect energy. Hence the indirect energy included contributions from transport, construction of all production facilities, production of all fuels and provision of all services such as steam and water. Using this very broad definition of indirect energy, Smith indicates that within the chemical industry the indirect energy takes a value of between 50% and 150% of the direct energy.

The major differences between Smith's approach and that discussed in the earlier part of this book lies in the way in which the contributors to the total energy requirement of a system are grouped together. In particular three differences may be highlighted.

(1) In this book, the energy associated with the production and delivery of fuels has been isolated as a separate identifiable component in Table 7.4 describing the energy requirement of the system. Smith in contrast combines this energy with that arising from other sources.

(2) The energy associated with transport operations is treated in this book as a separate identifiable sub-system within the production sequence. In consequence it will contribute to both fuel production energies and to energy content of fuels directly consumed in Table 7.4. In contrast Smith would combine all transport energies into his indirect energy value along with capital and the production of fuels and services.

(3) Services and capital energy in this book are again treated as separate sub-systems with corresponding contributions to all columns of Table 7.4, whereas Smith amalgamates these values into his indirect energy.

The effect of the two representations may be most easily seen by considering a simple example. Figure 8.5 shows a simple system consisting of two processing sub-systems A and B linked by a transport operation. For each of the processing sub-systems there are two additional sub-systems, one providing the necessary services and the other responsible for the construction of the plant. Hence within the overall system, seven sub-systems can be recognised and these are numbered as shown. Using Table 7.4, the energy associated with each sub-system will consist of a set of 16 numbers as shown by the array in each box. Assuming that these values have been normalised to a unit output of product from the overall system and using the convention developed earlier, the total energy requirement of the overall system will be an array of the form:

$$\begin{matrix} \Sigma A_i & \Sigma E_i & \Sigma K_i & \Sigma R_i \\ \Sigma B_i & \Sigma F_i & \Sigma N_i & \Sigma S_i \\ \Sigma C_i & \Sigma G_i & \Sigma P_i & \Sigma T_i \\ \Sigma D_i & \Sigma H_i & \Sigma Q_i & \Sigma V_i \end{matrix}$$

where for example:

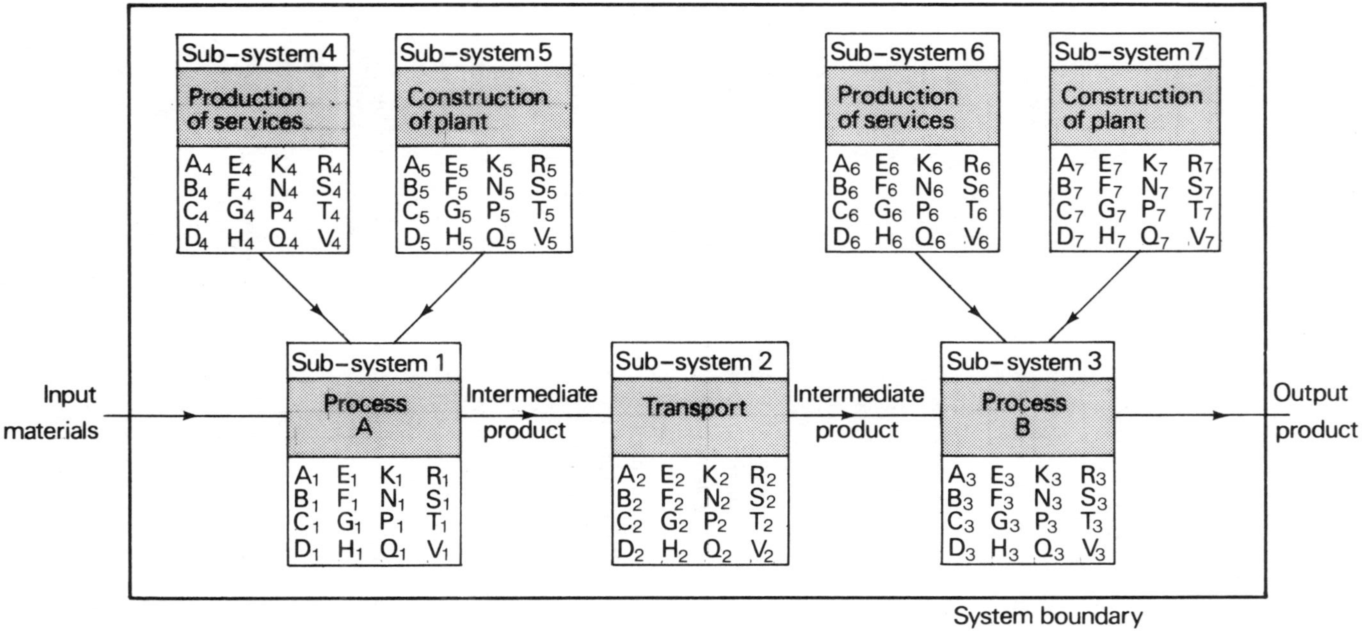

Figure 8.5 Hypothetical system containing two processing operations *A* and *B* linked by a transport sub-system. The array of numbers in each block represents the entries that would be made in Table 7.4 to describe the energy requirement of each sub-system when normalised with respect to the output of the overall system.

$$\Sigma H_i = H_1+H_2+H_3+H_4+H_5+H_6+H_7, \tag{8.1}$$

and so on. Using the Smith convention, the direct energy would be the sum of the direct energies associated with processes A and B alone and so would be (H_1+H_2). Compare this with Equation (8.1). Again using the Smith convention, the indirect energy will be:

$$(V_1-H_1)+V_2+(V_3-H_3)+V_4+V_5+V_6+V_7,$$

Note that the total energy requirement on the Smith convention is the sum of the direct and indirect energies and so would be:

$$V_1+V_2+V_3+V_4+V_5+V_6+V_7, \tag{8.2}$$

which is the same as ΣV_i in the array shown above as the set of numbers for Table 7.4. The two conventions therefore produce the same end result, but the detail within the calculation is arranged differently.

It should also be noted that Smith, in common with many other analysts, produced his results as a pair of numbers, corresponding to the direct and indirect energy. No detail of the breakdown of these numbers by fuel type was given, even though this detail must have been available in order to perform the original calculations. The unfortunate result of this form of representation is that the numerical values cannot be readily revised to accomodate different values for the fuel production efficiencies.

During the period between the 1968 World Energy Conference, where Smith presented his paper, and 1974, a number of researchers throughout the world took an increasing interest in energy studies of one form or another. In many respects these workers were operating in isolation and each was developing his own technique for dealing with problems of methodology as they arose. A significant development was the convening of a small conference in 1974 under the auspices of the International Federation of Institutes for Advanced Study (IFIAS) with the intention of producing guidelines that would be useful to researchers in this area. The results of this conference are now published [69]. It was recognised at the outset that there is always a danger in attempting to set up too rigid a framework for presenting results, particularly in the early days of the development of a relatively new field of endeavour, because this can act to inhibit new ideas and developments. Nevertheless a set of proposals were advanced. Some of these were novel, others simply formalised procedures which at the time were well established. Whether all of the proposals will be adopted by all workers in this area is a matter that only time can reveal.

One timely problem to which the conference addressed itself was that of nomenclature. Many of the terms in use at the time were open to ambiguous interpretation because, as already mentioned, they had been gleaned from other disciplines. For example, an expression such as 'energy cost' was frequently used to denote the energy requirements of a process and this could also be interpreted as the monetary cost of the energy. As a result, the term **energy requirement** was proposed and this is the term used in this book.

It was also recognised that because of the varying interests of energy analysts, some attempt should be made to distinguish between the various parameters that could be calculated to describe the performance of industrial processes. Three distinct definitions were proposed. The **gross energy requirement** (GER) was defined as **the total gross enthalpy that must be released at standard state from naturally occurring energy sources and consumed by the system in order to maintain the system in production**. This corresponds to parameter V in Table 7.4 and represents the total drain on available energy supplies needed to construct, maintain and operate a defined production system which is complete in the sense of Chapter 3. It should be noted that the original definition proposed by IFIAS referred to the *product* rather than the *system*, but, in view of the earlier comments on the inadequacy of defining a system in terms of a product only, the definition of gross energy requirement above has been modified. However, because the IFIAS definition specifies inputs materials as raw materials in the ground, ambiguity does not arise if the product alone is specified.

Because some products possess an intrinsic fuel value by virtue of their feedstock content (for example a polyethylene bottle), there is frequently a need to be able to state an energy requirement which excludes this energy content of the product. **Net energy requirement** (NER) was therefore defined as **the gross energy requirement less the energy content of the materials in the finished product**. NER therefore corresponds to parameter V in Table 7.4 less a proportion of parameter Q. Note that when the energy content of the finished product is the same as the feedstock requirement Q, the net energy requirement will be $(V-Q)$.

There is a certain lack of precision in the definition of NER because many workers include feedstock not as the energy content of the input materials but as the energy content of the output product, occasionally with a small allowance for materials losses. This technique certainly simplifies the calculations but leads to errors and uncertainties which may be significant (see Section 7.3). If the definition of NER is strictly adhered to, then the difference between GER and NER will represent the gross energy that can be derived from the combustion of the finished products of the system.

It was also recognised that much traditional energy analysis had been concerned with the consumption of fuels by the main processing operation within any system. It was therefore proposed to define a **process energy requirement** (PER) as **the sum of the direct and indirect energies associated with the fuel consumption of the main processing operation of any system** (see Figure 3.6). As already mentioned (see Section 8.5), the PER will most usually represent 90 to 95% of the GER and when the system is defined to include only the main processing operation, then PER will be the same as GER for that system. It is interesting to note that the difference (GER–PER) will be a measure of the completeness of a vertically extended system (see Chapter3).

The IFIAS Conference noted that the contributors to the energy requirement of any system are of different relative importance and so followed much the same type of argument as is outlined in Chapter 3 in arriving at an assignment of operations to a series of orders or levels. The only essential difference between the arrangement shown in Figure 3.6 and the scheme proposed by IFIAS is the assignment of the fuel supply industries alone to the second order. All of the materials supply sub-systems are therefore pushed to one higher order than shown in recognition of the fact that their contribution is, in general, lower than that of the indirect energy associated with the supply of direct energy to the main processing operation. This occupation of the second order by the single fuel supply sub-system allows a definition of a second order analysis as the process energy requirement of the system but does not affect the numerical value of the gross energy requirement.

One feature that the IFIAS proposals have in common with the earlier work of Smith [124] is that they do not separate out transport as an industrial operation in its own right. To some extent this is understandable because the convention is aimed primarily at extended industrial processing sequences where individual transport operations tend to be relatively small contributors to the total energy requirement. However, because transport operations represent direct energy consuming operations with their own capital and maintenance energies as well as direct and indirect energy inputs associated with their fuel consumptions, it is wiser to examine these operations as separate sub-systems rather than include them within processing steps.

One other proposal of the conference relates to flow chart construction. Flow charts are a useful tool in energy process analysis and it is certainly true that conventional chemical engineering representations are of only limited application. It was therefore proposed to introduce a new set of symbols based on the earlier published work of Berry and Fels [10]. The symbols proposed are shown in Figure 8.6. To show how these symbols may be used in flow chart construction, Figure 8.7 shows how the process illustrated by Figure 8.5 can be represented. To construct this flow chart we have made the following simplifying assumptions:

(a) all values refer to 1 kg of output,
(b) there are no materials losses in the process,
(c) the services referred to are steam only and both processes *A* and *B* each require 1 kg.

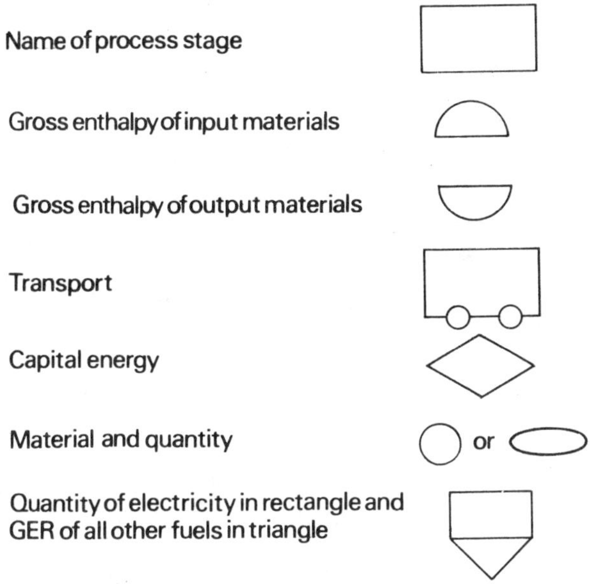

Figure 8.6 IFIAS flow chart symbols.

A comparison of Figures 8.5 and 8.7 show that much of the numerical data of Figure 8.5 is lost in the IFIAS representation. There are however two points to be remembered. First, all of the detail given in Figure 8.5 may not be available. This is especially true when a detailed analysis of specific plant is not possible and reliance must be placed on limited published information. Secondly, Figure 8.7 keeps electrical consumption separate from the consumption of other fuels. To calculate an energy requirement for the whole system it is necessary to take into account a value for the fuel production efficiency of electricity before it can be combined with the energy consumption of other fuels.

Despite these different methods of representing energy requirements, it should be noted that if the total energy requirement of any defined system is calculated by any of the different techniques, the answers will still be the same. In general, the only difference that occurs within the various methodologies is the way in which the different contributions to the total energy requirement are grouped together. It should also be noted that if the results of energy analysis are prepared in the form of Table 7.4, then they may be readily converted into any of the conventions reported here.

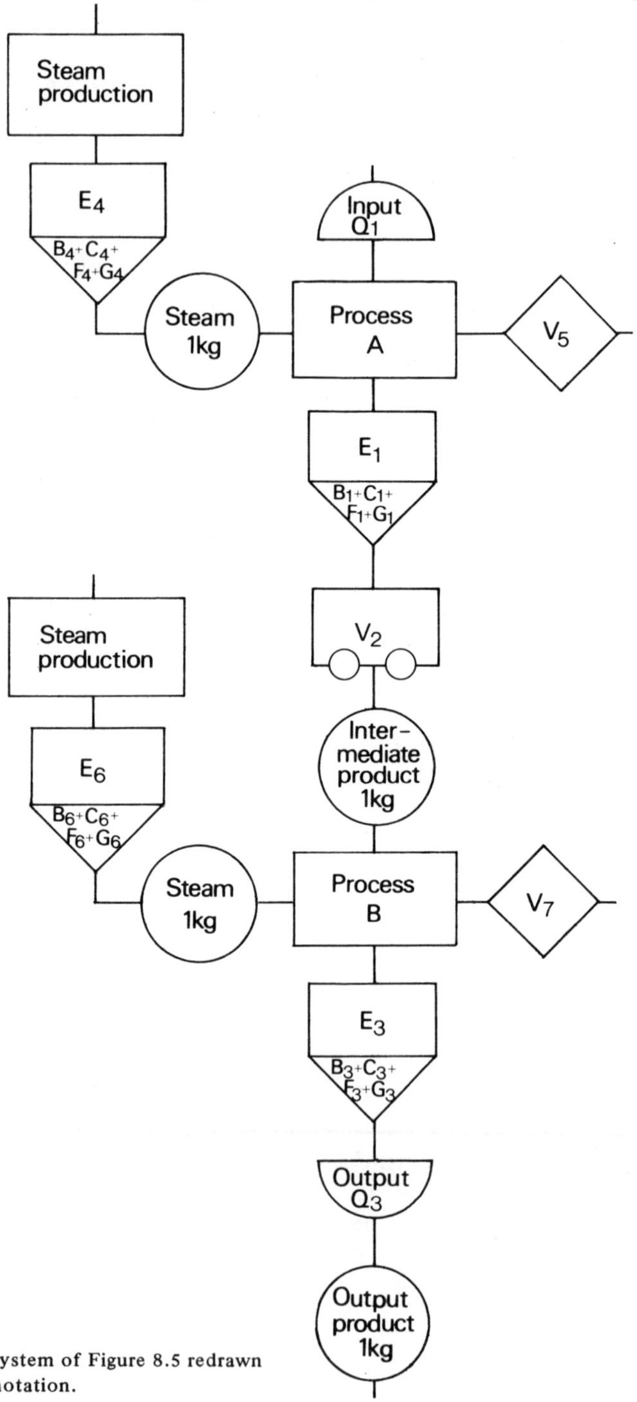

Figure 8.7 The system of Figure 8.5 redrawn using the IFIAS notation.

8.6 UNITS

Energy analysis is bedevilled by a variety of different units not only in the measure of energy but in relation to the simpler physical parameters such as mass, length and volume as well. A cursory examination of even a few published papers soon reveals the size of the problem. Within the pure sciences the use of S.I. units has now been adopted and the IFIAS Convention of 1974 recommended their use in energy analysis. This recommendation has been adopted in this book.

It must be recognised however that industry is far from standardised insofar as units are concerned. The use of imperial measures alongside metric units is not uncommon. There is moreover a tendency to use units very loosely so that the word **ton** for example may mean a long ton (2240 lb), a short ton (2000 lb) or a metric tonne (1000 kg). There is also the added complication that certain British and American units use the same words for different quantities, for example gallon.

As a result, the energy analyst must often perform a very large number of simple conversion calculations to standardise units and Appendix 1 gives a selection of the more common conversion factors needed.

For industrial analysis the most convenient unit of energy is the megajoule (MJ) and for energy normalised against output mass the most useful unit is the megajoule per kilogramme (MJ/kg). When very low energy requirements are encountered (see Section 7.4) it is often better to change the unit of mass to the tonne (t) rather than change the unit of energy to the kilojoule (kJ), although of course 1 kJ/kg = 1 MJ/tonne.

One area in which there is still a need for two units is in the measurement of transport distances. Both the U.K. and the U.S. retain the mile whereas the kilometre is widely used elsewhere. Hence in this book we have values in both units.

8.7 DATA BANK

Energy analysis has not yet reached the stage where complete and detailed analyses of all types of plant have been carried out. Yet it is quite clear from the preceding Chapters that because of the interdependence of various industries, a very large body of information is needed before any progress can be made. Indeed it is virtually impossible to derive fuel producing energies without some knowledge of the production energies for other commodities such as steel and copper.

It would moreover be an impossible task for any analyst to have to carry out a detailed energy analysis of every material input to any system in which he is interested.

Clearly then there is a need for a bank of data on which the analyst can rely for many of the inputs to his calculations. Such a collection of data has been supplied as Part 2 of this book. It must however be remembered that all of this information relates to published data and is therefore to a greater or lesser extent historical. Modern developments in technology mean that some of the values are expected to change with time.

In addition, because of the interdependence of the different industries, some of the values will have been derived using inaccurate data for some of their inputs. Such inaccuracies will gradually disappear as more detailed analyses or iterations are carried out, but in many cases the information given represents the 'best' values available even though they may involved some gross assumption. The sources of the data and the reliability are discussed in more detail in Part 2.

In general, because of the uncertain accuracy of much published data it is often wise to restrict its use in energy analysis to evaluation of higher order sub-systems. However, when published data are used to evaluate the energy of the main production process, great care must be taken to ensure that the system is properly defined because this is the major weakness of most published information.

8.8 APPLICATIONS OF ENERGY ANALYSIS

In these first eight Chapters we have attempted to outline some of the detail and background to the procedure of energy analysis. Using the ideas contained in these Chapters it is possible to define and analyse any system of interest. However, every system is unique and difficulties can occur in the most unexpected places. It is clearly not possible to give a set of guidelines to cover all eventualities, but to give some indication of the way in which energy analysis can be carried out and to highlight some of the more common problems, the remainder of this book is devoted to analysis of specific systems. At the same time it is possible to gain an insight into the ways in which the results of energy analysis can be used in practice.

Chapter 9 examines transport systems with a special emphasis on road transport. This element of processing pervades all aspects of industry and few systems do not involve at least one type of transport. From the analyst's viewpoint, road transport, that is the running of a lorry, represents a simple system which should in principle be readily amenable to detailed energy analysis. It also highlights a point made earlier, that a transport system is one in which the function implies no change in the chemical state of the materials and in which there are, in theory, no materials losses. The only change in the materials is a change in their geographical location and so the system function is an abstract operation.

One of the most immediate uses of energy analysis is in the direct examination of specific production plants. Such exercises provide the raw data for the component sub-systems of much larger industrial organisations and possess an intrinsic value for the plant manager himself. Chapter 10 therefore gives an analysis of a single factory to illustrate the nature of the available information, the level of detail that can be introduced and the way in which the results of the analysis can be used to suggest energy saving practices.

A practice which has been receiving considerable attention in recent years is recycling and this term is used here to include any feedback of materials from one part of the system to another, whether it be the reclamation of useable materials from refuse or the introduction of a returnable container system in some aspect of food and drink distribution. There can be no doubt that many potentially useful schemes will reclaim materials resources but the energy requirements of these schemes is in many instances unknown. Chapter 11 therefore considers the general principles of the energy analysis of recycling systems and shows how the results of energy analysis can be used as an aid in decision making.

Metals production represents a significant sector of any industrial economy and the products of the metals industry find applications in almost every other industry. Systems which result in the production of metals are usually large and complex and are seldom under the commercial control of a single company. Moreover, external pressures such as political considerations and strategic requirements often lead to the use of ores that would be regarded as uneconomic in a free market situation. The definition and analysis of such systems and the results produced are of fundamental importance to almost every energy analysis. Chapter 12 therefore considers some metal producing systems to highlight the nature of the problems that arise.

Chapter 9

Energy requirements of transport

9.1 FACTORS AFFECTING TRANSPORT ENERGIES

Transport is a vital element in most industrial processes. Frequently the energy requirements of transport represent a significant fraction of the total processing energy. Table 6.4 for example shows that the transport of crude oil from the Middle East to Western Europe contributes over one third of the energy needed to produce a commercial oil fuel from crude oil. Moreover, transport energy dominates in the distributive trades. Calculation of transport energy requirements is therefore an important exercise.

In general transport energy requirements arise from three main contributions.

(1) The energy content of the fuels consumed directly by the vehicle together with the associated fuel production energy (Table 6.13). This is usually directly proportional to the distance travelled, although it may be affected by factors such as loading, type of journey, age of vehicle and level of maintenance.

(2) The energy needed to construct and maintain the vehicle. This is the sum of the contributions from a number of sources. Some, such as tyres, vary directly with the journey distance whereas others, such as mechanical repairs, are independent of journey distance.

(3) The energy needed to provide facilities for the vehicle to carry out its journey. The construction and maintenance of roads and railway tracks fall into this category.

Information concerning the performance of all types of vehicles is available in a variety of forms. At one extreme, national statistics provide averages for all vehicles in a particular group. This is the only satisfactory source of information for large transport networks such as national railway systems, where detailed monitoring of individual journeys is seldom carried out and even when available is not necessarily representative of the overall network performance. Such aggregated data are however of only limited use in road transport systems, where the energy requirements are sensitive to a variety of factors such as vehicle size, loading, level of vehicle maintenance and type of journey involved.

In this latter case it is tempting to go to the other extreme and use performance data supplied by vehicle manufacturers. However, such information usually represents performances under ideal conditions which seldom occur in practice. It is reported for example that the fuel consumption of automobiles in Sweden is some 10% higher than the expected value and in the U.S.A. the difference is about 20% [116]. More reliable information can usually be obtained from fleet operators or from detailed trade statistics.

It is convenient at this point to consider the units used for expressing transport energy requirements. First, there are two units of distance in widespread use for transport operations, the kilometre and the mile. In this Chapter therefore, transport energy requirements have been expressed in both units for convenience. Secondly, the energy consumed per unit mass of product transported depends not only upon the energy requirement of the vehicle for that journey but also upon the load being carried. The most commonly used unit in statistics is energy per tonne-km (or per tonne-mile). This however assumes that loads are weight limited and this is not always so. A 20 tonne trailer delivering empty metal cans, for example, will only achieve a fraction of the maximum permitted weight loading before the dimensions of the load reach the maximum permitted legal limit. Similarly many vehicles do not carry full loads for a variety of reasons. In addition, many factors contributing to the total energy requirement depend not upon the load but upon distance travelled. Preventive maintenance of lorries for example is usually carried out after a fixed mileage. It is convenient therefore, particularly for road transport, to calculate energy per km (per mile) rather than energy per tonne-km (per tonne-mile).

The following Sections give data on the energy requirements of road, rail and sea transport. Of these, road transport is considered in the greatest detail because it is the most convenient and widely used form of transport and is also the least energy efficient.

9.2 ROAD TRANSPORT SYSTEMS

By far the greatest proportion of goods are transported by road vehicles both for long haul (trunking) movement and for local distribution. Road vehicles are available in a variety of sizes ranging from small petrol driven pick-ups to large diesel-powered articulated lorries with gross weights in excess of 32 tonne. Not surprisingly, the energy requirements of these different vehicles vary considerably.

The energy requirement of a road transport system can be considered as the sum of the energy requirements of a three component system. The main contributor is the fuel directly consumed by the vehicle on its journey. Then there are the two sub-systems responsible for (a) the construction and maintenance of the vehicle and (b) the construction and maintenance of the roads. The energy requirement associated with fuel consumption typically accounts for some 61% of the total. Construction and maintenance of the vehicles requires a further 32% and construction and maintenance of roads and other route facilities amounts to 7% [98].

9.3 FUEL REQUIREMENTS OF ROAD TRANSPORT

Fuel consumption of road vehicles is sensitive to a number of factors. Figure 9.1 for example shows the variation in the energy requirements of vehicles (that is energy content of fuel plus fuel production energy) in the U.K. with vehicle size, measured in terms of the maximum payload. An almost identical curve is obtained if vehicle size is measured in terms of gross vehicle weight. The most striking feature of Figure 9.1 is the rapid increase in energy requirement with lorry size. Two other features of Figure 9.1 are that petrol driven vehicles require consistently more energy than their diesel equivalents and that articulated vehicles, which employ a motor unit to pull a removable trailer, require more energy than rigid vehicles. However, the motor unit is frequently used in conjunction with a number of trailers, allowing considerable flexibility of operation.

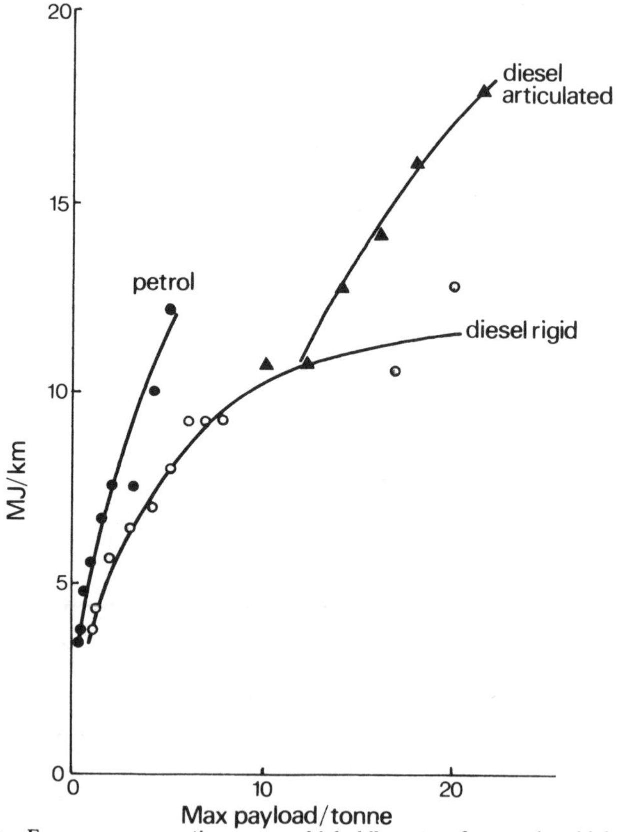

Figure 9.1 Energy consumption per vehicle-kilometre for road vehicles of different sizes.

Although Figure 9.1 seems to suggest that small vehicles are the more energy efficient because the energy per kilometre is lower than for larger vehicles, it must be remembered that the important parameter is the energy required to carry a unit mass of load over a distance of 1 km. The fuel consumption per kilometre must therefore be divided by the load carried and Figure 9.2 shows the energy per tonne-km for the same vehicles as plotted in Figure 9.1. It is now clear that larger vehicles are much more energy efficient than small vehicles and the apparently large difference between rigid and articulated vehicles emphasised in Figure 9.1 is much less apparent. The data presented in Figures 9.1 and 9.2 represent aggregated data for actual vehicle performances for a large number of vehicles in use in the U.K. and therefore include a wide range of different loadings and journey types. Figure 9.2 is of course somewhat artificial because the plotted values have been calculated from the data in Figure 9.1 assuming that full loads were being carried.

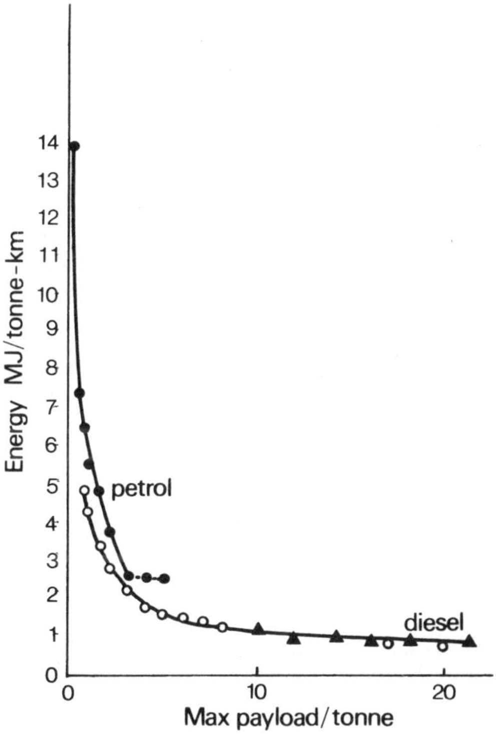

Figure 9.2 Energy per tonne-kilometre for road vehicles in the U.K.

The energy requirement per unit mass of load is extremely sensitive to the actual load carried. Suppose that an articulated lorry has a maximum payload of 20 tonne. From Figure 9.1, the total energy requirement per kilometre for this vehicle is approximately 17 MJ. When the vehicle is fully loaded, this energy requirement corresponds to:

$$17/20 = 0.85 \text{ MJ/tonne-km.}$$

If this same vehicle were carrying only 10 tonne, the energy per unit mass of load would be 1.7 MJ/tonne-km, assuming that the vehicle performance does not change as the load is reduced. Under such conditions the energy per unit mass then varies with loading as in Figure 9.3.

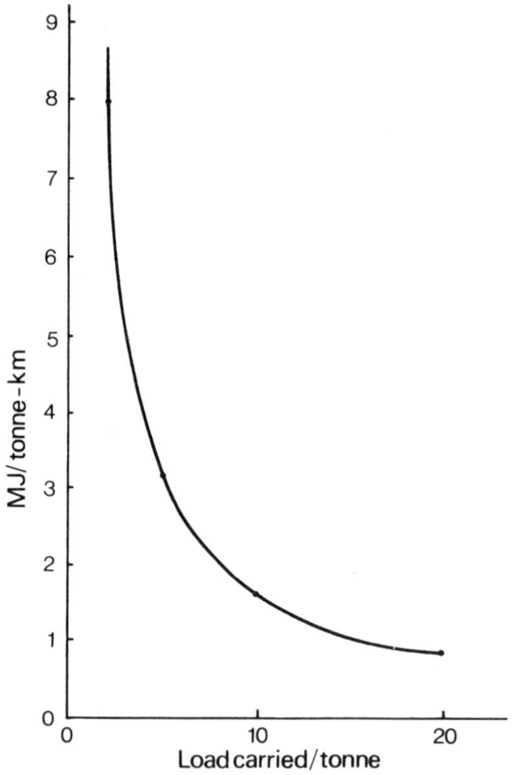

Figure 9.3 Energy per tonne-kilometre for different loads carried on a vehicle with a maximum payload of 20 tonne, assuming that the vehicle performance does not vary with vehicle loading.

Figure 9.3 assumes that vehicle performance is unaffected by load carried but this is not so [53]. Figure 9.4 for example shows the change in fuel energy requirement for three types of vehicle with changing load. Since energy requirement per kilometre for a lightly loaded vehicle is lower than for one containing the maximum load, neglect of this factor leads to an overestimation of energy per unit mass for sub-maximum loads.

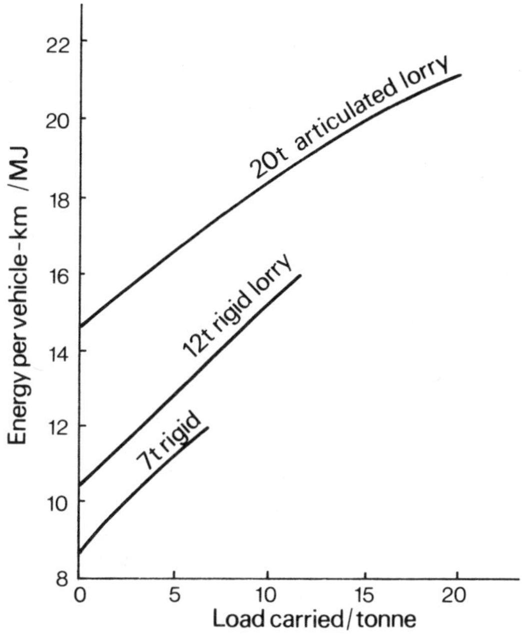

Figure 9.4 Changes in the performance of road vehicles as the loading is changed. The calculation is based on half rural and half urban driving.

If we modify the above example using Figure 9.4, then a 20 tonne payload lorry carrying only 10 tonne has a fuel energy requirement of 14.81 MJ per vehicle-km and not 17 MJ per vehicle-km as calculated previously. This leads to an energy requirement of 1.48 MJ/tonne-km as compared with 1.70 MJ/tonne-km when changes in vehicle performance are neglected. This factor is important and must be allowed for when calculating road transport energies.

Because the energy per tonne-km (or per tonne-mile) depends upon loading, and because loading is a variable factor, it is more convenient to express fuel requirements of vehicles in terms of fuel per vehicle-km. An added practical advantage of this is that fleet records are usually maintained in this form. If therefore a value can be assigned for the fuel or energy requirements per kilometre (or per mile) with full loading, the corresponding value for fuel or energy per tonne-km (or per tonne-mile) can be calculated using the multipliers given in Table 9.1 which has been computed to take into account the variations arising from Figures 9.3 and 9.4. Hence in the earlier example of a 20 tonne lorry with a performance of 17 MJ/vehicle-km, when this same lorry is only half-loaded the multiplier from Table 9.1 will be 0.087. So the energy requirement is:

Table 9.1

Multipliers for converting energy requirement per vehicle-km (vehicle mile) into energy per tonne-km (per tonne-mile) for different lorry loadings taking into account the change in vehicle performance with loading

Actual load (t) / max vehicle load (t)	1	2	3	4	5	6	7	8	9	10	11	12	13	14	15	16	17	18	19	20
1	1.000	–	–	–	–	–	–	–	–	–	–	–	–	–	–	–	–	–	–	–
2	0.876	0.500	–	–	–	–	–	–	–	–	–	–	–	–	–	–	–	–	–	–
3	0.831	0.457	0.333	–	–	–	–	–	–	–	–	–	–	–	–	–	–	–	–	–
4	0.802	0.434	0.313	0.250	–	–	–	–	–	–	–	–	–	–	–	–	–	–	–	–
5	0.783	0.418	0.300	0.239	0.200	–	–	–	–	–	–	–	–	–	–	–	–	–	–	–
6	0.769	0.408	0.290	0.229	0.191	0.167	–	–	–	–	–	–	–	–	–	–	–	–	–	–
7	0.758	0.399	0.283	0.223	0.185	0.162	0.143	–	–	–	–	–	–	–	–	–	–	–	–	–
8	0.750	0.394	0.277	0.217	0.180	0.157	0.139	0.125	–	–	–	–	–	–	–	–	–	–	–	–
9	0.744	0.388	0.273	0.213	0.177	0.154	0.136	0.122	0.111	–	–	–	–	–	–	–	–	–	–	–
10	0.737	0.384	0.269	0.210	0.174	0.150	0.133	0.120	0.109	0.100	–	–	–	–	–	–	–	–	–	–
11	0.732	0.381	0.266	0.207	0.171	0.148	0.130	0.118	0.107	0.098	0.091	–	–	–	–	–	–	–	–	–
12	0.728	0.378	0.263	0.204	0.169	0.145	0.128	0.115	0.105	0.096	0.089	0.083	–	–	–	–	–	–	–	–
13	0.724	0.375	0.261	0.202	0.167	0.143	0.125	0.114	0.103	0.095	0.088	0.082	0.077	–	–	–	–	–	–	–
14	0.721	0.374	0.259	0.200	0.165	0.141	0.124	0.112	0.101	0.094	0.086	0.081	0.076	0.071	–	–	–	–	–	–
15	0.719	0.372	0.257	0.199	0.164	0.139	0.122	0.110	0.100	0.092	0.085	0.079	0.075	0.070	0.067	–	–	–	–	–
16	0.717	0.371	0.256	0.197	0.163	0.138	0.121	0.109	0.099	0.091	0.084	0.078	0.074	0.069	0.066	0.063	–	–	–	–
17	0.715	0.370	0.254	0.196	0.161	0.137	0.120	0.108	0.098	0.090	0.083	0.077	0.073	0.068	0.065	0.062	0.059	–	–	–
18	0.714	0.369	0.253	0.195	0.160	0.136	0.119	0.107	0.097	0.089	0.082	0.076	0.072	0.067	0.064	0.061	0.058	0.056	–	–
19	0.714	0.368	0.252	0.194	0.159	0.136	0.118	0.106	0.096	0.088	0.082	0.076	0.071	0.067	0.064	0.060	0.058	0.055	0.053	
20	0.713	0.367	0.251	0.193	0.158	0.135	0.118	0.105	0.095	0.087	0.081	0.075	0.070	0.066	0.063	0.060	0.057	0.054	0.052	0.050

$$0.087 \times 17 = 1.48 \text{ MJ/tonne-km},$$

which agrees with the value calculated directly from Figure 9.4 by proportions.

The multipliers in Table 9.1 are a convenient way of summarising variations in performance with loading while calculating transport energy. Use of these multipliers however requires a knowledge of the fuel requirements of a vehicle at full load. Determination of this parameter in practice is difficult because fleet operators tend to maintain records for the overall performance of a vehicle, which will usually carry a variety of load sizes. Table 9.2 for example gives an analysis of a typical fleet of vehicles. The variation in fuel requirements between

Table 9.2

Fuel consumption of various lorries derived from direct analysis of a fleet operator. All fuel consumptions are stated in miles per imperial gallon of oil ful (petrol or diesel). Contract vehicles refer to those which are permanently used by a specific customer; general haulage vehicles carry a variety of loads as available. Numbers of vehicles in each group are shown in brackets. Vehicle tonnages refer to the size of the maximum payload in tonne. (Author's data).

Total fleet 10.49 mpg (989)	All articulated vehicles 7.76 mpg (557)	All contract articulated vehicles 8.00 mpg (178)	5-7t:	11.57 mpg	(1)
			10-11t:	11.45 mpg	(32)
			11-15t:	8.12 mpg	(6)
			15-18t:	8.13 mpg	(20)
			>18t:	7.01 mpg	(119)
		All general haulage articulated vehicles 7.64 mpg (379)	11-15t:	8.85 mpg	(37)
			15-18t:	9.37 mpg	(48)
			>18t:	7.21 mpg	(294)
	All rigid vehicles 14.02 mpg (432)	All contract rigid vehicles 14.52 mpg (343)	<1t:	18.26 mpg	(73)
			2-4t:	17.99 mpg	(44)
			4-5t:	14.63 mpg	(33)
			5-6t:	13.54 mpg	(14)
			6-8t:	13.20 mpg	(89)
			8-10t:	11.75 mpg	(53)
			10-12t:	11.00 mpg	(22)
			12-15t:	9.64 mpg	(15)
		All general haulage rigid vehicles 12.10 mpg (89)	4-5t:	13.77 mpg	(3)
			5-6t:	13.50 mpg	(1)
			6-8t:	12.87 mpg	(26)
			8-10t:	12.02 mpg	(21)
			10-12t:	11.50 mpg	(37)
			12-15t:	9.80 mpg	(1)

different vehicle sizes can be clearly seen as can the variation between different uses (that is general haulage or contract hire). For this fleet, the average loading of general haulage vehicles is estimated as approximately 67% of maximum load and that of contract vehicles placed somewhat lower at 55%, because such vehicles have less opportunity for picking up return loads.

Average fuel consumptions such as those shown in Table 9.2 may be readily converted to fuel consumptions under full loading using Table 9.1, providing that the average load is known or can be estimated. For example, a 20 tonne articulated general haulage vehicle gives 7.21 miles per gallon (Table 9.2) and this corresponds to a performance at an average 67% loading. From Table 9.1, the multiplier for this loading is 0.07 (that is corresponding to 13 tonne). If x is the fuel consumption at maximum loading then:

$$0.07\,x = \frac{7.21}{0.67 \times 20}.$$

Hence, $$x = 7.69 \text{ m.p.g.}$$

Using data from a variety of sources, Table 9.3 summarises typical energies required by various vehicle sizes under full load. When used in conjunction with Table 9.1, this Table allows calculation of the energy associated with the fuel consumed by a vehicle. For example, if a 14 tonne payload articulated lorry is carrying 6 tonne, Table 9.3 gives the total energy required as 15.38 MJ/km and Table 9.1 gives a multiplier of 0.141, so that the energy per tonne-km will be:

$$15.38 \times 0.141 = 2.17 \text{ MJ/tonne km.}$$

It is important to remember that calculations such as the above refer to a single journey. In practice the return journey must also be considered. From Figure 9.4, the multiplier to convert the performance under maximum load to that at zero load can be estimated as 0.7. Hence the energy required for the return journey will be:

$$15.38 \times 0.7 = 10.77 \text{ MJ/km.}$$

The load carried on the outward journey was 6 tonne and so the contribution due to the empty return load will be 10.77/6 = 1.79 MJ/tonne-km. Hence for the round trip, the energy for transport will be:

$$2.17 + 1.79 = 3.96 \text{ MJ/tonne-km.}$$

In both of these calculations, the detailed breakdown of energy into fuel energy content and fuel production energy can be readily determined from Table 9.3.

Table 9.3

Energy requirements for lorries arising from direct fuel consumption under full load. All values have been corrected to the energy values given in Table 6.13. U.K. values based on authors' data and published values derived from Commercial Motor (1977), Leach (1976) and Leach and Slesser (1973)

Lorry type (Maximum payload in tonne)	Oil fuel energy requirements MJ/km (MJ/mile)		
	Fuel production energy	Energy content of fuel	Total
Rigid vehicles			
<1	0.80 (1.29)	4.17 (6.71)	4.97 (8.00)
1-2	1.03 (1.66)	5.31 (8.54)	6.34 (10.20)
3	1.21 (1.95)	6.32 (10.17)	7.53 (12.12)
4	1.42 (2.28)	7.27 (11.70)	8.69 (13.98)
5-8	1.73 (2.78)	8.81 (14.18)	10.54 (16.96)
9	1.93 (3.11)	9.75 (15.69)	11.68 (18.80)
10-12	1.98 (3.19)	10.01 (16.11)	11.99 (19.30)
13-20	2.12 (3.41)	10.77 (17.33)	12.89 (20.74)
Articulated vehicles			
10	1.95 (3.14)	9.89 (15.91)	11.84 (19.05)
10-12	2.05 (3.30)	10.40 (16.73)	12.45 (20.03)
13-14	2.54 (4.09)	12.84 (20.66)	15.38 (24.75)
15-16	2.57 (4.14)	12.99 (20.90)	15.56 (25.04)
17-18	2.68 (4.13)	13.52 (21.75)	16.20 (25.88)
18	3.37 (5.42)	17.04 (27.42)	20.41 (32.84)

One further factor to be considered is the variation of fuel consumption with traffic conditions. Everall [53] has monitored the fuel consumption of vehicles in town driving, which incurs much stopping and starting, and rural driving, where the vehicle keeps moving at constant speed for relatively long periods. Typical performance characteristics are shown in Figure 9.5 for a 20 tonne payload vehicle and there is a marked difference between urban and rural performance. The values used to construct Figure 9.4 are based on the assumption that the vehicle spends half of its journey in town driving and the other half in country driving. In practice it is seldom possible to obtain detailed information on the precise type of running conditions for all lorries in a fleet and the best that can be done usually is to assume an average mix such as that used in Figure 9.4. In the absence of detailed performance characteristics, a reasonable assumption is that urban driving requires 20% more energy and rural driving 20% less energy than the average value.

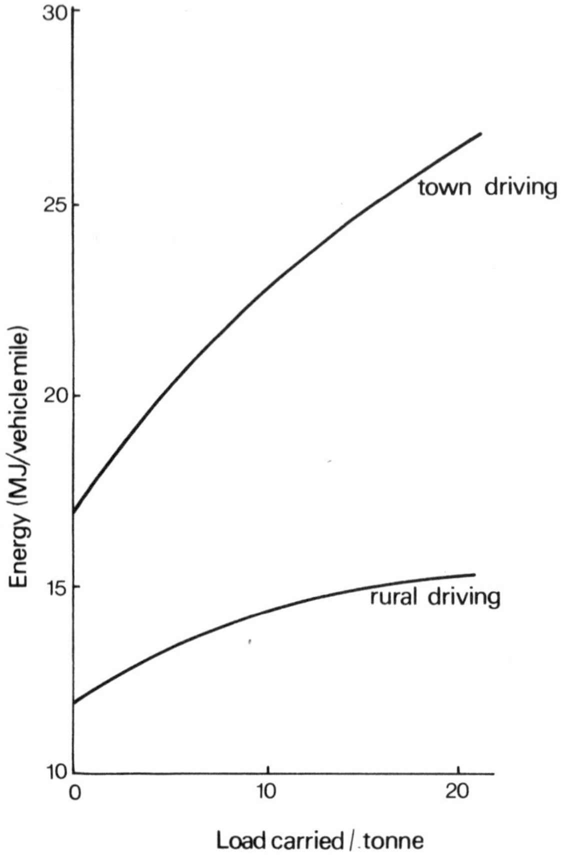

Figure 9.5 Changes in the performance of a 20 tonne articulated lorry under different driving conditions.

So far the performance characteristics of road vehicles have been considered in terms of the mass of the load carried, and for most purposes it is the most significant factor. However mass is governed by local regulations and is usually determined by the dimensions of the vehicle and trailer together with the permitted axle loadings. Most countries have regulations limiting the overall dimensions of lorries. In the E.E.C. countries for example, the maximum dimensions for a 20 tonne payload vehicle are length 15 m, width 2.5 m and height from road surface 4 m. After allowing for the size of the driver's cab and the height of the platform from the road, this permits a payload volume of the order of 110 m^3. Hence at an average load density of 182 kg/m^3, the payload will reach both the weight and volume limit at the same time. For loads of greater density, the mass restricts the load that can be carried, whereas the volume limits the maximum load for loads of lower density. Note that the term 'density' as used here is averaged over the whole of the packed load. For example, a typical lorry which is capable of carrying a 20 tonne payload might on average only hold about 5 tonne of empty metal cans before the volume limit is reached.

9.4 CONSTRUCTION AND MAINTENANCE OF ROAD VEHICLES

Within this group of operations are included (a) construction of the vehicle, (b) lubrication, (c) tyres, (d) garaging and (e) spares including batteries and paintwork. There are essentially two methods of obtaining values for the energies associated with these factors. The first is analysis of the available statistical data such as that carried out by Mortimer [98], in which case the items detailed above are usually reduced into two groups, vehicle construction and vehicle maintenance. The second approach is to calculate an energy requirement for each item in turn. In this latter approach, since the calculated energy will be eventually partitioned between either the number of vehicle miles or between the number of tonne-miles travelled, it is not necessary to achieve great accuracy in the initial energy calculations and notional values will usually suffice.

9.4.1 Vehicle construction

A variety of techniques have been used to estimate the energy needed to construct road vehicles. Makhijani and Lichtenberg [88] estimate the energy to manufacture a truck of mass 4.5 tonne as 3.204×10^5 MJ, assuming that the whole of the unladen weight of the vehicle is steel. This basic assumption is justified by arguing that approximately 80% of the mass of the vehicle is in fact steel. The energy requirement for the remaining 20% of materials, together with the fabrication energy requirement, are expected to be little different from the value derived by assuming an additional 20% steel. The load that can be carried by such a vehicle is not stated, but if similar to the vehicles in use in the U.K., it would be capable of carrying a payload of 12 tonne [41].

A higher value of 5.44×10^5 MJ for the construction energy requirement for trucks is quoted by Tuininga [135], although no details of vehicle size are given. However, from the stated fuel consumption, the performance is approximately that expected from a vehicle with a loading of about 12 tonne, that is the same as that quoted by Makhijani and Lichtenberg [88]. The marked difference in values is somewhat surprising since Tuininga [135] has excluded from his calculations the manufacturing energy of imported components. However, if we take a range of 3×10^5MJ to 6×10^5MJ as typical for a 12 tonne payload vehicle, then since the estimated lifetime of such a vehicle is 338,000 km (210,000 miles) [41], this corresponds to an energy of between 0.89 and 1.78 MJ/vehicle-km (1.43 to 2.86 MJ/vehicle-mile).

Irrespective of the disagreement between the two sets of results, the technique of Makhijani and Lichtenberg [88] illustrates an important point arising from the fact that vehicle lifetimes, measured as the distance travelled, tend to increase with the size of the vehicle [41]. Table 9.4 shows the calculations for lorries of different sizes based on published data. Note that although unladen vehicle weight increases by a factor of 10 over the range, the construction energy per kilometre only increases by a factor of 4. Moreover, if construction energy is calculated as a percentage of the direct fuel consumption of the vehicle then all the values lie in the relatively narrow range 4.2% to 9.9% with an average value of 7.5%. Because of the relatively small influence of vehicle size on this ratio, the use of the ratio as a means of calculating vehicle construction energies from fuel consumption energy requirements is a very attractive and simple technique. The problem is that of finding an acceptable value for the ratio.

Table 9.4
Construction energies of different lorry sizes using the technique of Makhijani and Lichtenberg (1972)

Max. load (t)	Unladen weight (t) (Note 1)	Lifetime $\times 10^3$ ml (Note 1)	Construction energy as if vehicle were 100% steel MJ (Note 2)	Construction energy per km (per ml) MJ	Total fuel energy requirement MJ/km (MJ/ml) (Note 3)	Construction energy as a % of the fuel energy
1	0.8	75	40000	0.33 (0.53)	4.97 (8.00)	6.6
2	1.75	90	87500	0.60 (0.97)	6.34 (10.20)	9.4
3	2.0	90	100000	0.69 (1.11)	7.53 (12.12)	9.2
4	2.5	90	125000	0.86 (1.39)	8.69 (13.98)	9.9
5	2.75	90	137500	0.95 (1.53)	10.54 (16.96)	9.0
6	3.0	150	150000	0.68 (1.00)	10.54 (16.96)	5.9
7	3.5	150	175000	0.72 (1.16)	10.54 (16.96)	6.8
8	3.75	150	187500	0.78 (1.25)	10.54 (16.96)	7.4
10	4.0	210	200000	0.59 (0.95)	11.99 (19.30)	4.9
12	4.5	210	225000	0.67 (1.07)	12.45 (20.03)	5.4
14	5.0	240	250000	0.65 (1.04)	15.38 (24.75)	4.2
16	6.5	240	325000	0.84 (1.35)	15.56 (25.04)	5.4
18	8.0	240	400000	1.04 (1.67)	16.20 (25.88)	6.4
20	10.0	240	500000	1.29 (2.08)	20.41 (32.84)	6.3

Notes
(1) Commercial Motor (1977).
(2) Based on a gross energy requirement for steel of 50 MJ/kg.
(3) From Table 9.3.

There is an alternative method of arriving at a value for this ratio which does not assume an energy requirement for steel. This is the use of national statistics to compare the energy required to construct vehicles in any one year with the fuel energy consumed by all goods vehicles during the same year. It assumes that any departure from the steady-state situation where vehicle production just replaces scrapped vehicles are sufficiently small to introduce no significant errors. Recorded values for the ratio of constructional energy/fuel energy are 12.8% for Holland [135], 14% for the U.K. [83], 16.7% for the U.S. [88] and 41.0% for the

U.K. [98]. The value of 12.8% is expected to be somewhat low since it excludes imported and capital goods used in the construction. Similarly, Mortimer [98] expresses some reservations about the validity of some U.K. data and the value of 41% may therefore be somewhat high, a point also noted by Tuininga [135]. Part of the discrepancy may arise from differing practices in the different countries but this does not explain the whole of the variation. A further cause may lie in the often-used assumption that the whole mass of a vehicle is steel when in fact some 20% of the mass is composed of materials such as copper, aluminium and plastics, which possess gross energy requirements well in excess of that for steel. In addition many of the machined parts, even though made of steel, incur an energy penalty because of the high electrical requirement in machining, coupled with the inevitable materials loss during production.

As a result of such considerations, a reasonable value for vehicle construction energy requirement can probably be calculated by assuming a value of 15% of the direct fuel energy requirement.

9.4.2 Statistical approach to vehicle maintenance

There are two reasons why the use of monetary statistical data is somewhat limiting in calculating vehicle maintenance energies. First, aggregation of data in national statistics makes it impossible to separate commodities such as tyres, which are essentially part of the running costs, from semi-variable and fixed costs such as spare mechanical parts and the provision of garage facilities. So far attempts to use national statistics have had to rely on simple apportionments of costs. Mortimer [98] for example has converted monetary values to energy assuming that an energy intensity (that is MJ/£) can be defined as the average for motor vehicle construction and rubber products.

Even when monetary data are available in a disaggregated form as in the published *Tables of Operating Costs* [41], there are still considerable uncertainties in the data since large fleet operators usually obtain significant discounts for component supplies. Moreover, maintenance costs usually include significant labour costs which are unrelated to energy use.

Despite these problems, Mortimer [98] has shown that the maintenance energy requirements for lorries is equivalent to 7% of the direct fuel energy requirement and this compares well with the 8% assumed by Leach [83]. In both cases, the energy refers to tyres and spares and does not include any contribution from the provision of garage facilities. Nonetheless, these values agree surprisingly well with the total maintenance energy for automobiles in the U.S.A. as 6% of the fuel energy consumption [115].

To see how these values compare with a more detailed analysis, it is useful to examine separately the major operations involved in maintenance.

9.4.3 Lubrication

The major contributor to the energy required for lubrication derives from the use of engine and sump oil. Although it might be expected that oil consumption will depend upon total distance travelled, this is only a minor effect. The major sources of oil consumption are oil changes during routine servicing and repairs, and leakage from engines. The incidence of breakdowns requiring major repairs and of the leakage of oil from engines are specific to individual vehicles and analysis of fleet records gives no evidence of any predictable pattern relating to vehicle size, make or age. This absence of well defined trends is reflected in published costs for vehicle maintenance [41], where a fixed allocation is usually made without reference to vehicle characteristics.

From our own data, a typical value for lubricating oil consumption is 885 km/litre (2500 miles/gallon). The use of other lubricants such as grease and axle oil is negligible in comparison. Taking the production of this oil as being similar to that of light fuel oil in Table 6.13, we obtain a total energy requirement of 0.055 MJ/vehicle-km, or 0.088 MJ/vehicle-mile, which is independent of vehicle size. When represented as a percentage of the direct fuel energy requirement, lubricant consumption varies from 0.3% for larger vehicles to 0.9% for small vehicles. An average value of 0.6% therefore seems reasonable.

9.4.4 Tyres

The calculation of an energy requirement for the consumption of tyres is complicated by a number of variables. However, a notional value for goods vehicles has been calculated in Table 9.5 using the following generalisations, which are based on information supplied by a number of U.K. tyre manufacturers.

(a) Tyre size is related to vehicle size. Adopting a 'worst case' approach, data is given for a tyre of mass 55 kg which would be suitable for use on vehicles with axle loads in excess of 5 tonne.

(b) Both cross-ply and radial-ply tyres are in use. In the U.K. in 1974, radial tyres held approximately 40% of the market for small sizes and 70% for larger sizes. A reasonable average value for all vehicles is 55% radial tyres.

(c) The exact composition of tyres varies from manufacturer to manufacturer and such information is highly confidential. However typical values are shown in Table 9.5.

(d) The retreading of commercial vehicle tyres is an important aspect of road transport economics and most vehicle operators employ a mix of new and retreaded tyres. The materials requirements for retreading are shown in Table 9.5.

(e) A typical purchase pattern of tyres by a fleet operator is one third new tyres and two thirds retreaded tyres.

Taking the above items into consideration, a typical average materials requirement can be calculated for a tyre used by a commercial fleet operator and this is shown in Table 9.5. An allowance of 5% loss of rubber compound in processing has been made. Note that because the casing may be used twice, the contribution from components such as steel beading will be less than that in a new tyre.

The gross energy requirement of this 'average' tyre has been calculated using a gross energy requirement for all polymers of 100 MJ/kg and a gross energy requirement for steel of 50 MJ/kg. More accurate values from Part 2 of this book can be used but, as can be easily demonstrated, they do not significantly affect the final conclusion. The energy requirement for assembling the tyres is based on a value of 200 MJ/kg of rubber compound moulded [17].

The lifetime of a retreaded tyre is lower than that of a new tyre and for all types of tyre this lifetime has been variously estimated as 50,000 to 100,000 km (30,000 to 60,000 miles). Again adopting a 'worst case' approach, the energy requirement for tyres from Table 9.5 is 0.20 MJ/km (0.32 MJ/mile). The tyre considered in Table 9.5 is essentially for the larger vehicle sizes and a comparison of the above energy requirements for tyres with the energy directly consumed as fuel (Table 9.4) shows that the tyre energy corresponds to an additional 1.1 to 1.8% of the fuel energy. A reasonable average is thought to be 1.5%.

Table 9.5

Data for tyre construction for road vehicles

Component	Mass in new tyre/kg		Mass in retread/kg		Composition of average tyre in use kg	G.E.R. of average tyre MJ
	radial	cross-ply	radial	cross-ply		
Rubber compound	42.1	46.5	8.0	10.8	27.6	2760
Nylon cord	–	6.0	–	–	1.4	140
Steel (cord plus beading)	12.9	2.5	–	–	4.2	210
Processing energy	–	–	–	–	–	6640
Gross energy requirement of average tyre						9750

9.4.5 Garaging

The essential component of garaging is the provision of facilities for servicing and repairing vehicles. Practices vary widely and the extraction of physical data from published statistics is virtually impossible. We have obtained garage data from a number of fleet operators who themselves maintain the whole of their fleet and typically consumption is of kerosine for space heating and electricity for lighting and powering machines. Usually the annual requirements of these two fuels are known and records are usually available of the total number of vehicle-km or vehicle-miles travelled in this same period by the vehicles serviced. Typical average values may therefore be calculated. These are:

kerosine: 0.082 litre/km (0.028 imp. gall/mile),
electricity: 0.022 kWh/km (0.036 kWh/mile).

Table 9.6 shows these fuel consumptions converted into energies using Table 6.13. For the whole group of vehicles to which these garage data belong, the average fuel energy requirement is 12.16 MJ/km (19.57 MJ/mile) so that the energy requirements of the garage represent 32% of the direct fuel energy requirement.

Table 9.6
Energy requirement per km (per mile) due to garaging

Fuel type	Fuel production energy (MJ)	Energy content of fuel (MJ)	Feedstock (MJ)	Total (MJ)
Electricity	0.25 (0.40)	0.08 (0.13)	nil	0.33 (0.53)
Oil fuels	0.57 (0.92)	2.99 (4.81)	nil	3.56 (5.73)
Other fuels	nil	nil	nil	nil
Totals/MJ	0.82 (1.32)	3.07 (4.94)	nil	3.89 (6.26)

9.4.6 Spares

Apart from tyres, which have been considered separately, the only significant large components likely to be needed in running a lorry fleet are replacements to engine and transmission. Engine and differential lifetimes are of the order of 300,000 miles and gearboxes last about 150,000 miles, both of which are well in excess of the expected lifetimes of many lorries. Inevitably there will be individual vehicles where these components do not last the expected mileage, but when the energy for their production is averaged over the total fleet mileage, the contribution to the gross energy requirement of any vehicle is negligible.

Suppose that a vehicle requires a replacement gear box of mass 125 kg after 150,000 miles. Assuming a gross energy requirement for steel of 50 MJ/kg, this corresponds to an energy for replacement of 0.04 MJ/mile or about 0.1% of the direct fuel energy requirement. Since in practice the demand for such replacements will be considerably less, this value represents an upper limit.

The batteries on a vehicle are usually of the lead-acid type and might typically weigh 30 kg. Assuming that the whole of the mass is lead with a gross energy requirement for production of 51 MJ/kg [88], the battery energy is 1530 MJ. This may be compared with a value of 1875 MJ obtained by TRRL [134].

Such a battery will last for 100,000 miles so that the energy requirement will be 0.01 MJ/mile or about 0.03% of the direct fuel energy requirement of the vehicle.

Most fleet lorries are repainted halfway through their lifetime (say 100,000 miles). If paint is assumed to have a gross energy requirement that is the same as that for the production of most polymers, and if it is assumed that 5 gallons of paint are used per vehicle, this gives rise to an energy requirement for painting of 0.02 MJ/mile or about 0.06% of the direct fuel energy consumption.

Hence combining the contributions from spares, batteries and paint, an overall value of 0.2% should be a reasonable estimate.

9.4.7 Summary of construction and maintenance energy requirements

From the discussion in the preceding Section, Table 9.7 summarises the energy typically required to keep a lorry on the road. If these values are compared with those available from statistical evidence (see Section 9.3.2) the values in Table 9.7 are considerably higher. The major reason for this is the very high contribution arising fron the garaging component. It should be remembered that the value for garaging shown here represents that for a well-maintained lorry fleet in which all operations are carried out in the same location. Frequently, when a transport operator does not operate a large number of vehicles, much of the maintenance work is carried out by outside contractors and may not be recorded in statistics as part of the transport operation. It must also be remembered that there are a large number of vehicles for which maintenance is the minimum needed to keep the vehicle in a 'legal' state and whilst such a state of affairs may be deplored, its existence in practice must be recognised. For these reasons, the value shown in Table 9.7 probably represents the national value that would exist if all vehicles were maintained to a high standard.

Table 9.7

Typical gross energy requirements for a 20 tonne payload vehicle

Contribution	MJ/vehicle km	MJ/vehicle ml
Fuel energy from Table 9.4	20.41	32.84 (67.0%)
Lubrication @ 0.6% of fuel energy	0.12	0.20 (0.4%)
Tyres @ 1.5% of fuel energy	0.31	0.49 (1.0%)
Garaging @ 32% of fuel energy	6.53	10.51 (21.4%)
Spares @ 0.2% of fuel energy	0.04	0.07 (0.1%)
Gross energy requirement excluding construction	27.41	44.11 (89.9%)
Construction energy @ 15% of fuel energy	3.06	4.93 (10.1%)
Gross energy requirement	30.47	49.04 (100.0%)

Table 9.8 gives the detailed breakdown of the energy requirements of different vehicle sizes using the data from the preceding sections.

9.5 ROAD TRANSPORT ROUTE FACILITIES

The operation of an efficient commercial road transport system requires a network of good roads together with ancilliary facilities such as parking and street lighting. It may therefore be argued that a proportion of the energy required to provide these facilities should be attributed to the energy requirement of commercial road transport. From statistical evidence, Mortimer [98] has estimated that the total energy requirement in 1968 for the provision of these facilities in the U.K. was 5.5×10^4 MJ, of which 52% was due to the construction and maintenance of roads and 42% was due to street lighting. The remainder arose from the construction of parking facilities, road cleaning and snow clearance.

The most difficult problem is deciding how to apportion this energy amongst the different road users because a number of partitioning parameters could be used, for example number of vehicles of each type, number of miles travelled and so on. Mortimer [98] concluded that the most appropriate method of partitioning was on the basis of vehicle loading, that is tonne-km (or tonne-miles) travelled by each class of vehicle. On this basis he showed that 61.6% of the total energy required to provide route facilities should be attributed to goods vehicles. An allowance can be made for this energy requirement by increasing the performance characteristics shown in Table 9.8 by an amount equal in magnitude to 11.4% of the energy consumed directly as vehicle fuel. Table 9.9 shows the revised energy requirements for road vehicles when an allowance is made for the provision of route facilities and it has been broken down into fuel types on the basis of 50% electricity and 50% oil fuels.

It should be noted that the inclusion of highway energies in transport energy requirements is not practised by all analysts. Leach [83] for example points out that a significant proportion of the highway energy quoted above is for the provision of street lighting, which is shared by pedestrians as well as road vehicles. Moreover, the apportionment of energy on the basis of tonne-mileage does not take into account an unknown number of other users such as cyclists, who do not appear in national statistics. There are therefore equally plausible arguments for and against including road construction and maintenance energies within the transport energy requirement. Whichever procedure is adopted it is important to ensure that the type of analysis is explicitly stated in the system definition.

Table 9.8

Energy requirements per vehicle-km (per vehicle-mile) for different size vehicles. Values include direct fuel consumption, construction and maintenance energies

Vehicle type maximum payload in tonne	Electricity MJ/km (MJ/ml)		Oil fuels MJ/km (MJ/ml)			Other fuels MJ/km (MJ/ml)		Total energy MJ/km (MJ/ml)
	Fuel production energy	Energy content	Fuel production energy	Energy content	Feedstock	Fuel production energy	Energy content	
Rigid vehicles								
<1	0.40 (0.64)	0.12 (0.19)	1.37 (2.20)	7.20 (11.58)	0.01 (0.02)	0.08 (0.13)	0.52 (0.84)	9.70 (15.60)
1–2	0.44 (0.71)	0.14 (0.23)	1.61 (2.59)	8.34 (13.42)	0.01 (0.02)	0.11 (0.18)	0.66 (1.06)	11.31 (18.21)
3	0.47 (0.76)	0.15 (0.24)	1.79 (2.88)	9.36 (15.06)	0.02 (0.03)	0.13 (0.21)	0.79 (1.27)	12.71 (20.45)
4	0.51 (0.82)	0.16 (0.26)	2.00 (3.22)	10.31 (16.59)	0.02 (0.03)	0.15 (0.24)	0.91 (1.46)	14.06 (22.62)
5–8	0.56 (0.90)	0.18 (0.29)	2.31 (3.72)	11.87 (19.10)	0.02 (0.03)	0.18 (0.29)	1.10 (1.77)	16.22 (25.65)
9	0.59 (0.95)	0.19 (0.31)	2.51 (4.04)	12.82 (20.63)	0.03 (0.05)	0.20 (0.32)	1.22 (1.96)	17.56 (28.26)
10–12	0.60 (0.97)	0.19 (0.31)	2.56 (4.12)	13.08 (21.05)	0.03 (0.05)	0.20 (0.32)	1.25 (2.01)	17.91 (28.83)
13–20	0.63 (1.01)	0.20 (0.32)	2.70 (4.34)	13.84 (22.27)	0.03 (0.05)	0.22 (0.35)	1.35 (2.17)	18.97 (30.51)
Articulated vehicles								
<10	0.60 (0.97)	0.19 (0.31)	2.53 (4.07)	12.96 (20.85)	0.03 (0.05)	0.20 (0.32)	1.24 (2.00)	17.75 (28.57)
10–12	0.62 (1.00)	0.20 (0.32)	2.63 (4.23)	13.47 (21.67)	0.03 (0.05)	0.21 (0.34)	1.30 (2.09)	18.46 (29.70)
13–14	0.70 (1.13)	0.23 (0.37)	3.13 (5.04)	15.93 (25.63)	0.03 (0.05)	0.26 (0.42)	1.61 (2.59)	21.89 (35.23)
15–16	0.71 (1.14)	0.23 (0.37)	3.16 (5.08)	16.08 (25.87)	0.03 (0.05)	0.26 (0.42)	1.62 (2.61)	22.09 (35.54)
17–18	0.73 (1.17)	0.23 (0.37)	3.27 (5.26)	16.61 (26.73)	0.04 (0.06)	0.27 (0.43)	1.69 (2.72)	22.84 (36.74)
18	0.85 (1.37)	0.27 (0.43)	3.96 (6.37)	20.16 (32.44)	0.05 (0.08)	0.35 (0.56)	2.13 (3.43)	27.77 (44.68)

Table 9.9

Energy requirements per vehicle-km (per vehicle-mile) for different size goods vehicles. Values include direct vehicle fuel consumption, vehicle construction and maintenance energies as well as an allowance for the construction and maintenance of the route facilities.

Vehicle type maximum payload in tonne	Electricity MJ/km (MJ/ml)		Oil fuels MJ/km (MJ/ml)			Other fuels MJ/km (MJ/ml)		Total energy MJ/km (MJ/ml)
	Fuel production energy	Energy content	Fuel production energy	Energy content	Feedstock	Fuel production energy	Energy content	
Rigid vehicles								
<1	0.62 (1.00)	0.19 (0.31)	1.42 (2.28)	7.44 (11.97)	0.01 (0.02)	0.08 (0.13)	0.52 (0.84)	10.28 (16.55)
1–2	0.71 (1.14)	0.23 (0.37)	1.67 (2.69)	8.64 (13.90)	0.01 (0.02)	0.11 (0.18)	0.66 (1.06)	12.03 (19.36)
3	0.80 (1.29)	0.25 (0.40)	1.86 (2.99)	9.72 (15.64)	0.02 (0.03)	0.13 (0.21)	0.79 (1.27)	13.57 (21.83)
4	0.89 (1.43)	0.28 (0.45)	2.08 (3.35)	10.73 (17.26)	0.02 (0.03)	0.15 (0.24)	0.91 (1.46)	15.06 (24.22)
5–8	1.02 (1.64)	0.32 (0.51)	2.41 (3.88)	12.37 (19.90)	0.02 (0.03)	0.18 (0.29)	1.10 (1.77)	17.42 (28.02)
9	1.00 (1.16)	0.35 (0.56)	2.62 (4.22)	13.38 (21.53)	0.03 (0.05)	0.20 (0.32)	1.22 (1.96)	18.80 (30.25)
10–12	1.02 (1.64)	0.36 (0.58)	2.68 (4.31)	13.65 (21.96)	0.03 (0.05)	0.20 (0.32)	1.25 (2.01)	19.19 (30.87)
13–20	1.19 (1.91)	0.38 (0.61)	2.83 (4.55)	14.45 (23.25)	0.03 (0.05)	0.22 (0.35)	1.35 (2.17)	20.45 (32.89)
Articulated vehicles								
<10	1.15 (1.85)	0.35 (0.56)	2.65 (4.26)	13.52 (21.75)	0.03 (0.05)	0.20 (0.32)	1.24 (2.00)	19.14 (30.79)
11–12	1.16 (1.87)	0.37 (0.60)	2.75 (4.42)	14.06 (22.62)	0.03 (0.05)	0.21 (0.34)	1.30 (2.09)	19.88 (31.99)
13–14	1.37 (2.20)	0.44 (0.71)	3.28 (5.28)	16.66 (26.81)	0.03 (0.05)	0.26 (0.42)	1.61 (2.59)	23.65 (38.06)
15–16	1.39 (2.24)	0.44 (0.71)	3.31 (5.33)	16.82 (27.06)	0.03 (0.05)	0.26 (0.42)	1.62 (2.61)	23.87 (38.42)
17–18	1.44 (2.32)	0.45 (0.72)	3.43 (5.52)	17.38 (27.96)	0.04 (0.06)	0.27 (0.43)	1.69 (2.72)	24.70 (39.73)
18	1.74 (2.80)	0.55 (0.88)	4.16 (6.69)	21.13 (34.00)	0.05 (0.08)	0.35 (0.56)	2.13 (3.43)	30.11 (48.44)

9.6 RAILWAY TRANSPORT

Unlike road transport, commercial railway networks are usually under the control of the State or very large companies with interchange of traffic occurring between neighbouring networks. Goods fed into a railway system may follow a variety of routes to a given destination because the aim is to move maximum train loads (up to 1,000 tonne in the U.K.). Consequently the most useful data relates to the operation of the whole network apportioned between the volume of different types of traffic using the network. Much of the available data is of this type [38, 98], although limited data are available for specific journeys [84]. As for road transport, examination of the energy required for rail traffic can be considered under the three headings of direct fuel energy, vehicle construction and maintenance, and route construction and maintenance.

9.6.1 Fuel requirements of rail transport

Information on the performance of rail transport is usually produced in terms of energy per tonne-km (or tonne-mile) and may be calculated either as an average for the whole of a network or as a specific value for a particular type of load or journey. Care must therefore be taken to identify the precise nature of the base from which any quoted values are derived. For example, it would be misleading to assume that the average energy consumption of a high-speed freight train over a major trunking route is representative of the network as a whole.

In general there are three main factors which introduce variations into the energy performance of motor units. These are speed, loading and the mix of motor units, that is steam, diesel or electric. Clemow [38] has reported the effect of average train speed on the fuel energy requirement and showed that for a heavy freight train, a fuel energy requirement of 0.045 MJ/tonne-km at 48 km/hr (30 mph) rose to a requirement of 0.08 MJ/tonne-km at 113 km/hr (70 mph) with the change obeying an almost linear relationship. A similar linear relationship exists for all types of train (for example, passenger, freightliner, and so on). The importance of this is that if rail performances are calculated from the performance of a single train over a specific journey [84] then some account must be taken of the average speed of the train.

The second factor influencing energy performance, namely loading, gives rise to effects very similar to those observed in road vehicles. First, in a manner similar to that demonstrated in Figure 9.3, the load moved will influence the numerical value for the energy per tonne-km (per tonne-ml). Secondly, loading will also influence the energy requirements per vehicle-km but in a somewhat different manner to that for road transport. Clemow [38] has pointed out that the traction energy supplied to a train is used to overcome inertia, friction, wind resistance and the effect of gradients. Of these wind resistance is expected to be fairly low and will in any case be taken into account when considering the effect of speed. On average, it is expected that the effect of gradients will be evened out on long journeys so that they will not affect network averages significantly.

Friction is expected to be low but what there is, will depend upon the number of trucks and the load carried by each. However, to a first approximation, rolling friction between rigid surfaces may be regarded as sensibly constant once the train is in motion.

By far the most significant influence of loading is the effect on the inertia of the train. Moreover, because of the low friction between the rails and the wheels, the engine must usually exert a tractive force for a long period to attain the required speed. Consequently the number of times the train must start from rest significantly affects the energy requirement. Clemow [38] estimates that a freight train making no unnecessary stops between start and finish of a journey may require approximately 0.10 MJ/tonne-km, whereas the same train making two stops on the same journey would require approximately 0.14 MJ/tonne-km.

The effect of the mix of different motor units is somewhat less clear. The railway system in the U.K. currently enploys both diesel and electric trains, but in general the energy requirements of the two types of traction are thought to be approximately equivalent. For train operation, an input of 3.1 kWh of electricity at the feeder station is equivalent to the consumption of 1 litre of diesel oil. From Table 6.13, these consumptions are equivalent to 46.50 MJ and 45.16 MJ. In consequence, the co-existence of two different types of motor unit does not introduce any serious complicating factors into the calculation of rail performance energies.

Table 9.10 shows the energy requirements of rail transport as reported by a number of workers. Although the spread of values shown is 0.33 to 1.41 MJ/tonne-km, the range of values that are strictly comparable is much narrower and lies between 0.58 to 0.80 MJ/tonne-km. The three lower values may be omitted because these represent estimated values for fully loaded trains or for trains working in circuit, that is well defined journeys with full loads in one direction and empty return loads. The highest value of 1.41 MJ/tonne-km has been omitted from this average range because it is derived from 1968 statistics when the railway system was undergoing modernisation.

It is however wrong to assume that *all* rail traffic possesses the same gross traction energy requirement as freight is shown to have in Table 9.11. Based on these data, the overall average for rail transport, weighted in terms of the distances travelled by the different types of traffic, is 1.58 MJ/tonne-km or per passenger-km. The high value of this average compared with the value for freight only is due to the relatively low performance of passenger trains, which travel long distances and/or stop frequently. For the purposes of industrial energy analysis however, a typical value of 0.66 MJ/tonne-km (1.06 MJ/tonne-mile) is more appropriate. Where however detailed performance data is available for specific types of journey, such as circuit working of the trains transporting high tonnages of raw materials on a regular basis between two points, then the more detailed data should be used.

Table 9.10

Energy requirement for rail freight arising from fuel directly consumed for traction

Energy requirement		Source	Remarks
MJ/tonne-km	MJ/tonne-mile		
0.33	0.53	(1)	Average for fully loaded trains
0.36	0.58	(2)	Fully loaded diesel powered
0.48	0.77	(1)	Part loaded diesel i.e. circuit working
0.58	0.93	(3)	Average loading for 1972
0.68	1.09	(4)	Fully loaded diesel powered
0.76	1.22	(5)	U.S. data for 1968
0.77	1.24	(4)	Fully loaded electric powered
0.80	1.29	(6)	Average loading 1974 data
1.41	2.27	(3)	Average loading 1968 data

Sources
(1) British Rail private communication (1975).
(2) Leach, G. (1975).
(3) Mortimer, N. D. (1974(b)).
(4) Leach, G. and Slesser, M. (1973).
(5) Hirst, E. (1968).
(6) Clemow, C. J. (1975).

Table 9.11

Gross traction energy requirement for different types of rail traffic in the U.K. [after Clemow (1975)]

Type of traffic	Estimated distance travelled in km $\times 10^6$	Gross energy requirement in MJ/tonne-km or MJ/passenger-km
Intercity passenger	88	1.0
Provincial passenger	87	1.6
London & SE passenger	127	1.5
Freight	114	0.8
Parcels	25	7.5

9.6.2 Capital and maintenance energy of rail transport

It is difficult to separate the consumption of the fuels and materials used in the construction and maintenance of the rolling stock from those used in the construction and maintenance of the track. However, since construction and maintenance of both rolling stock and track are usually carried out by the same operator, their separate identification is of no great significance for energy analysis and a composite value will suffice.

The most detailed published breakdown of the energy requirements of a railway network are those given by Clemow [38] for the U.K. railway network in 1973. He showed that the maintenance of rolling stock and track involved a gross energy requirement of some 3.1×10^{10} MJ compared with a gross energy requirement for traction of 6.5×10^{10} MJ. Of the gross energy requirement for maintenance, approximately 37.5% arose from the consumption of fuels, 56.3% was due to civil engineering operations and the remaining 6.2% was due to mechanical and electrical engineering operations.

The major problem is how to apportion this energy requirement between the different types of traffic. Perhaps the simplest method is to adopt the procedure employed for road vehicles and partition on the basis of the number of miles travelled by each type of traffic. Again using the data of Clemow [38], Table 9.12 shows this apportionment. Since freight traffic travelled a total of 1.14×10^8 km, this corresponds to a maintenance energy requirement of 70.3 MJ/km and the gross traction energy requirement for freight, averaged over all freight carried in this same period, is 161.4 MJ/km. Thus maintenance energy represents 43.5% of the traction energy requirement.

An alternative way of apportioning maintenance energy is to calculate the total maintenance energy as a proportion of the total traction energy. Again

Table 9.12

Apportionment of maintenance energy on the basis of the number of miles travelled by different types of rail traffic

Traffic type	Percentage of total miles travelled in 1973	Maintenance energy apportioned on the basis of mileage MJ $\times 10^{10}$
Intercity passenger	19.95	0.618
Provincial passenger	19.73	0.612
London & SE passenger	28.80	0.893
Freight	25.85	0.801
Parcels	5.67	0.176
Totals	100.00	3.100

using the data of Clemow [38], this gives a maintenance energy for freight transport of 47.1% of the gross traction energy requirement, which is close to that calculated by apportioning on the basis of distance.

A reasonable average value for maintenance energy requirement for all types of ancilliary operation, averaged over all types of *freight* traffic, would therefore appear to be about 45% of the gross traction energy requirement. This value may be compared with the much lower values of 28% estimated by Mortimer [98] for the 1968 operation of the U.K. network.

9.7 SEA TRANSPORT

Cargo ships vary widely in size, ranging from very small cargo ships carrying perhaps 5000 tonne to supertankers carrying in excess of 250,000 tonne. Not surprisingly therefore a wide variation in performance characteristics is expected and this is reflected in the gross energy requirement of this form of transport.

As with road and rail transport, the energy requirement of shipping arises from three main contributions, direct fuel consumption, construction of the vessel and maintenance of the vessel. Each of these contributors is examined in more detail in the following sub-sections.

However, before considering reported values for energy requirements it is useful to note the terminology of shipping because it will help in the extraction of data from some published work.

(1) Distance are usually expressed in **nautical miles** and 1 British nautical mile × 1.1515 statute miles = 1.0006 international nautical miles.

(2) Speeds are usually measured in **knots** where 1 knot is 1 international nautical mile per hour = 1.1508 statute miles per hour.

(3) The weight of a vessel may be described in a variety of units of which six are in common use.

   (a) **Cargo weight** is the weight of the maximum cargo load that the vessel may carry.

   (b) **Deadweight tonnage** is the weight of the cargo, fuel and stores that the vessel may carry.

   (c) **Displacement tonnage** is the mass of water displaced when the vessel is fully loaded.

   (d) **Gross tonnage** (or gross register tonnage) is calculated from the volume of all enclosed space on the vessel by allowing 1 ton (2240 lb) for every 100 cu.ft.

   (e) **Net tonnage** (or net register tonnage) is calculated from the enclosed space on the ship which has an earning capacity, by allowing 1 ton (2240 lb) for every 100 cu.ft.

   (f) **Steelweight tonnage** (or light deadweight tonnage) is the weight of the vessel when empty.

(4) All masses are calculated in terms of 2240 lb = 1 ton unless otherwise specifically stated.

9.7.1 Fuel requirement of sea transport

The two main factors affecting the fuel requirement of a ship are its size and its speed. For a ship of displacement d travelling at speed v, the fuel consumption varies as $d^{\frac{2}{3}}$ and v^3 [83]. As an example of the sensitivity of fuel consumption to speed variations, it has been estimated that a 215,000 tanker could halve its fuel consumption if it were to reduce its speed from a normal 15 knots to about 9.5 knots. An alternative relation between speed and fuel consumption has been demonstrated by Mortimer [99] who found an almost linear relationship between the two parameters.

Typical values for the gross energy requirements of the fuels consumed by cargo vessels and by oil tankers are shown in Figure 9.6. From the plotted results, an average value can be deduced graphically and these average values may be used as typical performance characteristics for all types of ships. They are shown in Table 9.13.

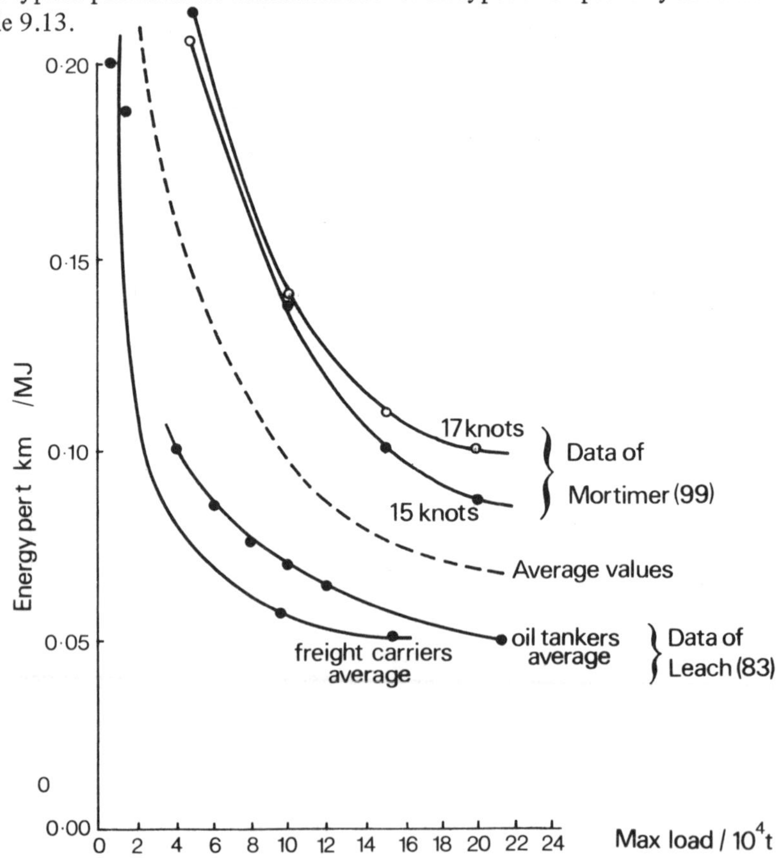

Figure 9.6 Energy requirements of ships assumed to be fully loaded. Full circles are based on the data of Leach (1975), open circles are based on the data of Mortimer (1974(b)).

9.7.2 Capital and maintenance energy of sea transport

The energy requirements arising from the construction and maintenance of a vessel represent an addition of about 11% on the gross fuel energy requirement for a 50,000 ton vessel, rising to about 14% for a 200,000 ton vessel [99]. A reasonable overall average would therefore appear to be 12% for all vessel sizes and the effect of introducing this factor into performance energies is shown in Table 9.13.

Table 9.13

Average values for the energy requirements of ship transport

Cargo mass in tonne	Gross energy requirement		Gross energy requirement including a 12% allowance for the construction and maintenance of the ship	
	MJ/tonne-km	MJ/tonne-mile	MJ/tonne-km	MJ/tonne-mile
20,000	0.208	0.335	0.233	0.375
30,000	0.190	0.306	0.213	0.343
40,000	0.170	0.274	0.190	0.307
50,000	0.152	0.245	0.170	0.274
60,000	0.137	0.220	0.153	0.246
70,000	0.125	0.201	0.140	0.225
80,000	0.115	0.185	0.129	0.207
90,000	0.107	0.172	0.120	0.193
100,000	0.100	0.161	0.112	0.180
110,000	0.094	0.151	0.105	0.169
120,000	0.088	0.142	0.099	0.159
140,000	0.084	0.135	0.094	0.151
150,000	0.080	0.129	0.090	0.144
160,000	0.078	0.126	0.087	0.141
170,000	0.076	0.122	0.085	0.137
180,000	0.073	0.117	0.082	0.131
190,000	0.072	0.116	0.081	0.130
200,000	0.071	0.114	0.079	0.128
250,000	0.069	0.111	0.077	0.124

Chapter 10

Energy analysis of a factory

10.1 ENERGY ANALYSIS AND FACTORIES

So far the concepts underlying industrial energy analysis have been illustrated with examples chosen to highlight specific points and any unnecessary detail has been eliminated. Such examples are seldom however encountered in practice because the unit for which information is available is frequently the factory, a physical system whose boundaries only occasionally coincide with the system boundaries defined by the analyst. Consequently the detail of the factory operation must be taken into account from the beginning. In addition, managements of factories attempt to optimise the use of men and machines so that performance is monitored primarily to check this aspect of operation. Therefore data available may not be in the form needed by the analyst. The purpose of the present Chapter is to examine a real factory, to illustrate the type of problems that may arise and to show how these problems may be overcome.

The ease with which a factory can be analysed depends to a great extent upon the degree of cooperation of the management in supplying information in a usable form. Such cooperation usually can only be obtained when the potential benefits of the analysis become evident to the supplier of the data. So what are these benefits?

Few factories have any precise idea of the energy consumption of different production areas and in the absence of detailed internal monitoring, the energy efficiencies of different operations is also usually unknown. Knowledge of the consumption in the different areas of a factory is useful for a number of reasons and of these four stand out.

(a) Because the cost of fuels has risen markedly in recent years, energy saving measures are being sought by most factory managers. Apart from the more obvious improvements in the 'good housekeeping' aspects of factory practice, most effective savings are usually to be made in the high energy-consuming areas. These areas can be identified by energy analysis.

(b) If new or modified production lines are proposed, the cost of the energy needed to run them must be compared with that of the existing process. Only by having a detailed breakdown of fuel consumption in the different areas of a factory can such a comparison be made and this is provided by energy analysis.

(c) If an operator wishes to compare his energy efficiency with that of a competitor, or indeed with that of another factory within the same company, it is important to compare like with like. Only by identifying the energy requirements for different operational areas of the factory can a satisfactory comparison be achieved. Without a detailed analysis, it is usually not possible to rectify anything but gross differences.

(d) Cost accountants need a method of apportioning fuel costs between different factory cost-centres and detailed energy analysis provides a realistic basis for this to be done.

A second reason for factories to take an interest in energy analysis is the increasing importance of environmental impact analysis. This is essentially a wide-ranging examination of the influence of an industry on the neighbourhood, and in some cases on the nation. Such analyses seek to identify the benefits and drawbacks of processes producing specific commodities in a particular way, examining not only the main production process but also the influence of all industries needed to support the production of the commodities. Factory managers involved in such analyses are therefore interested not only in the operation of their own factories but of the supporting industries. In other words, they have a direct interest in the gross energy requirement of the product. This technique is widely used in the United States and is gradually being adopted elsewhere.

Irrespective of whether the energy analyst is asked to calculate an energy requirement for a system producing only one product from a factory or for a system covering all products, it is usually necessary to analyse the complete factory. This arises because production sequences within a factory are usually interlinked. Even when this is not so, the consumption of fuels and services by a particular production sequence is seldom monitored separately so that the requirements of all operations must be examined and the total energy requirement partitioned between them. The way in which this is done is crucial to the accuracy of the final results and considerable thought must be given to the most satisfactory choice of partitioning parameter.

One recurrent problem in all factory analyses is the form of the available data. Although goods and services are invoiced in terms of physical quantities delivered, records are usually maintained in terms of monetary value. It is tempting to convert back to physical quantities using an assumed average price, but this is fraught with difficulties even when seemingly reasonable assumptions are made. A simple example shows this. The electricity bills for a U.K. factory in two successive months were £5518.37 and £4319.35. If we assume an average price for electricity over the two months, then apparently some 22% less electricity was used during the second month. Inspection of the detailed electricity accounts shows that the electricity consumptions for the two months were 482,600 kWh and 344,800 kWh, that is electricity consumption in the second month was over 28.5% less. The origin of the discrepancy can be easily discovered by a closer examination of the way the prices were calculated.

Month 1

Fixed charge		=	590.15
Daytime rate;	231000 units @ 0.54p	=	1247.40
Night rate;	251600 units @ 0.422p	=	1061.75
Fuel cost adjustment;			
	482600 units @ 0.5427p	=	2619.07
	Total	=	£5518.37
Average cost per unit in this month		=	1.41p

Month 2

Fixed charge		=	590.15
Daytime rate;	231000 units @ 0.54p	=	1247.40
Night rate;	113800 units @ 0.422p	=	480.24
Fuel cost adjustment,			
	344800 units @ 0.5805p	=	2001.56
	Total	=	£4319.35
Average cost per unit in this month		=	1.25p

Without the detailed account it is difficult if not impossible to derive a sensible average cost per unit of electricity in this instance, and even the detailed account for a single month is of little help because the proportion of the fixed charge attributed to each unit of electricity depends upon the number of units consumed. In addition, the fuel cost adjustment, which is included to allow a charge to be made for the variation in fuel prices from month to month, changes frequently.

Similar problems occur in the conversion of all financial data into physical quantities, and the time required to carry out such conversions is usually no less than that needed to trace the information back to the original invoices, a procedure which may also eliminate many other errors. In general therefore, financial data should always be regarded as second-best.

The following Sections of this Chapter describe analysis of a typical factory. The example is based on a real operation but the presentation has been simplified in the following respects.

First, numerical data have been converted to S.I. units. A wide variety of units of measurement are used in industry. Mixtures of metric and imperial units are not uncommon and many of the more traditional industries still use units unique to themselves. For example, the metal can industry uses the 'basis box' as a measure of the steel shipped to can makers, referring to area rather than weight so that 1 basis box is 31,360 square inches. Consequently the energy analyst must frequently devote considerable time to standardising units.

Secondly, numerical data in the present example have been aggregated for the appropriate period. The performance of a factory can seldom be gauged from data relating to a single week or month. Most companies are invoiced monthly and averages of 6 to 12 months are needed to smooth out erratic deliveries and performance, as well as seasonal variations such as those occurring in the demand for space heating. Most operators maintain weekly production sheets and the aggregation of these into longer periods is a time consuming exercise. However, close inspection of such production records often highlights problems such as power cuts, annual holidays, strikes or shortages of supplies which interrupt production and lead to unrepresentative performance characteristics.

Finally, some of the numerical values have been rounded off to simplify presentation so that a number of the results and tables may appear to be inconsistent. More detailed calculations do eliminate the apparent discrepancies.

10.2 OUTLINE DESCRIPTION OF THE FACTORY

The factory analysed is a trade injection moulding factory in which five products labelled *A, B, C, D* and *E* are manufactured using four injection moulding machines referred to by the letters *W, X, Y* and *Z*. Figure 10.1 shows the products made on each machine. Products from machines *X* and *Y* have sprues that must be removed whereas products from machines *W* and *Z* use hot runner tools and do not have sprues. After sprue removal, moulded products are visually inspected and imperfect components rejected. Products *A* and *B* then pass to the paint shop for finishing and after a further inspection are sent to the packing area. Products *C* and *D* pass to an assembly area where they are joined and after a further inspection they too are sent to the packing area. Product *E* receives no further treatment after moulding and is sent directly to the packing area after inspection. All of the products *A, B* and *E* together with the assembled components $(C+D)$ are packed in batches of 50 in polyethylene sacks ready for despatch. The operations within the factory are shown in Figure 10.1.

Data has been assembled for a 12 month period and when aggregated can be summarised in the form of Tables 10.1 and 10.2. During this same period the factory consumed the following fuels:

2.5×10^6 kWh of electricity,
20,000 imperial gallons of kerosine,
110,000 therms of natural gas,
2,000 imperial gallons of petrol.

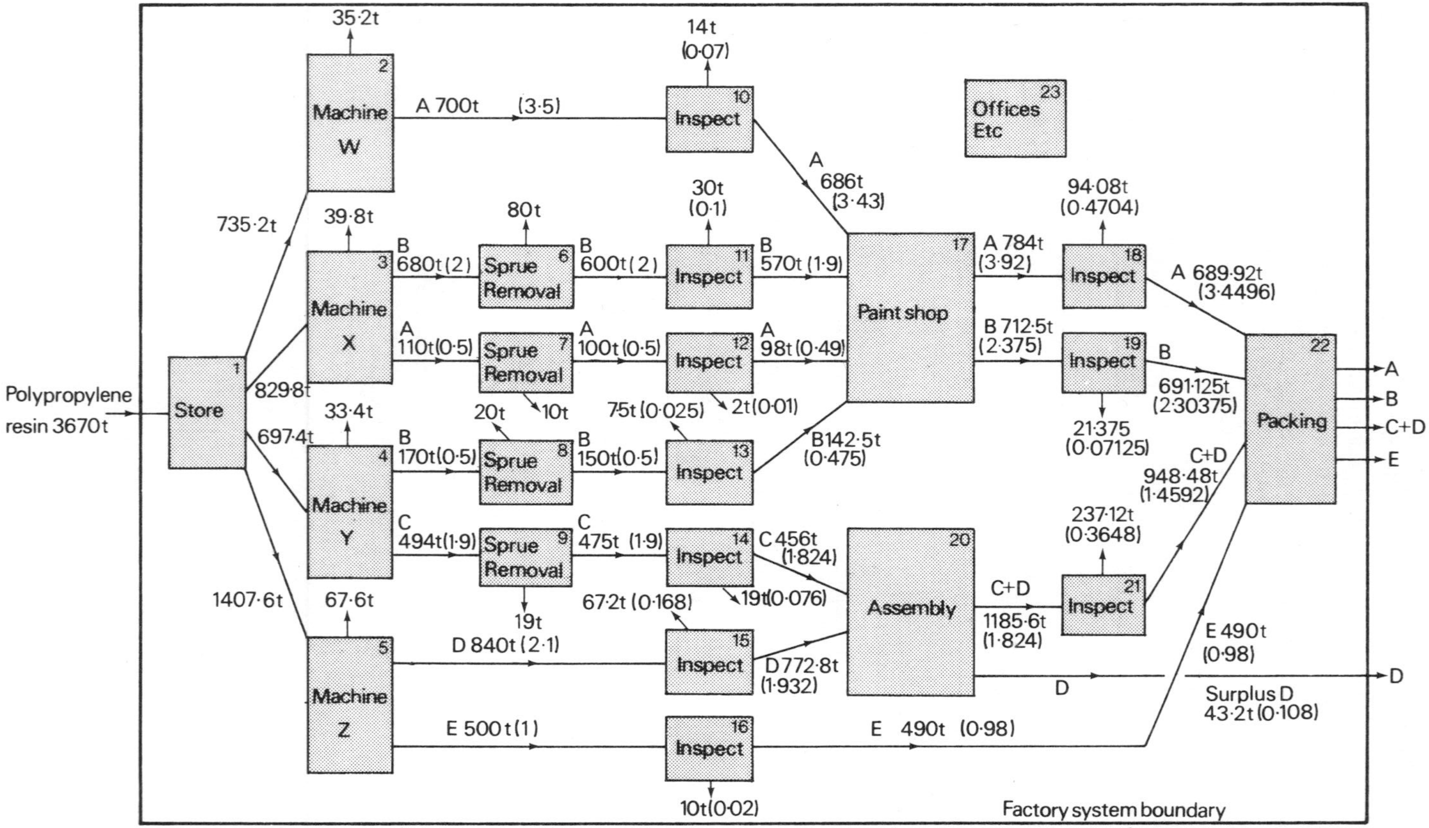

Figure 10.1 Flow diagram of the factory described in Section 10.2 showing the flow of polypropylene mass and the products *A*, *B*, *C*, *D* and *E*. The quantitative flows have been calculated from Tables 10.1 and 10.2 and may be identified as follows; a flow shown as *A* 700 t (3.5) means that the flow consists of product *A*, the total mass flowing in a 12 month period is 700 tonne and this corresponds to 3.5 million components.

Table 10.1
Product specification and reject production

Product	Finished mass (kg)	Sprue mass (kg)	Rejects after moulding as a percentage of the number moulded	Rejects after painting as a percentage of the number painted	Rejects after assembly as a percentage of the number assembled
A	0.20	0.02†	2	12	—
B	0.30	0.04	5	3	—
C	0.25	0.01	4	—	2
D	0.40	—	8	—	2
E	0.50	—	2	—	—

†Machine *X* only

Table 10.2
Moulding production record for a 12 month period

Product	Machine	Number moulded $\times 10^6$
A	*W*	3.5
B	*X*	2.0
A	*X*	0.5
B	*Y*	0.5
C	*Y*	1.9
D	*Z*	2.1
E	*Z*	1.0

In the same period, 3,670 tonne of polypropylene were fed to the moulding area of the factory.

Using the data in Tables 10.1 and 10.2, the materials flows in the factory may be calculated by applying the law of conservation of mass to each of the sub-systems of Figure 10.1. These mass flows, together with the appropriate numbers of components, are shown in Figure 10.1 and they highlight three potential problems for the analyst.

(1) In the period studied, there is greater production of component *D* than of component *C*. Since *C* and *D* are jointed in the assembly area, there must be an additional output representing this surplus of component *D*. These components will be held over to form part of the subsequent year's production and so must not be regarded as waste.

(2) The total mass of moulded components and sprues is 3494 t, yet the factory records a polymer input of 3670 t. The difference of 176 t (that is 4.8% of input) therefore represents an apparently unaccountable loss. This is a common feature found in most factories when considering mass balances. Assuming no mistakes in stock records, this apparent loss of polymer within an injection moulding factory usually results from the use of plastic to purge machines or from the production of badly formed components which do not even reach the first inspection stage and occur when a machine fault is developing. Similarly, waste produced when a new machine tool is being tested will also contribute to this unaccounted loss.

(3) All inspection and sprue removal operations are responsible for producing waste and the quantities of waste are shown in Figure 10.1 as side-flows from the relevant operations. Some of this waste may be reclaimed for re-use, a possibility that is discussed in more detail later.

10.3 GROSS ENERGY ANALYSIS OF THE FACTORY

The data of Tables 10.1 and 10.2 and the calculated materials flows of Figure 10.1 are often the only information readily available. It is quite common for materials flows to be known in great detail while the fuel consumption is only available in outline. In such a situation it is possible to prepare a very approximate gross analysis and the necessary data for such an analysis are summarised in Figure 10.2.

Some care is needed in interpreting the meaning of the system shown by Figure 10.2 and the essential features are best summarised by listing the sub-systems included and excluded.

Sub-systems included

(a) All manufacturing operations within the factory which convert polypropylene resin to saleable or usable products.

(b) All administrative functions within the factory offices.

(c) All sales functions carried out off-site and for which the factory system supplies the fuel (that is the petrol for the salesmens' cars).

(d) The production and delivery of all fuels used. This is automatically taken into account if the energy requirement is presented in the form of Table 7.4.

Sub-systems excluded

(a) Capital energy used in the construction of the buildings and of machinery used within the factory.

(b) The production and delivery of polypropylene.

(c) The production, packing and delivery of ancilliary materials such as plastic sacks used in the packaging area of the factory.

(d) The provision of services such as water.

(e) The production and delivery of paints and solvents used in the paintshop.

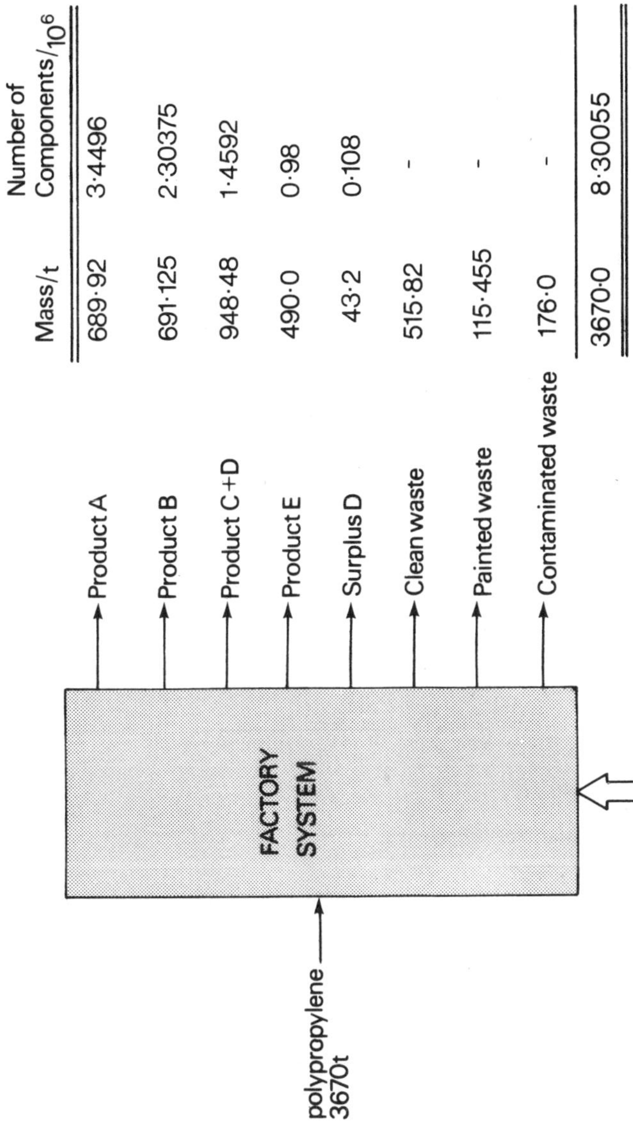

	Mass/t	Number of Components/10^6
Product A	689·92	3·4496
Product B	691·125	2·30375
Product C+D	948·48	1·4592
Product E	490·0	0·98
Surplus D	43·2	0·108
Clean waste	515·82	-
Painted waste	115·455	-
Contaminated waste	176·0	-
	3670·0	8·30055

Figure 10.2 Summary of the gross data from Figure 10.1 for the factory system.

With the system defined in this way, the numerical analysis can now be performed. The simplest and least accurate analysis is that derived by normalising the fuel inputs with respect to the number of saleable or usable components produced. This is the best that can be done when the only available data are the overall fuel consumption and the total number of components produced. On this basis the normalised fuel requirements per component are:

0.301 kWh of electricity,
0.0024 imperial gallons of kerosine,
0.013 therms of natural gas,
0.0002 imperial gallons of petrol.

Converting these into units of energy using Table 6.13, the energy requirement for the production of components using the defined system will be as shown in Table 10.3. This energy requirement is not expected to be particularly accurate. Even without a detailed knowledge of the processing, there is seldom any justification for assuming that all products from a factory are processed in the same way so that precisely the same fuel requirements apply. Such an analysis does however possess one useful property; it establishes an *average* energy requirement for the production of components. More accurate values for each of the products should lie around this value and any set of numerical values for energy requirement must yield Table 10.3 as the average when weighted by the number produced.

Table 10.3

Energy required to produce one saleable component from the factory system of Figure 10.2 when the fuel inputs are normalised with respect to the number of saleable components produced

Fuel type	Fuel production energy (MJ)	Energy content of fuels (MJ)	Feedstock energy (MJ)	Total energy (MJ)
Electricity	3.43	1.08	nil	4.51
Oil fuels	0.09	0.43	nil	0.52
Other fuels	0.20	1.37	nil	1.57
Totals/MJ	3.72	2.88	nil	6.60

More realistic energy requirements are likely to be obtained by assuming that the fuel requirements for processing depend in some way on the product mass. The most common assumption is that energy is directly proportional to product mass. If therefore the fuel requirements are partitioned between products using the masses of the saleable or usable outputs, then for a total saleable mass of 2862.725 t (Figure 10.2), the fuel requirements per kilogramme of saleable output are:

0.873 kWh of electricity,
0.007 imperial gallons of kerosine,
0.038 therms of natural gas,
0.0007 imperial gallons of petrol.

Converting into units of energy using Table 6.13, these fuel requirements lead to Table 10.4. The energy requirement per component may be readily calculated by multiplying each of the entries in Table 10.4 by the component mass. Hence the energy requirement of the system producing component A will be as shown in Table 10.5. Repeating this procedure for the other products gives the system energy requirements of 5.74 MJ for component B, 12.48 MJ for the jointed component $(C+D)$, 9.59 MJ for component E and 7.67 MJ for the surplus components D. When an average energy requirement per component is calculated, weighted by the number produced, a value of 6.61 MJ/component is obtained. The difference between this value and that in Table 10.3 arises from the arithmetical rounding of values.

By assuming a processing energy requirement proportional to component mass, the energy requirements obtained for the systems producing the different components are markedly different. This demonstrates that the simple analysis based solely on the numbers produced is not sufficiently accurate and that a more detailed analysis is required.

Before considering such an analysis it is worthwhile considering the effect of including the surplus D components with the saleable output. They represent 1.5% of the mass output and 1.3% of the number output, and if omitted corresponding errors would be introduced in Tables 10.3 and 10.4. However by including them in the analysis, their energy requirement may be carried into the following year.

Table 10.4

Energy required to produce 1 kg of saleable output from the system of Figure 10.2 when fuel inputs are normalised with respect to the total saleable mass

Fuel type	Fuel production energy (MJ)	Energy content of fuels (MJ)	Feedstock energy (MJ)	Total energy (MJ)
Electricity	9.95	3.14	nil	13.09
Oil fuels	0.24	1.27	nil	1.51
Other fuels	0.57	4.01	nil	4.58
Totals/MJ	10.76	8.42	nil	19.18

Table 10.5

Energy required to produce one component A of mass 0.2 kg using the data in Table 10.4

Fuel type	Fuel production energy (MJ)	Energy content of fuels (MJ)	Feedstock energy (MJ)	Total energy (MJ)
Electricity	1.99	0.63	nil	2.62
Oil fuels	0.05	0.25	nil	0.30
Other fuels	0.11	0.80	nil	0.91
Totals/MJ	2.15	1.68	nil	3.83

10.4 DETAILED ANALYSIS OF THE FACTORY

A more detailed analysis than that described in the preceding Section requires more precise information not only concerning the consumption of fuels in specific areas of the factory but regarding the procedures employed within the factory as well.

Table 10.6 gives the available fuel consumption data and is typical of that available for most factories. Some fuels, such as petrol and kerosine in this instance, are known to be used exclusively in specific areas of the factory. Others, such as electricity and natural gas, are only available in a partially broken-down form depending upon the amount of internal metering carried on within the factory.

Available fuel consumption data often falls into two types. First, and usually the most significant, is the fuel supply to areas of the factory which carry out specific processes. In Table 10.6, the fuels to the paintshop and to the moulding and assembly area are of this type. Such fuel consumption is usually dependent in some way on the throughput of materials and is analogous to the direct costs of an accountant. The second type, providing for example heating and lighting, are ancilliary to the main production process. Such fuel consumption is usually independent of the materials throughput and is analogous to the accountant's fixed costs. This distinction is important because fuels depending upon throughput can usually be partitioned between producing systems using some easily identified parameter related to materials throughput. On the other hand, partitioning of the ancilliary fuels is often a matter of judgement and alternative partitioning schemes are often equally valid.

Table 10.6
Metered or known fuel consumptions in different factory areas

	Electricity (kWh)	Kerosine (gall)	Natural gas (therm)	Petrol (gall)
Salesmen's cars	nil	nil	nil	2,000
Office heating	nil	nil	10,000	nil
Factory heating excluding the paintshop	nil	nil	100,000	nil
Lighting of factory and offices but excluding the paintshop	0.1×10^6	nil	nil	nil
Paintshop - all operations	0.4×10^6	20,000	nil	nil
Machines and assembly	2.0×10^6	nil	nil	nil
Totals	2.5×10^6	20,000	110,000	2,000

The essential purpose of this more detailed analysis is to subdivide the factory's operation into the four subsidiary systems *A*, *B*, (*C*+*D*) and *E* as shown in Figure 10.3 and then to evaluate the energy requirement of each of the sub-systems represented. As discussed in Chapter 8, when an energy requirement has been calculated for each of these single product systems and normalised with respect to the output, the resulting value can reasonably be quoted as the energy required to produce the components within the factory system.

The problem is however twofold. First, few of the energy consuming sub-systems of Figure 10.3 coincide with physical operations within the factory; moulding of product *A* for example arises from the operation of two separate machines. Moreover, even when the sub-systems in Figure 10.3 are related to existing physical operations, they may still not coincide with systems for which fuel consumptions are measured.

The effect of this is shown in Figure 10.4. This is a simplified form of Figure 10.3 and assumes that storage, sprue removal, inspection and packing do not involve any fuel consumption. The heating and lighting of all these areas is accomodated within the 'ancilliaries' sub-system. These simplifications leave four subsidiary systems *A*, *B*, (*C*+*D*) and *E* and only 14 sub-systems. Superimposed on Figure 10.4 are a set of physical system boundaries corresponding to identifiable

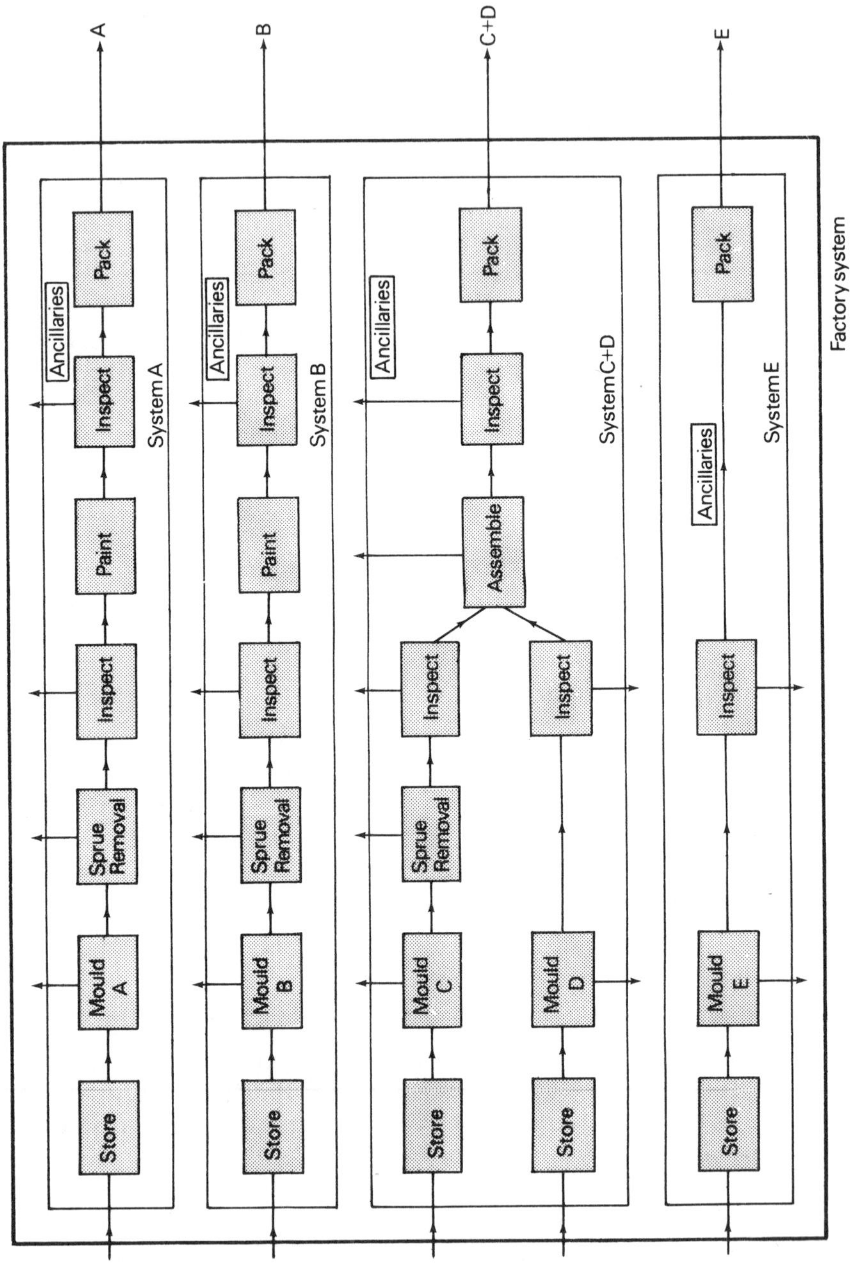

Figure 10.3 The factory system of Figure 10.1 simplified to show the four subsidiary systems contained within the overall system such that each subsidiary system produces a single component type.

machines or areas of the factory and, as can be seen, only moulding machines *A* and the assembly area lie totally within the boundary of any of the subsidiary systems. Also shown in Figure 10.4 are the three systems for which fuel consumption is monitored. None of these fuel systems lies wholly within any boundary of the subsidiary systems *A*, *B*, (*C*+*D*) or *E*.

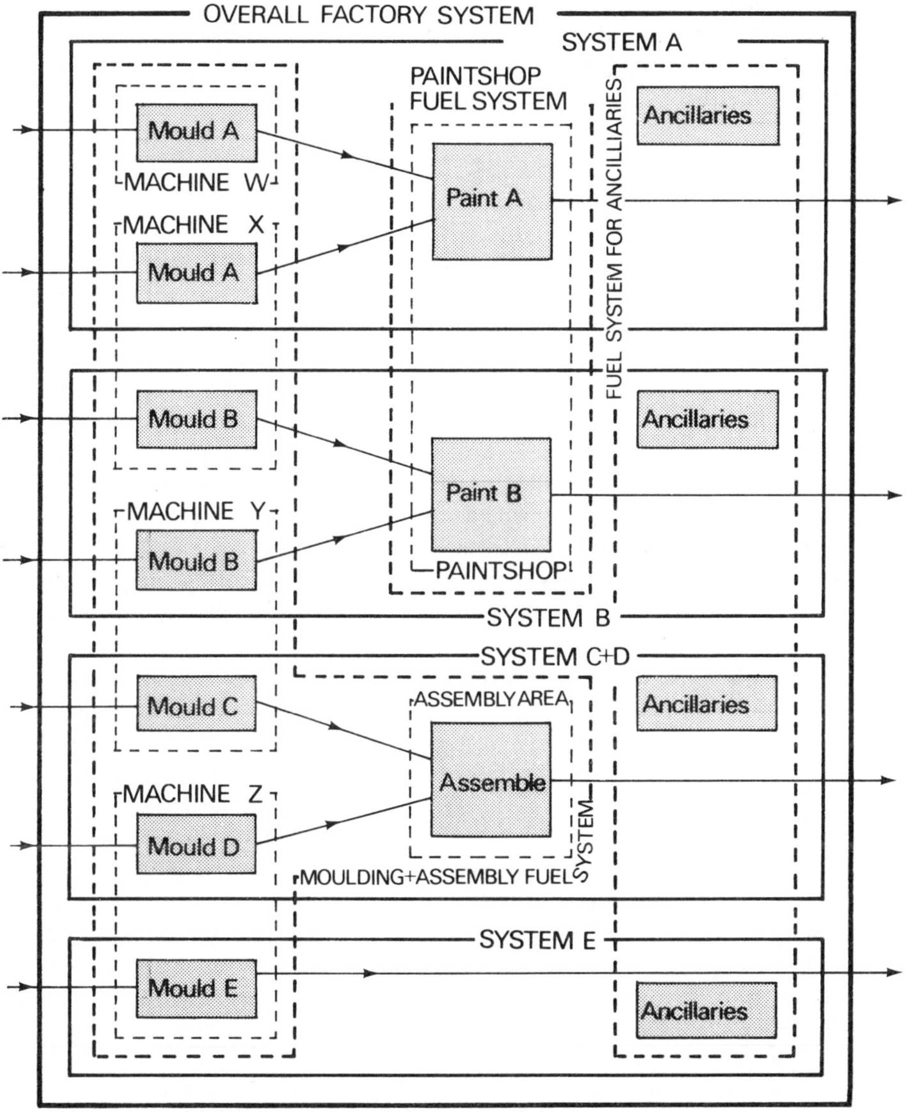

Figure 10.4 Diagrammatic representation of the different system boundaries that can be defined for the factory system of Figure 10.1. These are the overall factory system, the subsidiary systems defined in Figure 10.3, the physical component systems and the systems for which the fuel consumptions are known.

Any errors introduced by this simplification will arise from the use of incorrect partitioning parameters for fuels consumed by the absent sub-systems. These are however expected to be small because of the low fuel consumption of the absent operations.

10.4.1 Partitioning ancilliary energies

Ancilliary energy use arises principally from the use of petrol in salesmens' cars, natural gas used in office and factory heating and electricity used for lighting offices and factory. From Table 10.6, the total requirement of these ancilliary fuels is:

0.1×10^6 kWh of electricity,
2,000 imperial gallons of petrol,
110,000 therms of natural gas.

To calculate the fuel requirements of sub-systems it is importamt to recognise the distinction between a partitioning parameter and a normalising parameter. A **partitioning parameter is applied when the fuel consumption of a system must be apportioned between component sub-systems.** It is usually based on some physical quantity chosen so that when the partitioning has been carried out, the fuel consumption attributed to each sub-system represents as closely as possible the fuel that was actually used by that sub-system in processing the materials in the defined manner. As a result, a partitioning parameter may be related to the inputs, outputs or some intermediate flow of materials within the sub-system and may be mass or volume or surface area or any one of a host of possible properties thought to be the major influence on the processing energy requirement. Hence choice of a suitable partitioning parameter usually requires detailed knowledge of the internal working of the system or sub-system.

In contrast, a **normalising parameter is simply chosen to eliminate the dependence of the fuel or energy requirement of a system on the throughput of that system.** It is almost always related to the output of the system but may occasionally be referred to the input; the two are of course related by the materials conversion efficiency (see Section 2.2). In general therefore, there is no reason why the property chosen to assign a partitioning parameter should be the same as that used in normalisation, though occasionally it is the same.

Returning to the present problem, the simplest method of partitioning the ancilliary energy amongst the four subsidiary systems *A*, *B*, $(C+D)$ and *E* is on the basis of the number of saleable components produced by the overall factory system. For this purpose, the product formed by joining components *C* and *D* is regarded as a single product. Furthermore we have excluded production of surplus components *D*. Assuming that these stay within the factory and are used during the following year, they will then attract a contribution from that year's ancilliary energy use.

Using data from Figure 10.2, the fuel consumption per 100 components irrespective of type will be:

1.221 kWh of electricity,
0.024 imperial gallons of pertol,
1.343 therms of natural gas.

The corresponding energy requirement may therefore be calculated using Table 6.13 and this is shown in Table 10.7. In this instance, the partitioning and normalising parameters are the same.

Table 10.7

Energy required to produce 100 saleable components of any type from the factory system of Figure 10.1 due to the use of fuels in ancilliary operations

Fuel type	Fuel production energy (MJ)	Energy content of fuels (MJ)	Feedstock energy (MJ)	Total energy (MJ)
Electricity	13.92	4.40	nil	18.32
Oil fuels	0.75	3.92	nil	4.67
Other fuels	20.23	141.61	nil	161.84
Totals/MJ	34.90	149.93	nil	184.83

It can be argued with some justification that the operation of the factory is more closely described by the number of components moulded rather than by the number that are actually sold. Using the numerical data for Figure 10.1, partitioning factors for the different components may be calculated in terms of the relative numbers of components moulded as shown in Table 10.8. When the total ancilliary fuel consumption is multiplied by the partitioning factor for each product, the result is the fuel attributed to the whole production of that component. If these are now normalised with respect to the saleable number produced and converted into units of energy using Table 6.13, the energy requirements per unit product due to ancilliary fuel supplies will be as shown in Table 10.9.

There are now two features to note. First, unlike the earlier method of apportionment, where the partitioning and normalising factors were the same, the energy requirements for the different products are different. This will always be the result of using different factors for partitioning and normalising except when the process has a materials conversion efficiency of 100%. Secondly, the value for the product $(C+D)$ has been found by summing the fuel requirements for the separate components C and D. By apportioning in this way, that proportion of the ancilliary energy attributed to the production of surplus components D has been included in the analysis.

Table 10.8
Calculation of partitioning factors and hence ancilliary fuel requirements based on the number of components moulded

Component	Number moulded $\times 10^6$	Partitioning factor = number moulded as a fraction of the total number moulded	Fuel requirements: Electricity (kWh)	Petrol (gall)	Natural gas (therms)
A	4.0	0.348	34,800	696	38,280
B	2.5	0.217	21,700	434	23,870
C	1.9	0.165	16,500	330	18,150
D	2.1	0.183	18,300	366	20,130
E	1.0	0.087	8,700	174	9,570
Total	11.5	1.000	100,000	2,000	110,000

Table 10.9
Energy requirement per 100 saleable components due to ancilliary fuel consumption when partitioning is based on the relative numbers of components moulded

Component	Electricity/MJ: Production and delivery	Energy content	Oil fuels/MJ: Production and delivery	Energy content	Other fuels/MJ: Production and delivery	Energy content	Total energy (MJ)
A	11.50	3.63	0.63	3.29	16.71	117.01	152.77
B	10.74	3.39	0.59	3.07	15.60	109.24	142.63
C+D	27.19	8.59	1.49	7.80	39.50	276.57	361.14
E	10.12	3.20	0.55	2.90	14.71	102.96	134.44

The partitioning of these ancilliary fuels relies heavily on the analyst's judgement, because it is possible to argue for the use of a variety of different parameters such as the mass moulded or the mass output. At the same time, arguments could be proposed for partitioning the salesmens' petrol on the basis of output and the heating and lighting on the basis of mass moulded. In this instance we can be guided by the very approximate analysis of Section 10.3, which indicates that the most significant contributor to the total system energy requirement arises from the electricity used in moulding. If as a result of more detailed analysis, the ancilliary energy is found to be particularly important in some instances, then we can return and consider alternative methods of apportionment to see if they introduce any significant differences.

10.4.2 Partitioning energy in the paintshop

The paintshop's total fuel requirement of 0.4×10^6 kWh of electricity and 20,000 imperial gallons of kerosine must be partitioned between products *A* and *B* (Figure 10.4). Again, in the absence of more detailed information the relative numbers of components entering this area of the factory is probably the best partitioning parameter. From Figure 10.1, it is clear that 62.3% of the throughput is component *A* and 37.7% is component *B*. Hence the fuel attributable to product *A* is:

0.249×10^6 kWh of electricity,
12,460 imperial gallons of kerosine.

This fuel consumption is now normalised with respect to the output *from the paintshop* and when converted to units of energy using Table 6.13, the energy requirement is as shown in Table 10.10. A similar calculation for product *B* will give precisely the same result. In fact it is unnecessary to perform this calculation because it is clear from the materials flows of Figure 10.1 that this process has a materials conversion efficiency of 100% and so, although the partitioning parameter is defined as being different to the normalising parameter, they are in fact equivalent. Hence the energy requirements of the different systems producing both these products will be the same (see Section 10.4.1).

Table 10.10

Energy requirement attributable to 100 components *A* or *B* leaving the paintshop. Partitioning has been carried out on the basis of the number of individual components entering the operation

Fuel type	Fuel production energy (MJ)	Energy content of fuels (MJ)	Feedstock energy (MJ)	Total energy (MJ)
Electricity	72.44	22.87	nil	95.31
Oil fuels	10.06	52.81	nil	62.87
Other fuels	nil	nil	nil	nil
Totals/MJ	82.50	75.68	nil	158.18

If we examine actual practice within this factory, this method of partitioning can be shown to be unsatisfactory here, illustrating the need for detailed knowledge of the sub-systems studied. Products for painting are laid out on wire mesh trays adequately spaced so that the paint spray produces a satisfactory coating. After spraying a conveyor belt moves the trays into a hot air blast which vaporises the solvent. This is then recovered by an extraction system. The number of components that can be loaded onto a tray depends upon the geometry of the moulding and it is found that a tray will hold 40 of component *A* but only 20 of component *B*. The rate of working of this section of the factory is governed by the rate at which it can process a tray-load and this implies that the most satisfactory partitioning parameter would be the relative number of tray-loads passed rather than the number of components. For the inputs shown in Figure 10.1, the number of tray-loads of component *A* will be 98,000 and the number of tray-loads of component *B* will be 118,750. Partitioning on the basis of these parameters gives a fuel consumption for product *A* of:

0.1808×10^6 kWh of electricity,
9.040 imperial gallons of kerosine.

Table 10.11

Energy requirement attributable to 100 components *A* leaving the paintshop when energy within the paintshop has been partitioned on the basis of the number of tray-loads passed

Fuel type	Fuel production energy (MJ)	Energy content of fuels (MJ)	Feedstock energy (MJ)	Total energy (MJ)
Electricity	52.58	16.60	nil	69.18
Oil fuels	7.29	38.29	nil	45.58
Other fuels	nil	nil	nil	nil
Totals/MJ	59.87	54.89	nil	114.76

Table 10.12

Energy requirement attributable to 100 components *B* leaving the paintshop when energy within the paintshop is partitioned on the basis of the number of tray-loads passed

Fuel type	Fuel production energy (MJ)	Energy content of fuels (MJ)	Feedstock energy (MJ)	Total energy (MJ)
Electricity	105.21	33.22	nil	138.43
Oil fuels	14.55	76.39	nil	90.94
Other fuels	nil	nil	nil	nil
Totals/MJ	119.76	109.61	nil	229.37

Again, after normalising with respect to the output number of components from the paintshop and converting into units of energy using Table 6.13, the energy requirement of product *A* is as shown in Table 10.11. In the same way, a value for product *B* may be calculated, and this is shown in Table 10.12.

10.4.3 Partitioning energy in the moulding and assembly area

From Figure 10.4, it is clear that partitioning of energy in the moulding and assembly area is a two-stage process. First, the fuels are partitioned between the four moulding *machines* and the assembly area. Then the energy of each machine is partitioned between the systems producing different products from the machines. To do this requires further detail concerning both the electrical performance of the machines during the period under examination and the times for which they were working. Such data are shown in Table 10.13.

Table 10.13
Moulding machine performance characteristics and operation in the year studied

Machine	Product moulded	Moulding current (A)	Idling current (A)	Warmup current (A)	Moulding time (hr)	Idling time (hr)	Warmup time (hr)
W	*A*	400	80	150	2700	420	260
X	*B*	600	200	250	1500	900	195
X	*A*	600	200	250	500	220	65
Y	*B*	600	200	250	500	220	68
Y	*C*	600	200	250	1900	500	192
Z	*D*	800	200	250	1300	250	130
Z	*E*	800	200	250	1300	270	130

Note that the performance of the machines is divided into three groups: moulding, idling and warm-up. The **moulding current** is that drawn when the machine is in normal production and moulding components. When the machine develops a fault, it remains connected to the supply with heaters operating until the fault is rectified. This is termed **idling** and the current drawn during this mode of operation will be less than the moulding current because the hydraulic part of the machine is inoperative. This factory operates for a 12 hour period, five days per week. One hour prior to working, the machines are switched on to allow them to **warm-up** to operating temperature. The currents drawn during this period are as shown.

The currents drawn by the machines under the different loading conditions are characteristcs of the machines and in this instance were determined by direct measurement. Therefore they represent the best available on machine performance. In the absence of such information, a second best approach is to assume that the power drawn by the machines is proportional to the motor ratings. The machine operating times given in Table 10.13 were derived from the weekly production sheets held by the factory to monitor production.

From the data in Table 10.13 it is possible to calculate the relative proportions of electricity consumption in the moulding and assembly area going to each machine and to each component-producing system using the product of current and time as the partitioning parameter (Table 10.14). The justification for this is simply that the energy drawn by any electrical component will be proportional to this product. It is possible to carry out both stages of the partitioning at the same time because of the level of detail in Table 10.13.

Table 10.14

Partitioning of electricity in the moulding and assembly area using the product of current and time as partitioning parameter

Machine	Product	Current × time when moulding (10^3 Ah)	Current × time when idling (10^3 Ah)	Current × time during warmup (10^3 Ah)	Total current × time (10^3 Ah)	Total current × time as % of overall total	Electricity (10^6 kWh)
W	*A*	1080	33.6	39	1152.6	15.56	0.312
X	*B*	900	180	48.75	1128.75	15.24	0.305
X	*A*	300	100	16.25	416.25	5.62	0.112
Y	*B*	300	100	17	417	5.63	0.112
Y	*C*	1140	380	48	1568	21.17	0.423
Z	*D*	1040	260	32.5	1332.5	17.99	0.360
Z	*E*	1040	260	32.5	1332.5	17.99	0.360
Assembly	*C*+*D*	60	–	–	60	0.8	0.016
Totals		5860	1313.6	234	7407.6	100.0	2.000

The partitioned electricity requirements of the sub-systems may now be normalised with respect to the output from each of the sub-systems measured as the number of components. This is shown in Table 10.15.

Table 10.15

Calculation of the normalised energy consumption per 100 components output from each sub-system in the moulding and assembly area of the factory

Machine	Product	Electricity consumption 10^6 kWh	Number of components produced × 10^6	Electricity per 100 components output kWh	Electrical energy (MJ/100 components): Production and delivery	Energy content	Total energy
W	*A*	0.312	3.5	8.91	101.57	32.08	133.65
X	*B*	0.305	2.0	15.25	173.85	54.90	228.75
X	*A*	0.112	0.5	22.40	255.36	80.64	336.00
Y	*B*	0.112	0.5	22.40	255.36	80.64	336.00
Y	*C*	0.423	1.9	22.26	253.76	80.14	333.90
Z	*D*	0.360	2.1	17.14	195.40	61.70	257.10
Z	*E*	0.360	1.0	36.00	410.40	129.60	540.00
Assembly	*C*+*D*	0.016	1.824	0.88	10.03	3.17	13.20

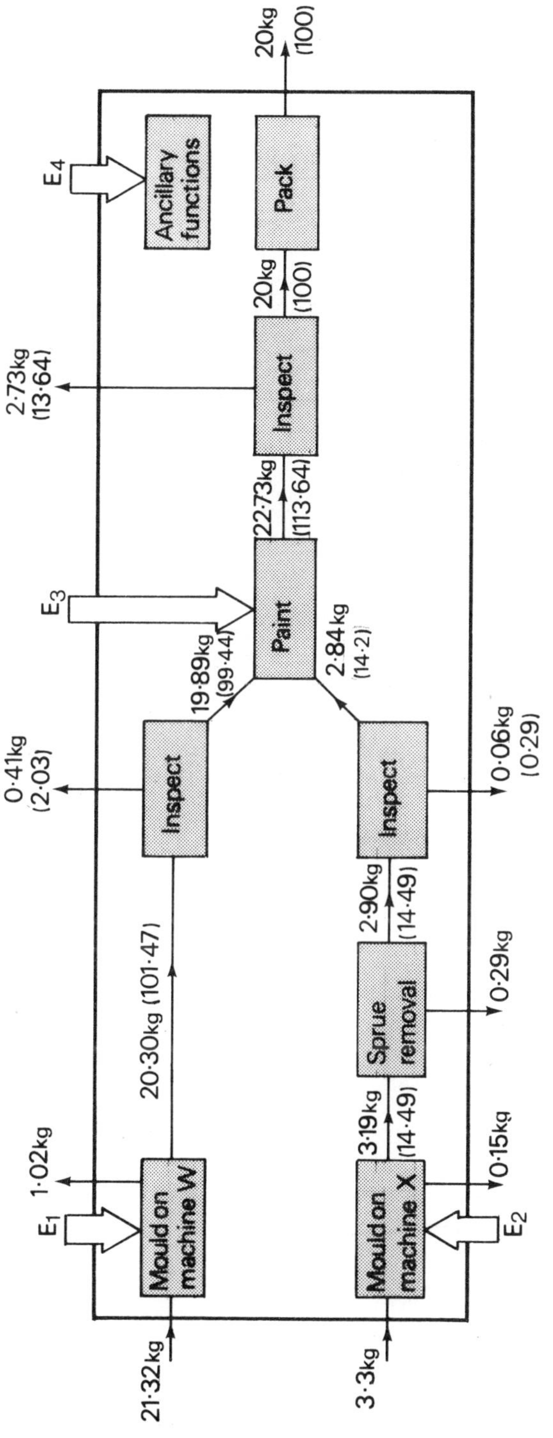

Figure 10.5 Flow diagram for the subsidiary system producing components *A* showing the detailed flows of materials and the points at which fuels are fed into the system.

10.4.4 Total energy requirements for product manufacture

The energy requirements of the individual sub-systems in Figure 10.4 have been calculated in Sections 10.4.1 to 10.4.3. The remaining problem is to combine these to yield the energy requirements of the individual subsidiary systems *A*, *B*, (*C*+*D*) and *E*.

Consider the production of component *A*. The detailed system flow diagram is shown in Figure 10.5 and the materials flows have been calculated from Table 10.1 for an overall system output of 100 components. It has been assumed that the proportions derived from machines *W* and *X* are in the ratios of the total productions from these machines. The overall materials conversion efficiency for producing component *A* is 81.1%. Figure 10.5 shows four inputs of energy to this system, labelled E_1, E_2, E_3 and E_4. Energy requirement E_1 is given in Table 10.15 as 133.65 MJ per 100 components. Hence the energy associated with the production of 101.47 components will be 133.65 × 101.47/100 = 135.6 MJ. Similar calculations are possible for the other energy requirements and when combined they give an overall system energy requirement of 472.38 MJ as shown in Table 10.16.

Table 10.16
Total energy requirement of the system producing 100 components *A*

Operation	Electricity/MJ		Oil fuels/MJ		Other fuels/MJ		Total energy
	Production and delivery	Energy content	Production and delivery	Energy content	Production and delivery	Energy content	
Machine *W* (Table 10.15)	103.06	32.55	nil	nil	nil	nil	135.61
Machine *X* (Table 10.15)	40.73	12.86	nil	nil	nil	nil	53.59
Painting (Table 10.11)	59.74	18.86	8.29	43.52	nil	nil	130.41
Ancilliaries (Table 10.9)	11.50	3.63	0.63	3.29	16.71	117.01	152.77
Totals/MJ	215.03	67.90	8.92	46.81	16.71	117.01	472.38

By repeating this procedure, energy requirements may also be calculated for the systems producing products *B*, (*C*+*D*) and *E* and these are shown in Table 10.17.

Table 10.17
Total energy requirements of the systems producing 100 components

Component produced and overall materials conversion efficiency	Sub-system	Electricity/MJ		Oil fuels/MJ		Other fuels/MJ		Total energy MJ
		Production and delivery	Energy content	Production and delivery	Energy content	Production and delivery	Energy content	
A	Machine *W* (Table 10.15)	103.06	32.55	–	–	–	–	135.61
	Machine *X* (Table 10.15)	37.01	11.66	–	–	–	–	48.69
81.1%	Painting (Table 10.11)	59.74	18.86	8.29	43.52	–	–	130.41
	Ancilliary (Table 10.9)	11.50	3.63	0.63	3.29	16.71	117.01	152.77
	Total for component *A*	211.31	66.72	8.92	46.81	16.71	117.01	467.48
B	Machine *X* (Table 10.15)	150.92	47.66	–	–	–	–	198.58
	Machine *Y* (Table 10.15)	55.44	17.51	–	–	–	–	72.95
77.0%	Painting (Table 10.12)	108.46	34.25	15.00	78.75	–	–	236.46
	Ancilliary (Table 10.9)	10.74	3.39	0.59	3.07	15.60	109.24	142.63
	Total for component *B*	325.56	102.81	15.59	81.82	15.60	109.24	650.62
C+*D*	Machine *Y* (Table 10.15)	269.72	85.18	–	–	–	–	354.90
	Machine *Z* (Table 10.15)	203.44	64.25	–	–	–	–	267.69
89.3%	Assembly (Table 10.15)	10.23	3.23	–	–	–	–	13.46
	Ancilliary (Table 10.9)	27.19	8.59	1.49	7.80	39.50	276.57	361.14
	Total for component *C*+*D*	510.58	161.25	1.49	7.80	39.50	276.57	997.19
E	Machine *Z* (Table 10.15)	418.78	132.24	–	–	–	–	551.02
93.3%	Ancilliary (Table 10.9)	10.12	3.20	0.55	2.90	14.71	102.96	134.44
	Total for component *E*	428.90	135.44	0.55	2.90	14.71	102.96	685.46

10.5 ENERGY ANALYSIS AND ENERGY SAVINGS

One of the main aims of energy analysis is the search for improvements in practices that will reduce the system energy requirement. The detailed factory analysis of Section 10.4 can be used to identify some of the ways in which energy consumption can be reduced. Note that we are primarily concerned with minimising energy requirement and no consideration is given here to the economic consequences. These must be evaluated separately once the energy analyst has proposed possible energy-saving schemes.

10.5.1 Reducing ancilliary fuels

From Table 10.17, it is clear that the fuels used in the ancilliary functions such as heating and lighting represent a significant contribution to the energy requirement of the system. Using Table 10.17 the proportions can be calculated as:

Product A: 32%,
Product B: 22%,
Product $(C+D)$: 36%,
Product E: 20%.

Moreover, it is also clear from Table 10.9 that the natural gas used for heating is by far the greatest contributor (88%) to this component of the energy requirement. Since natural gas has one of the highest production and delivery efficiencies (89.5% from Table 6.13), fuel substitution is unlikely to lead to any significant energy savings. Ways must therefore be found to reduce the direct consumption of fuel if the contribution of space heating to the ancilliary energy is to be reduced.

The provision of acceptable working conditions for the work-force whilst at the same time minimising the use of fuel is a persistent problem in industry [132]. In the factory reported here, local fire regulations demand the provision of two large access doors into the moulding area and these doors are usually kept open to allow ready access by fork lift trucks and delivery vehicles. Not surprisingly therefore a significant proportion of the heat within the factory is carried away in the air stream passing through the factory.

The installation of flexible double doors at the two entrances would permit the necessary access whilst at the same time reducing the heating fuel requirement of the factory area. It is estimated that up to 50% of the fuel energy could be saved by this installation, so reducing the factory consumption of natural gas from 110,000 therms to a minimum of 55,000 therms.

An alternative solution might be the use of electrically operated infra-red spot heaters [132]. Although this incurs the penalty of using electricity, with its accompanying low efficiency of production and distribution, a much decreased consumption of direct energy is possible because only small areas of the factory are heated. The total rated electrical heating that could be installed to give the same total energy requirements as the existing natural gas heating is 283 kW for factory and offices.

If infra-red spot heaters were installed, a typical scheme might be:

4 machines @ 3 kW	=	12 kW
4 sprue removal stations @ 3 kW	=	12 kW
4 mould inspection stations @ 3 kW	=	12 kW
Assembly area @ 20 kW	=	20 kW
2 Assembly inspection stations @ 3 kW	=	6 kW
4 Packing stations @ 6 kW	=	24 kW
Store @ 40 kW total	=	40 kW
10 Offices @ 3 kW	=	30 kW
Total	=	156 kW

The energy consumption (that is energy content plus fuel production energy) with such a scheme is approximately the same as would be achieved by the installation of double doors while retaining natural gas heating. If both double doors and spot heaters were installed, the electrical requirement would decrease below the level indicated because the ambient temperatures of those areas not directly heated would probably reduce the demand for such intense spot heating.

More detailed calculations, as well as costing, would be required if these proposals were to be considered further. However, implementation of *either* scheme would reduce the total energy requirements of the systems producing the products by a maximum of:

Product A: 16%,
Product B: 11%,
Product $(C+D)$: 18%,
Product E: 10%.

10.5.2 Reducing idling time

The idling time represents approximately 22% of the total available machine time (Table 10.13) and idling is responsible for approximately 18% of the direct electrical consumption of the moulding machines (Table 10.14). Idling is wasteful of energy because it represents unproductive operation of the machines. It is inevitable that even in the best ordered moulding factories some idling will occur because machine and tool faults always develop.

There are however two ways in which the energy arising from idling can be minimised. The first is to ensure that the discovery and rectification of faults is prompt. Frequently there is advance warning of faults developing in that factors such as rejection rate at inspection start to rise some time before the fault becomes obvious on the machine itself. Preventive maintenance outside of the normal moulding shift also helps reduce idling time without reducing the time available for moulding.

The second way in which energy may be saved is to switch off the machine when it becomes obvious that rectification of the fault requires a prolonged period. For example, if machine *W* idles for 1.875 hours it will consume the same electrical energy as if it were switched off and subsequently allowed to warm-up from cold. Consequently if any repairs are likely to take longer than 1.875 hours the machine should not be allowed to idle. For the other machines in the factory this cut-off time is only 1.25 hours. It is estimated that idling times and energies could be halved by adopting these strategies.

10.5.3 Continuous working

The use of discrete working periods leads to the need for energy consumption in the daily warm-up periods. In this factory, the warm-up energy contributes approximately 3% to the direct energy consumed by the moulding machines and this could be eliminated by continuous working.

Obviously the possibility of continuous working as opposed to single shift working depends upon a number of factors quite unrelated to energy, such as the availability of labour and the demand for the increased output. Nevertheless, in this factory as in many others, continuous working can result in energy savings although it does represent a significant departure from the established working pattern. It should also be noted that in this instance the establishment of continuous working is unlikely to be prompted by the small energy saving likely to result.

10.5.4 Matching machine and product

One striking feature of Table 10.15 is that the energy requirement associated with products is markedly dependent on the performance of the particular moulding machine used. For instance, the energy required to produce component *A* on machine *X* is 2.5 times the energy needed to produce this same component on machine *W*. Similarly the energy to produce component *B* on machine *Y* is 1.5 times greater than the energy needed to produce it on machine *X*.

It is seldom possible to predict these different performances in advance but once data such as those given in Table 10.15 are available, it would seem sensible to suggest that components are made on the machines giving the best performance. In this instance there would be no difficulty in transferring all of the production of component *B* to machine *X*. On the other hand, it would not be possible to transfer all production of component *A* to machine *W* because *W* is already working full time on component *A*. However some additional moulding time would become available on machine *Y* if component *B* were transferred to machine *X* and using this time to produce component *A* could lead to improvements in fuel demand. Similar proposals cannot be made for the other components because the operating data for their production on different machines are not available.

Table 10.18
Effect on energy requirement for 100 components *A* of introducing energy saving measures

Energy contributor	Electricity/MJ		Oil fuels/MJ		Other fuels/MJ		Total energy
	Production and delivery	Energy content	Production and delivery	Energy content	Production and delivery	Energy content	
Production of 100 components *A* (Table 10.17)	211.31	66.72	8.92	46.81	16.71	117.01	467.48
Installation of double doors	–	–	–	–	−7.60	−53.19	−60.79
Eliminating all idling time	−25.51	−8.06	–	–	–	–	−33.57
Eliminating all warmup time	−4.55	−1.44	–	–	–	–	−5.99
Reducing loss at painting to 3%	−13.53	−4.27	–	–	–	–	−17.80
Producing all components on machine *W*	−19.34	−6.11	–	–	–	–	−25.45
Substituting natural gas for paintshop kerosine	–	–	−8.31	−43.59	+6.23	+43.59	−2.08
Total energy required to produce 100 components of *A* if all of the above measures are introduced	148.38	46.84	0.61	3.22	15.34	107.41	321.80

10.5.5 Improving materials conversion efficiency

Table 10.17 gives the overall materials conversion efficiency for each of the systems producing the various components. All of the polymer input to any system will pass through the moulding machine and if the moulded components are subsequently rejected, the fuel consumed in their production is wasted. Elimination of materials losses from whatever cause inevitably therefore results in a reduction of the energy requirement for the overall system.

For component *A* for example, the most significant materials loss arises from the 12% reject production in the painting process. If this could be reduced to the same level as that achieved for component *B* (that is 3%), it would represent a 9% saving in the moulding energy requirement for component *A*.

Following from the arguments in Section 10.5.2, if rejection at the inspection stage immediately following moulding can be shown to be an indication of the development of machine faults, then prompt remedial treatment will result in a decrease in the losses at this stage of production.

10.5.6 Overall effect of energy saving measures

The precise energy savings that might be achieved by implementing any of the measures described in the preceding Sections can only be determined by performing more detailed calculations than those included in the text. In essence the procedure is to recalculate the energy requirement of the modified system using the best estimates available for the proposed operations. Such calculations require techniques no different from those already introduced and since the details will obviously vary from system to system they are not presented here. It is however useful for illustration purposes to give an estimate of the effect of implementing these measures on the energy requirement of one of the subsidiary systems producing one of the components, and this is shown in Table 10.18. Some of the proposed changes, such as the installation of double doors to reduce factory heating, are more realistic and more easily implemented than others. However the Table gives some idea of the way that the different measures influence the total energy requirement of the system producing component *A* and a total maximum saving of 31% is indicated. Similar tables may be constructed for the other production systems.

Detailed calculations apart, the importance of the discussion in Section 10.5 is the way in which the detailed energy analysis of the factory system allows a ready identification of those areas where energy savings are possible and also helps to indicate ways of making savings.

10.6 Expanding the factory system

The system shown in Figure 10.1 includes only those operations taking place within the physical boundaries of the factory, together with the production and delivery of the fuels. However, the system can be made more complete in the sense discussed in Chapters 3 and 8 by including those operations needed to support this factory and which trace the materials supplies back to the extraction of materials from the earth. Some of the components of such a system are shown in Figure 10.6, where each sub-system requires analysis similar to that described in Section 10.4. We shall not attempt to carry out the detailed analysis for each of the sub-systems but instead rely on data from Part 2 of this book and Chapter 9. Consider each sub-system in Figure 10.6 in turn.

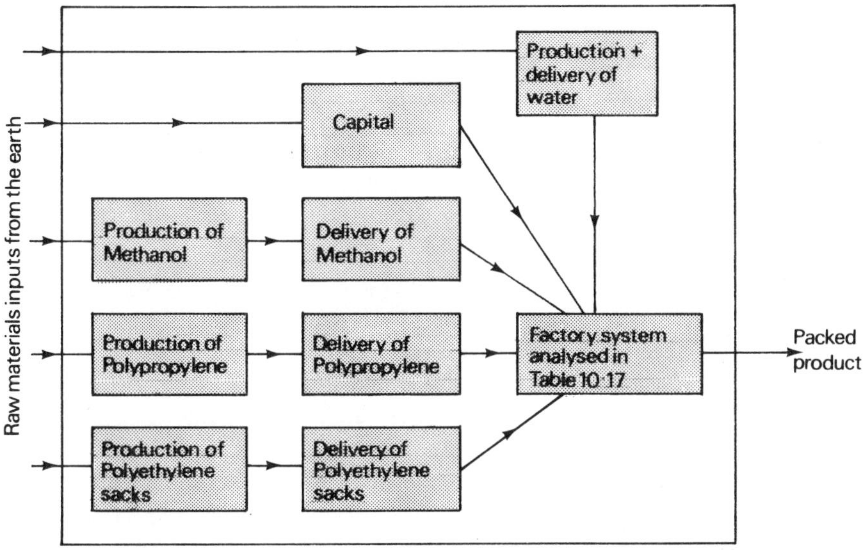

Figure 10.6 A simple extension of the factory system showing some of the additional sub-systems that may be included to produce a more complete analysis of the sequence of operations needed to produce the moulded components from raw materials in the ground.

Water consumption during the period examined was 0.7×10^6 imperial gallons. This was mainly for domestic consumption because the water supply to the moulding machines was on a recirculatory system, a feature common to most industrial processes. Partitioning over the saleable output leads to a water use of 8.5 imperial gallons per 100 components irrespective of type. The energy required to produce and supply 1 imperial gallon of water is 0.042 MJ (Part 2) so that the energy attributable to 100 components will be 0.36 MJ.

The capital energy is more difficult to evaluate. The most accurate form of analysis is to obtain data concerning the construction of the building from a quantity survey and then perform a detailed analysis of the type described in Section 10.4. However, for most purposes we may calculate a reasonable figure using arguments similar to those developed in Section 8.5, where a typical value for a factory of this type might be 14,000 MJ/m^2 of floor area. With a floor area of 2000 m^2, the total energy would be 28×10^6 MJ. If this is amortised over 30 years and then partitioned amongst the saleable products, it leads to an energy requirement per 100 components of 11.4 MJ.

Methanol is employed as a solvent in the paintshop and a reclamation scheme operates so that the annual consumption of methanol is only 20 t. This is apportioned over products *A* and *B* only, using the method described in Section 10.4.2, and when a production energy requirement from Part 2 of 55.22 MJ/kg is used, the total energy attributed to 100 components *A* is 14.36 MJ.

The delivery of methanol draws on transport data from Table 9.9. If we assume a round-trip distance of 100 miles on a 20 tonne payload lorry carrying full loads, the total annual transport energy for methanol is 4844 MJ. This is again partitioned between products *A* and *B* using the method of Section 10.4.2 to give a total energy requirement per 100 components *A* of 0.06 MJ.

The energy associated with the production of polypropylene from Part 2 is 110.16 MJ/kg. The production of 100 components *A* requires a polypropylene input of 24.66 kg (Figure 10.1) so that the total energy required for the polypropylene production for 100 components *A* will be 2716.54 MJ.

Again the delivery energy of polypropylene may be calculated from Table 9.9 and assuming a round-trip distance of 100 miles on fully loaded 20 tonne lorries, the energy per 100 components *A* will be 5.97 MJ.

The polyethylene sacks used to pack the finished product are each of mass 0.2 kg and hold 50 components. From Part 2, the energy to produce 1 kg of polyethylene is 92.32 MJ and assuming that the energy required to convert the polymer resin into sacks in negligible compared with the energy to produce the resin, the energy associated with 100 components from this factory will be 36.93 MJ.

Again, calculating transport energy for the sacks using 20 tonne lorries on a 100 mile round-trip gives an energy requirement per 100 components of 0.10 MJ.

The detailed energy requirement for product *A* is shown in Table 10.19 and similar tables could be prepared for the other components produced by this factory. It is now clear that within this larger system the factory contributes only about 15% of the total energy requirement. By far the most dominant contributor is the energy needed to produce the polypropylene resin which represents about 83% of the system energy requirement.

Because of the significance of the polypropylene in determining the energy requirement of the system shown in Figure 10.6, the greatest energy savings are likely to be produced *in this system* if the demand for polypropylene can be reduced. We know however from Figure 10.1 that the materials conversion efficiency for the production of component *A* is only 81.1% and the waste is currently discarded. Figure 10.5 however shows that of the total input of polypropylene to the system producing component *A*, 0.76 kg can be recovered in a clean state suitable for recycling. The remaining waste is either contaminated

Table 10.19
Energy requirements per 100 components A arising from the more complete system

Contributor	Electricity/MJ		Oil fuels/MJ		Other fuels/MJ			Total energy
	Production and delivery	Energy content	Production and delivery	Energy content	Feedstock energy	Production and delivery	Energy content	
Factory energy (Table 10.17)	211.31	66.72	8.92	46.81	–	16.71	117.01	467.48
Water	0.27	0.09	–	–	–	–	–	0.36
Capital								11.40
Methanol production	1.17	0.37	0.94	4.60	7.28	–	–	14.36
Methanol delivery	neg	neg	0.01	0.05	neg	neg	neg	0.06
Polypropylene production	217.99	68.31	188.65	979.00	1262.59	–	–	2716.54
Polypropylene delivery	0.36	0.11	0.82	4.18	0.01	0.07	0.42	5.97
Polyethylene sack production	6.64	2.10	1.73	8.46	18.00	–	–	36.93
Polyethylene sack delivery	0.01	neg	0.01	0.07	neg	neg	0.01	0.10
							Total	3253.20

with paint or with oil and dirt and so cannot be readily recycled. If the recyclable material is reclaimed, this represents a 3.1% saving in materials and an energy saving of 84.21 MJ in Table 10.19. This concept of recycling is of such importance that it merits a detailed examination. This is the purpose of the next Chapter.

Chapter 11
Recycling

11.1 GENERAL IDEAS OF RECYCLING

Recycling is the process whereby the materials output from one part of a production sequence is fed back to some earlier part of the same sequence. A variety of terms has arisen in this area of technology because of the different types of recycling practised. When materials such as waste off-cuts or waste products appearing within a factory are fed back to some earlier stage, the term **recycling** is generally used. However, when a product is used repeatedly as a matter of common practice, as in the case of a returnable glass bottle, the term **re-use** is applied. When we are considering outputs such as domestic waste, which are currently disposed of as useless but which contain valuable materials that could be recycled or re-used if they could be separated, the term **reclamation** is often used. Despite these differences in terminology, all such processes involve feedback loops of the type shown in Figure 11.1 and for simplicity in this Chapter, this type of operation is referred to as recycling irrespective of whether it involves re-use, recycling or reclamation.

In its simplest form, recycling can be represented as in Figures 11.1 and 11.2. Figure 11.2 shows a processing system, *S*, containing five sub-systems *A, B, C, D* and *E* forming a linear sequence of operations. For simplicity, assume that no materials losses occur but that sub-system *D* exhibits two outputs, one of which passes on to the final processing sub-system, *E*, while the other passes out

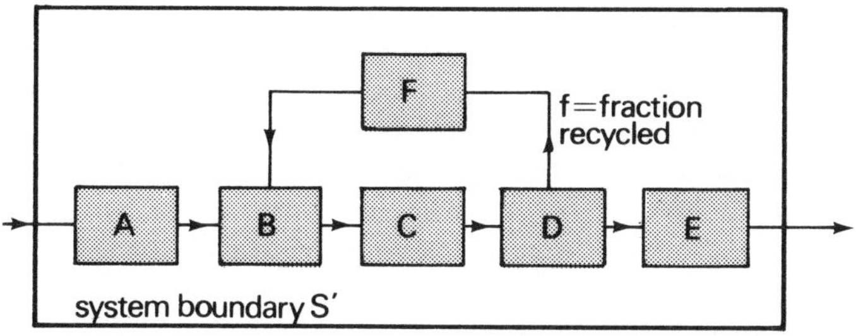

Figure 11.1 Simple system containing a recycling loop.

of the system via operation *G*. Now suppose that the output of sub-system *G* is suitable as an input for sub-system *B*. It would therefore be possible to recycle this output as shown by the system *S′* of Figure 11.1. Here the output from *D* is fed back to *B* via a new sub-system, *F*, which replaces *G* in Figure 11.1.

A simple example of this type of operation is the recycling of readily identifiable industrial waste. For example, if the main processing sequence represents the production of aluminium cans such that sub-system *D* represents the manufacture of cans fron aluminium sheet, then output P_2 in Figure 11.2 represents off-cuts from the sheet. Without recycling, sub-system *G* represents the operation needed to dispose of these off-cuts. With recycling, sub-system *F* represents those operations needed to collect, process and transport the off-cuts from the can-making factory back to the aluminium sheet manufacturer represented by sub-system *B*.

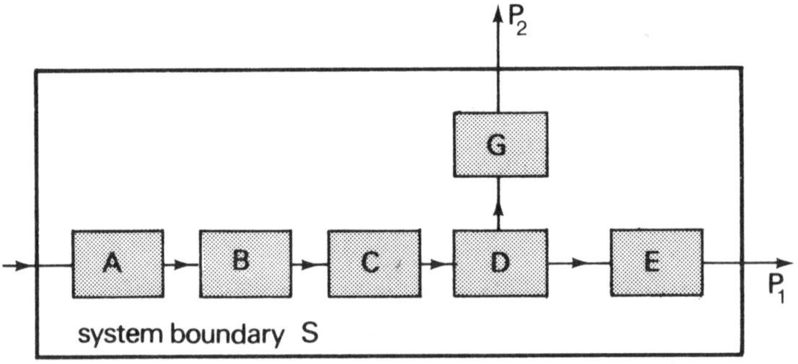

Figure 11.2 Simple processing system without feedback.

Although simple recycling schemes such as Figure 11.1 occur widely in industrial processing, many practical recycling schemes are considerably more complex and give rise to networks of inter-connecting sub-systems. Figure 11.3 for example shows in a simplified manner the principal flows of materials in the steel industry. The network of Figure 11.3 is amenable to numerical analysis although it is clearly more complex than systems such as that of Figure 11.1. It is however important to analyse the *complete* system rather than attempt extensive simplification by omitting some of the operations. The reason for this is that the system energy can be significantly affected by altering apparently minor flows within the network, as will be seen later in Chapter 12.

Chapter 2 showed that one of the most convenient ways of representing the energy requirement of a system is to normalise the energy consumption of a system with respect to the output mass of product. In this way the energy requirement may be rendered independent of materials throughput and, as discussed in Chapter 8, when the system produces a single output the normalised energy requirement can be reasonably held to be the energy needed to produce that product using the defined system. The process shown in Figure 11.2 is however somewhat more complicated because, although there is only a single saleable product, there are in fact two outputs, P_1 and P_2, one of which is waste. The energy requirement of this system can only be satisfactorily normalised using the mass of the useful or saleable product, P_1, as the normalising parameter. However the normalised energy requirement of the system must now be more carefully specified as the energy requirement per unit output of product P_1, and not simply as the system energy requirement per unit output, because this *could* be taken to imply the product P_2 even though P_2 is not used in normalisation. In this instance therefore, lack of precision in specifying the result can lead to ambiguity in interpretation.

The only output from the system defined in Figure 11.3 is waste. The mass of this waste is seldom known because it is extremely difficult if not impossible to measure. Moreover, even if it could be measured, it would be of little practical use as a normalising parameter for the energy requirement of the system. The flow parameter of real interest is the flow of steel products, *M,* from conversion to consumer and ideally therefore the energy requirement of the system should be normalised with respect to this flow. Such a proposal however represents a departure from the simple definition given in Chapter 2 because it involves normalising with respect to a materials flow that does not cross the systems boundary.

This type of problem always arises when the consumer forms part of the overall system. In some instances such systems can be re-drawn so that the consumer is excluded, though in the example of Figure 11.3 the steel production system would be changed by such an alteration. The modified system would have finished steel products as output but there would be *two* inputs iron ore concentrate *and* scrap steel.

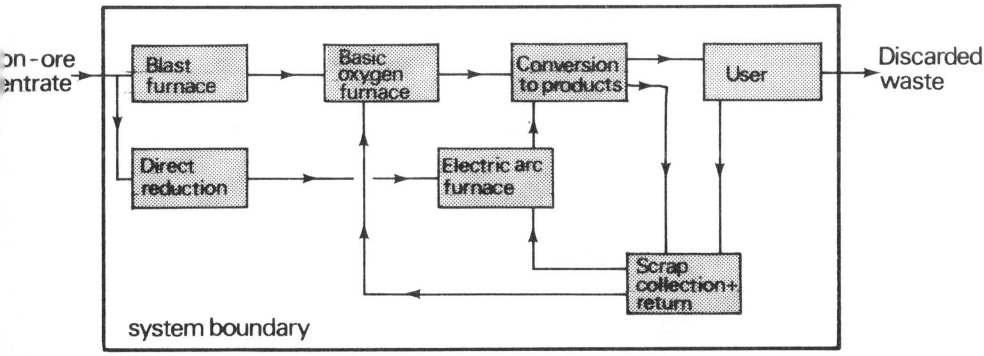

Figure 11.3 Simplified flows of materials in the steel industry.

Whilst it is possible to eliminate the consumer from the steel system of Figure 11.3, situations do occur where this becomes impossible. Consider for example Figure 11.4, which shows the flow of materials in the returnable glass bottle system. In this instance, the consumer represents an integral and essential component of the whole system. Without his participation the system will not operate and it would be extremely difficult to devise a system boundary that excluded the consumer whilst at the same time retaining the function of the system. In this instance the system must be analysed as it stands and the way in which this can be done is shown later.

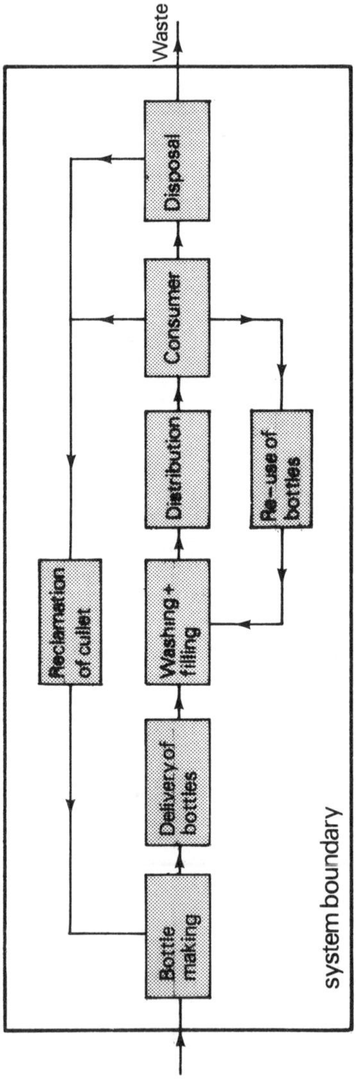

Figure 11.4 Simplified description of flows of materials in a returnable glass bottle delivery system.

The following Sections of this Chapter examine the techniques that may be used to analyse different recycling systems. However, it is most important to recognise that **at no point are energy requirements assigned to products**. Although this has been repeatedly stressed throughout the book, it assumes even greater importance in recycling networks where there may be flows of many different identifiable commercial products. The only unique energy parameter is **the energy requirement of the system** and it is impossible to partition this amongst each of the internal flows in the network.

The immediate benefit of recycling is the decreased demand for raw materials inputs to the system (that is inputs to operation A in Figure 11.1). However the energy benefits, if any, are not immediately apparent because they arise from the relative magnitudes of a number of contributing systems. For example, suppose that it was decided to recycle paper waste from household refuse. The most noticeable benefit would be a reduction in the demand for virgin pulp by the paper manufacturer because a proportion of his requirements would be met by recycled paper. There would therefore be an energy saving in the production of virgin pulp. However, in order to satisfy this demand, the waste paper must be sorted from refuse, cleaned and turned into pulp. At the same time, some of the energy formerly used in disposing of the waste paper (for example in transport and landfill operations) will also be saved. Hence the net increase or decrease in energy of such a recycling scheme requires the careful evaluation of the relative magnitudes of each of these different contributions before it is possible to give an answer to the problem. The situation is even less simple when network systems such as Figure 11.3 are involved.

Although here we are primarily interested in energy considerations, it should be noted that the energy requirement of recycling in no way predicts whether a specific recycling scheme will be *economically* viable, because the fuel requirements represent only one contribution to the overall cost of the process. Consideration of the energy associated with recycling does however provide an additional objective criterion by which to judge the value of potential recycling schemes and as such is an important tool in decision making.

11.2 ANALYSIS OF A SIMPLE RECYCLING SYSTEM

Simple recycling systems of the type shown in Figure 11.1 are commonly encountered in practice and they are readily described quantitatively. Figure 11.1 is identical to Figure 2.19 and so using Equation (2.46) the system energy per unit output of product P_1 is E'_s, which can be written as:

$$E'_s = (E_A+E_B+E_C+E_D+E_E) + \frac{f}{1-f}(E_B+E_C+E_D+E_F)\,. \quad (11.1)$$

Here E_A represents the energy requirement per unit output from the sub-system

A, and so on, and f represents the fraction of input to sub-system D that passes into the recycling loop. Note that this derivation assumes no materials losses in any of the sub-systems.

To consider whether recycling results in energy savings, the system energy of Equation (11.1) must be compared with that of an equivalent system in which recycling is absent. Figure 11.2 is such a system. For mass m fed into the system of Figure 11.2, the energy required to process this input will be:

$$mE_A + mE_B + mE_C + mE_D + m(1-f)E_E + mfE_G \,. \qquad (11.2)$$

The useful product emerging from sub-system E is of mass $m(1-f)$ and so the energy requirement of the system for a unit output of this product can be calculated by dividing expression (11.2) throughout by the factor $m(1-f)$. Hence:

$$E_s = \frac{1}{1-f}(E_A+E_B+E_C+E_D) + E_E + \frac{f}{1-f}.E_G \,. \qquad (11.3)$$

The energy *saving* produced by recycling is therefore given by the value of ΔE, where:

$$\Delta E = E_s - E_s'$$

$$= \frac{f}{1-f}(E_A+E_G-E_F) \,. \qquad (11.4)$$

If ΔE is positive, energy savings will result from recycling. If ΔE is negative, there will be a net loss of energy upon recycling.

The important feature of Equation (11.4) is that the energy saving depends upon three factors E_A, E_G and E_F. The contribution from E_A arises because recycling reduces the demand for processing raw materials up to the point at which they enter sub-system B. In the recycling of paper example mentioned earlier, this represents the reduced demand for virgin pulp by the paper manufacturer. Component E_G arises from a reduction in the demand for disposal facilities. In the paper example this would represent any energy used in disposing of paper in domestic refuse. Component E_F represents the energy needed to recycle the materials and the negative sign indicates a consumption of energy. In the paper example this would correspond to the energy needed to sort, clean, de-ink, pulp and deliver the paper so that it can be fed back into the paper-making process. Equation (11.4) therefore predicts that for energy saving to occur in recycling, the quantity $(E_A+E_G-E_F)$ must always be greater than zero. When

$$E_A + E_G - E_F = 0, \qquad (11.5)$$

the recycling scheme does not save energy but it does of course save materials and so may still be worthwhile.

The method of analysis given in Chapter 2 and leading to Equation (11.1) is rigorous and relatively straightforward when dealing with simple systems containing only one feedback loop. Many practical systems however involve multiple feedback loops (Figure 11.4) or networks (Figure 11.5) and in such cases the method of analysis becomes extremely cumbersome. It is however possible to carry out analysis in a much simpler manner. To demonstrate the technique, consider Figure 11.5. This is identical to the system S' of Figure 11.1 but shows the steady-state materials flows that would be achieved in the various branches of the network when a mass m per unit time is fed into the system. The mass flows are calculated using the requirement that the mass of materials entering a sub-system must equal the mass leaving the sub-system. At sub-system D, the parameter a represents the fraction of the mass m that passes into the feedback loop.

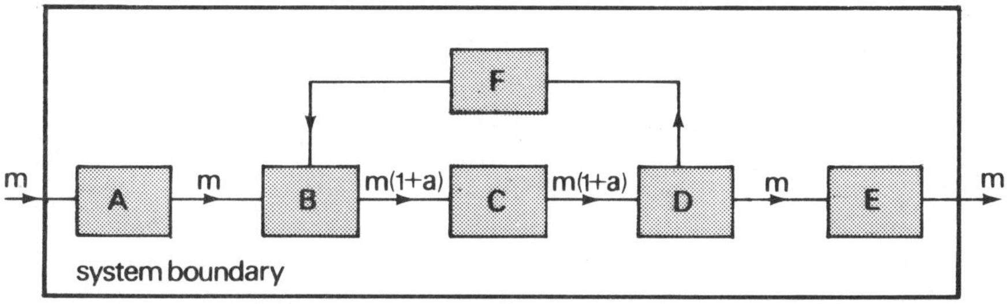

Figure 11.5 Analysis of a simple recycling system. m represents the input materials mass to the system and a represents the fraction of m that passes into the feedback loop.

For each sub-system, an energy E_i may be defined as the energy required to produce unit output of all products from that sub-system. Thus E_A will be the energy required to yield unit output from sub-system A and so on. The total system energy will therefore be the sum of the energies required by each sub-system in order to yield a system output of mass m. Each of the sub-system energies may be calculated as the product of E_i and the mass output of the i'th sub-system. Hence, writing E_s as the system energy requirement per unit mass output from the system, Figure 11.5 gives:

$$mE_s = mE_A + m(1+a)E_B + m(1+a)E_C + m(1+a)E_D + mE_E + amE_F \qquad (11.6)$$

$$= m(E_A + E_B + E_C + E_D + E_E) + ma(E_B + E_C + E_D + E_F). \qquad (11.7)$$

Note that parameter a is not the same as f in Figure 11.1; f is the fraction of the input to sub-system D that passes into the feedback loop. The two parameters are therefore related by the expression:

$$f = \frac{am}{(1+a)m}. \tag{11.8}$$

Hence,

$$a = \frac{f}{1-f}. \tag{11.9}$$

If Equation (11.9) is now substituted into Equation (11.7), the resultant value of E_s is identical to that of Equation (11.1) demonstrating the validity of the technique in this example. It can in fact be shown to be valid in all cases and is really a simplified version of the more rigorous derivation given in Chapter 2.

11.3 IN-HOUSE RECYCLING OF COPPER WASTE

One of the simplest forms of recycling and one that is widely practised in industry as a matter of good housekeeping is the internal recycling of waste produced during manufacturing processes, Such recycling frequently involves a single feedback loop of the type shown in Figure 11.2 and a typical example would be the recycling of waste metal produced in a casting shop.

Figure 11.6 shows a simple example of such recycling for the production of copper castings. The first sub-system covers all operations needed to manufacture copper from ore in the ground and deliver it to the casting shop. The three remaining sub-systems represent the operations carried out in the casting shop itself, that is melting the copper metal, pouring it into the mould and finally removing any unwanted metal in the finishing stage. Since all metal removed in the finishing stage is well characterised and is easily collected, recycling back to the melting stage is a simple matter. In this instance, assume that the return operation is carried out manually so that no energy is expended in the recycling. This corresponds in Figure 11.2 to setting $E_F = 0$.

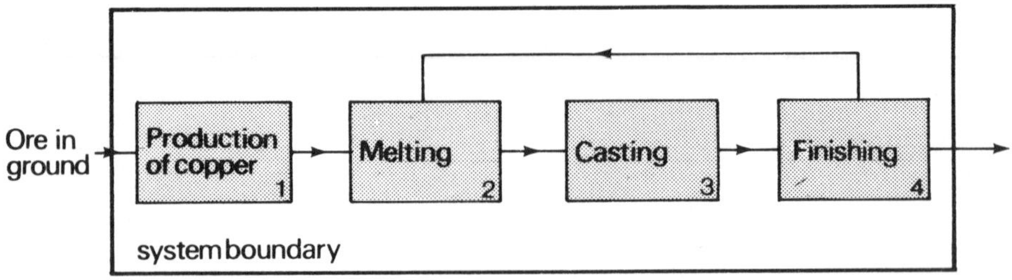

Figure 11.6 System needed to produce copper castings with internal recycling of the waste copper from the finishing process.

If the sub-systems are numbered as shown, then by analogy with Equation (11.1) the system energy requirement per unit output from this system will be E'_s given by:

$$E'_s = (E_1 + E_2 + E_3 + E_4) + \frac{f}{1-f}(E_2 + E_3 + E_4), \quad (11.10)$$

where f is the fraction of the material entering the finishing operation that is recycled.

If recycling were not practised but the waste from the finishing stage were disposed of instead, again with no energy commitment, then the system energy requirement per unit output of castings would be E_s given, by analogy with Equation (11.3), by:

$$E_s = \frac{1}{1-f}(E_1 + E_2 + E_3 + E_4). \quad (11.11)$$

We can give notional values for the savings in energy that might be produced by recycling, using the data from Part 2 of this book and assuming that $f = 0.1$, that is that 10% of the mass of a casting from the mould is removed during finishing.

From Part 2, the energy required to produce 1 kg of electrolytic copper from ore in the ground is 114.33 MJ and the energy required to produce 1 kg of liquid copper from cathode copper is 5.41 MJ. Assume that no energy is involved in the casting (pouring) operation or in the finishing operation. Hence:

$$\begin{aligned} E_1 &= 114.33 \text{ MJ}, \\ E_2 &= 5.41 \text{ MJ}, \\ E_3 &= 0, \\ E_4 &= 0, \\ f &= 0.1 \end{aligned}$$

Therefore:

$$\begin{aligned} E_s &= 133.04 \text{ MJ}, \\ E'_s &= 120.34 \text{ MJ}. \end{aligned}$$

The net energy saving is therefore 12.70 MJ per kg of output castings from the system. Note that this saving arises from the reduced demand for copper by the melting process. The energy used in the melting stage will be the same whether recycling is practised or not because 1.1 kg of copper must be melted to produce 1 kg of finished casting. If perfect castings could be produced, that is castings requiring no finishing whatsoever, the energy per kilogramme of output would be reduced to 119.74 MJ.

11.4 RECYCLING CHARACTERISED INDUSTRIAL SCRAP

The relative simplicity of the calculation in Section 11.3 arises because of the ease with which the recycling can be accomplished in practice. Frequently however, industrial scrap, although well characterised, is generated at a point remote from primary manufacture and so the return loop requires the expenditure of energy. In the manufacture of aluminium cans from aluminium sheet, for example, about 23% scrap sheet is produced because the cans are drawn from circular blanks punched from rectangular sheet. Scrap must be transported from the can-making factory back to the aluminium sheet manufacturer, melted and cast into ingot and then converted back to sheet. A typical arrangement of operations is shown in Figure 11.7. Transport has been included in this example. Although the transport energy requirement is relatively insignificant in this instance, this is not always so.

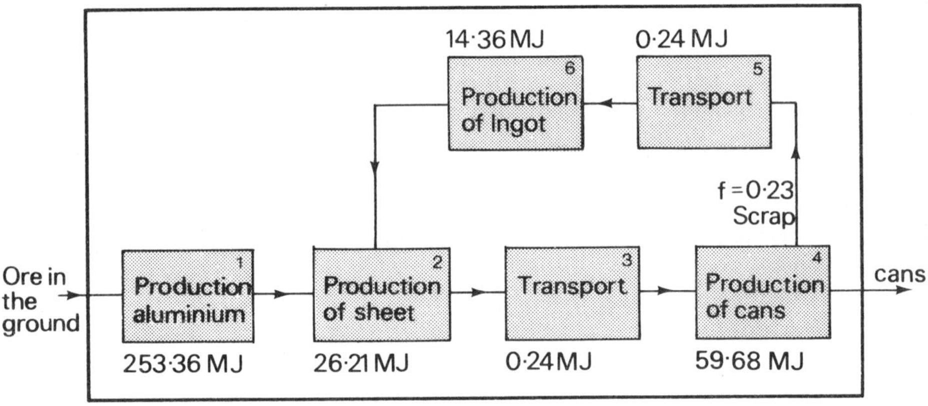

Figure 11.7 System required to produce aluminium cans from ore in the ground with recycling of scrap aluminium sheet from the can-maker. The energies shown are those for unit output mass from the relevant sub-system.

Numbering the sub-systems as shown in Figure 11.7, the system energy requirement of the operations shown will be E'_s, which by analogy with Equation (11.1) is given by:

$$E'_s = (E_1 + E_2 + E_3 + E_4) + \frac{f}{1-f}(E_2 + E_3 + E_4 + E_5 + E_6). \tag{11.12}$$

Substituting the values in Figure 11.7 into this Equation gives:

$$E'_s = 369.58 \text{ MJ per kg of cans produced.}$$

Without recycling, the system energy, E_s, will be given by:

$$E_s = \frac{1}{1-f}(E_1 + E_2 + E_3 + E_4) \tag{11.13}$$
$$= 440.90 \text{ MJ per kg of cans produced,}$$

by analogy with Equation (11.3).

In this instance it is clear that the combination of the very high energy needed to produce primary aluminium together with the high level of inevitable scrap during can production results in a significant energy saving by recycling. This incentive is also reflected in the economics of the system.

One of the main arguments in favour of the use of aluminium cans in beverage distribution compared with other containers is that the can may be readily recycled after use by the consumer since it is composed of a single material. We can therefore consider the impact of this argument on the energy needed to produce cans by setting up the sequence of operations as shown in Figure 11.8.

Following the procedure outlined in Section 11.2, a mass m (of alumunium as ore) is fed into the system, a mass $a.m$ passes into the industrial recycling loop and a mass $b.m$ passes into the post-consumer recycling loop. The masses flowing between each of the sub-systems may now be calculated to give a mass balance for each sub-system as shown. Note that with this system the demand for aluminium as ore is simply that needed to replace the mass m that is lost from the end of the system, assuming no other processing losses.

The total energy requirement of this system in the steady state is E_s, given by

$$mE_s = m(E_1 + (1+a+b)(E_2 + E_3 + E_4) + (1+b)(E_5 + E_6) + a(E_7 + E_8) + b(E_9 + E_{10})). \tag{11.14}$$

We are now faced with the problem mentioned earlier, namely that of choosing a normalising parameter. In this instance the parameter of importance is the mass of cans passing from the can manufacturer to the consumer. This is $(1+b)m$. Normalising with respect to this parameter therefore gives an energy per kg of cans as:

$$\frac{E_1}{(1+b)} + \frac{(1+a+b)}{(1+b)}(E_2 + E_3 + E_4) + (E_5 + E_6) + \frac{a}{1+b}(E_7 + E_8) + \frac{b}{1+b}(E_9 + E_{10}). \tag{11.15}$$

To evaluate a notional value for the use of cans in such a system, the same energies have been used in Figure 11.8 as in Figure 11.7. The additional operations have been assigned energies in the following way. All operations from the

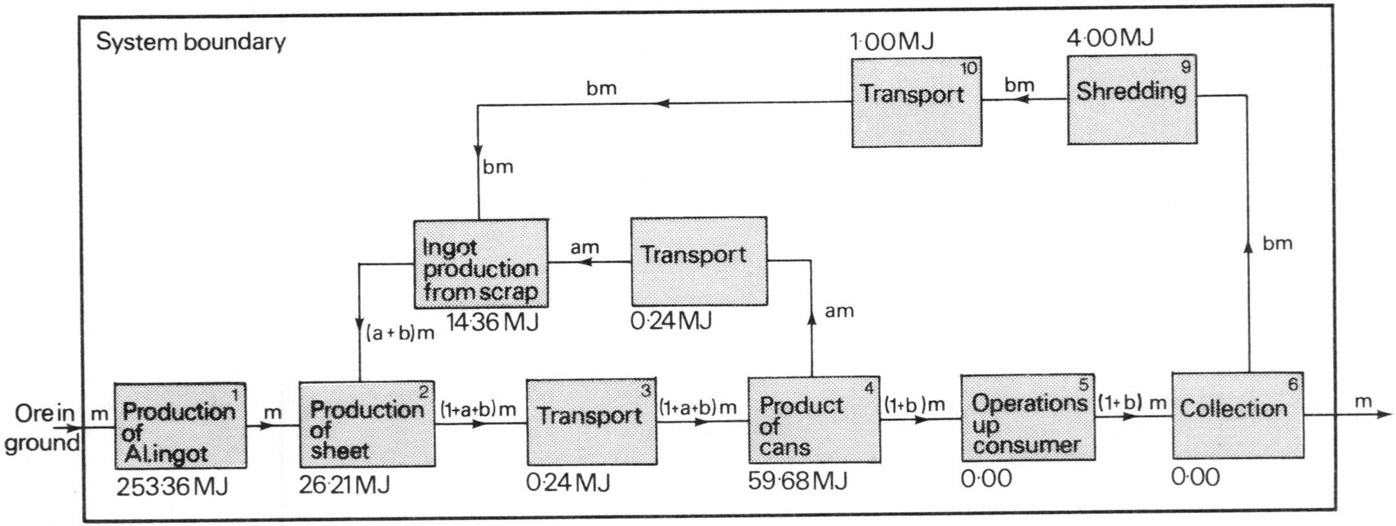

Figure 11.8 Production of aluminium cans showing post-consumer reclamation in addition to recycling of industrial scrap.

can-maker to the consumer have been set equal to zero because these operations are more closely associated with the contents of the can rather than with the can itself. The collection of used cans has also been set equal to zero because it is assumed that the consumer will return them to some central point. Both of these assumptions are arguable in the real case but they have been used here merely to simplify the calculation. The only significant operation carried out on the collected cans is shredding to improve their packing density for subsequent transportation. A variety of techniques are used for shredding cans but a typical value of 4.00 MJ/kg has been chosen. Transport energy is expected to be greater in returning the collected cans to the ingot producer because of the greater geographical distribution of post-consumer scrap. For convenience a value of 1.00 MJ/kg has been used. The values of the energies to be associated with a unit mass of output from each sub-system is therefore as shown in Figure 11.8.

The only other factors to which values must be assigned are the parameters a and b in Figure 11.8. The fraction of the aluminium sheet that is returned as scrap by the can manufacturer is $a/(1+a+b)$ and this can be set equal to 0.23 as in Figure 11.7 because this represents an inevitable loss due to the geometry of the shapes that must be cut from the sheet. Hence

$$\frac{a}{1+a+b} = 0.23. \tag{11.16}$$

The fraction of the cans reclaimed as post-consumer scrap will be $b/(1+b)$ and the numerical value of this fraction will depend upon the efficiency of the reclamation programme. For simplicity at this stage, assume that 50% of all cans are reclaimed:

$$\frac{b}{1+b} = 0.5. \tag{11.17}$$

Hence from Equations (11.16) and (11.17),

$$a = 0.6,$$
$$b = 1.$$

Substituting these values in Equation (11.15) and using the values of energy from Figure 11.8 gives an energy per kg of cans as 242.53 MJ, a value some 34% lower than when no post-consumer reclamation is carried out.

In a recycling system such as this, the parameter of crucial importance is the value of b, which is directly related to the fraction of the post-consumer cans that are reclaimed. The effect on the energy of cans of increasing the reclamation rate can be readily calculated using Equation (11.15) and for the data of Figure

11.8 the result can be shown graphically as in Figure 11.9. Here the dramatic decrease in the energy can be clearly seen as the reclamation rate increases. In this instance there would therefore be a good energy incentive to reclaim aluminium cans from post-consumer scrap.

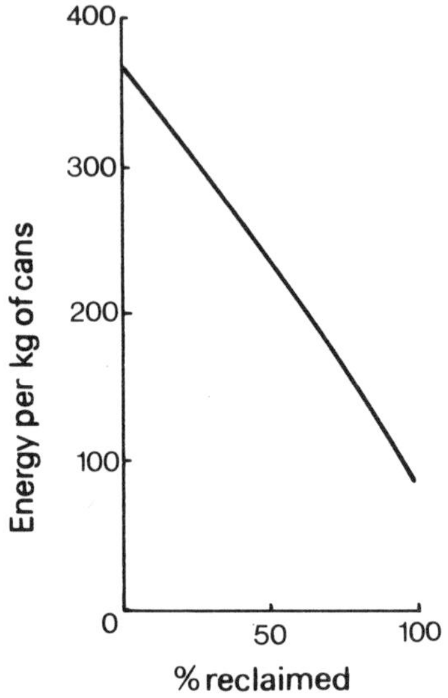

Figure 11.9 Energy of the aluminium can system as a function of the percentage of cans reclaimed from post-consumer scrap.

It must be remembered that the numerical values used in this Section, although based on published data, are introduced by way of illustration. The precise savings in energy likely to be achieved will of course depend upon the exact procedures and operations used. In general, as far as the aluminium can is concerned the very high energy needed to produce primary aluminium means that virtually all reclamation schemes that permit secondary aluminium production will be favourable from an energy viewpoint. This is not however true for all commodities.

11.5 RECYCLING PLASTICS

The crucial difference between plastics and metals from an energy point of view is that the total energy required to produce a polymer resin contains a contribution due to the fuel value of the feedstock (see Chapter 7). For example, approximately half of the energy associated with polyolefin production is feedstock energy and a significant proportion of this is reclaimable as available energy when plastic waste is generated.

Recycling schemes which involve the repeated use of plastics as materials can be treated in exactly the same way as the recycling of metals (see Section 11.2). In such cases, any energy savings will arise in part from a decreased demand for processing energy and in part from a decreased demand for feedstock energy.

Unlike the aluminium can example however, there are few instances in which post-consumer plastics can be reclaimed and used as materials. The principal reason for this is that the composition of commercial plastics is uncertain even when the base polymer can be identified, because of the wide variety of fillers and so on that are used.

The most significant contribution that plastics may make in the future to energy conservation measures is their use as fuels. Suppose therefore that a system such as that shown in Figure 11.10 were set up. the first operation is that needed to produce the polymer resin from crude oil. The second operation subsumes all of these operations used to convert the polymer resin to useful artifacts, to allow their use by the consumer and to dispose of the artifact after use up to the point at which it appears in the solid waste stream that is to be used as a fuel. The third operation is that in which the polymer is used as a fuel. What then is the energy requirement of this system?

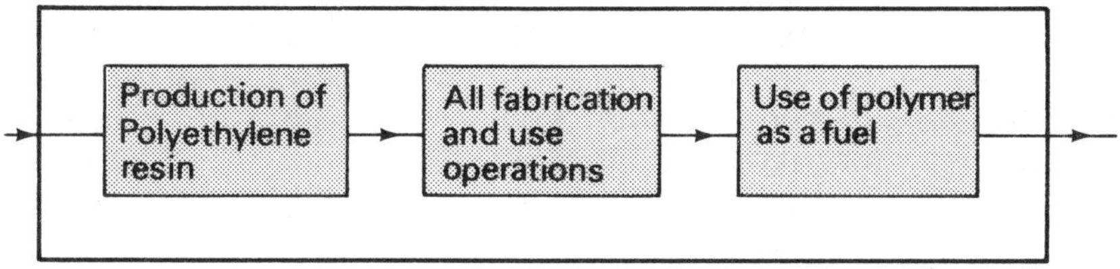

Figure 11.10 System which uses polyethylene as both a material and as a fuel.

For each of the sub-systems, a Table of energies is prepared as shown by Tables 11.1, 11.2 and 11.3. In this instance Table 11.1 is taken from Part 2 as the energy to produce 1 kg of low density polyethylene. Table 11.2 is a random collection of numbers which we will take to represent the conversion and use energy, and Table 11.3 has been calculated as 90% of the feedstock energy of Table. 11.1. Note that not all of the feedstock energy is available for reclamation because it represents the energy content of the *input* materials to operation (1) of Figure 11.10. Note also that Table 11.3 will be negative because energy is being generated. The total energy requirement of the system shown in Figure 11.10 will therefore be the sum of Tables 11.1 to 11.3 and is shown in Table 11.4.

Table 11.1
Energy required to produce 1 kg of low density polyethylene from crude oil

Fuel type	Fuel production and delivery /MJ	Energy content of fuel /MJ	Feedstock energy /MJ	Total energy /MJ
Electricity	16.59	5.24	nil	21.83
Oil fuels	4.33	21.16	45.00	70.49
Other fuels	nil	nil	nil	nil
Totals/MJ	20.92	26.40	45.00	92.32

Table 11.2
Energy required to convert and use polymer (random numbers)

Fuel type	Fuel production and delivery /MJ	Energy content of fuel /MJ	Feedstock energy /MJ	Total energy /MJ
Electricity	3.00	1.00	nil	4.00
Oil fuels	1.00	5.00	nil	6.00
Other fuels	1.00	7.00	nil	8.00
Totals/MJ	5.00	13.00	nil	18.00

Table 11.3
Energy required to use plastic as a fuel

Fuel type	Fuel production and delivery /MJ	Energy content of fuel /MJ	Feedstock energy /MJ	Total energy /MJ
Electricity	nil	nil	nil	nil
Oil fuels	nil	nil	−40.5	−40.5
Other fuels	nil	nil	nil	nil
Totals/MJ	nil	nil	−40.5	−40.5

Table 11.4
Total energy of the system shown in Figure 11.10

Fuel type	Fuel production and delivery /MJ	Energy content of fuel /MJ	Feedstock energy /MJ	Total energy /MJ
Electricity	19.59	6.24	nil	25.83
Oil fuels	5.33	26.16	4.50	35.99
Other fuels	1.00	7.00	nil	8.00
Totals/MJ	25.92	39.40	4.50	69.82

Chapter 12
Metals production

12.1 METAL PRODUCTION SYSTEMS

The production of metals and their conversion to a saleable form is of fundamental importance to all industry-based economies. These industries are large consumers of energy and a significant proportion of national energy resources in the developed countries is devoted to them. For example, the engineering and metals industries currently use about 21% of the total energy (on a heat supplied basis) consumed in the U.K. Not surprisingly therefore, evaluation of the energy required by systems that produce metals and metal products is an important function of energy analysis.

Calculation of an energy requirement for a metal-producing system is however complicated by a number of factors, and of these three are of special importance because they may significantly affect the energy required by any processing system. These are the low materials conversion efficiency of some stages of production, particularly mining and mineral processing, the existence of alternative processing routes and the development in recent years of increasingly complex processing networks for production. As a result, it is impossible to quote a single energy requirement for the production of a metal ingot from ore in the ground. Indeed the energy required to produce 1 kg of steel from a given starting material probably possesses as many values as there are processing plants carrying out the operation. Hence Part 2 gives a large number of values for the energy requirements of most metals operations. The purpose of the present Chapter is to show how these different values arise and to present guidelines for use in interpreting data, Not only from Part 2 but from other published sources as well. As will be seen however, much of the published information is singularly unhelpful because of the imprecision with which the systems have been defined.

At the same time as illustrating the origin of differences in energy requirements, we shall also demonstrate how the more complex production networks can be set up and analysed, as well as giving an indication of the way in which the data of Part 2 may be used.

In this Chapter, a number of specific calculations are carried out. It is important to recognise that these numerical calculations are intended by way of illustration only; they must not be regarded as definitive and applicable to all similar plants. They are based on published data, often for specific plants or processes and significant variations do exist from plant to plant even when different plants are carrying out apparently similar operations. Moreover, because our intention in producing the calculations is primarily to illustrate the technique, we have simplified some of the systems for convenience and a more detailed analysis would be necessary to produce more accurate and relevant results.

12.2 THE EFFECT OF MATERIALS CONVERSION EFFICIENCY

Section 2.2 introduced the concept of a materials conversion efficiency, and examined briefly the significance of this concept in energy analysis. Materials conversion efficiency, C, is defined as:

$$C = \frac{\text{Mass of saleable output}}{\text{Mass of input}}, \qquad (12.1)$$

but it was noted that some care is needed in interpreting numerical values of C since it can be calculated in a number of ways. For example, in a mineral processing plant the aim is to increase the concentration of desired mineral compared with waste rock, and the input mass could be specified in terms of the total mass of the mineral plus waste rock, or in terms of the mineral alone, or in terms of the metal content. Choice of different parameters will lead to different values for materials conversion efficiency and the effect of using different definitions can be illustrated by a simple hypothetical example shown in Figure 12.1. Here it is assumed that a metal ore contains 90% by mass of waste rock and 10% metal oxide. After passing through the mineral concentration and metal extraction plants, 6.5 kg of saleable metal are produced. The output of 6.5 kg of metal is easily identified, but the input could be 100 kg (total mass), 10 kg (mineral input) or 8 kg (metal content of input), which will give materials conversion efficiencies for this process of 0.065, 0.65 or 0.8125 respectively. All are describing the same process, yet each numerical value describes a different facet of the process and confusion is inevitable unless the basis of the calculation is carefully specified.

Specifying this parameter requires an understanding of the nature of the different definitions. From the simple example above, it is clear that conversion efficiency in terms of overall mass inputs and outputs is really a composite parameter. In the mineral processing stage for an ore containing 1% of desired mineral, the process would be working perfectly if this 1% of mineral alone could be made to appear in the output. The remaining 99% would be discarded as waste. The materials conversion efficiency measured in terms of the total masses of inputs and output would therefore be only 0.01 or 1%. On the other hand, materials conversion efficiency could be measured in terms of the mineral content of the input rather than the total mass of input materials of all types. Then this mineral content would also be present in the output and the process would have a materials conversion efficiency measured in this way of 100%. These differences arise because only one component of the input is valuable and worth concentrating.

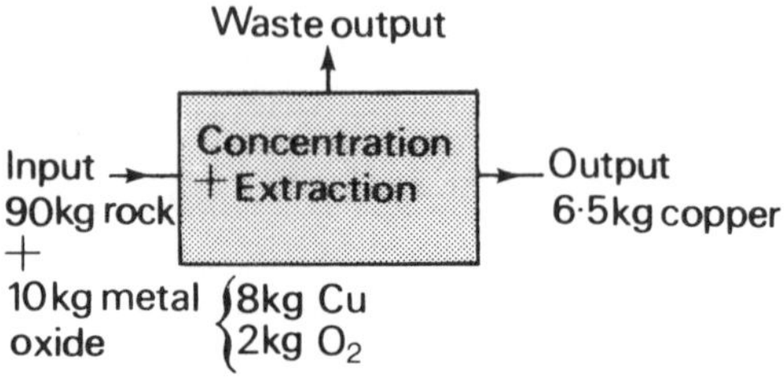

Figure 12.1 Hypothetical metal extraction process from ore.

A very similar situation exists during extraction. If copper, for example, were extracted from a pure copper oxide, CuO, then only 80% of the mass of the input materials would be copper while the remaining 20% would be oxygen. In terms of the total masses of inputs and outputs, the conversion efficiency of such a process would at best be 80%. However an alternative conversion efficiency based on the copper content of the input and output materials could be defined again and in the best possible case this would be 100% even though the gross value is only 80%. The difference between the two efficiencies arises in this instance from the different stoichiometric compositions of the input and output materials.

In addition to the above losses, there are also genuine losses of the desired output occurring during processing. These arise inevitably because few operations process materials with an efficiency of 100%. The materials loss is not necessarily due to bad practice but may be an inherent property of the process itself, For example, the use of slags in metal extraction allows many impurities in the metal to be taken into the slag and so removed. At the same time however the slag must also necessarily take up a small quantity of metal and this metal loss into the slag represents a genuine but unavoidable materials loss.

In general therefore, an overall materials conversion efficiency may be broken down into three more specific parameters.

(1) **Genuine losses** in which a proportion of the desired output is lost either by bad practice or as a result of inherent properties of the processing system.
(2) **Concentration losses** in which the desired output is present in the input materials, usually in a very dispersed form such as mineral in a run-of-mine ore. This would also include the overburden that must be removed in mining operations.

(3) **Stoichiometric losses** in which the desired output is present in the input but is 'diluted' by chemical bonding to some other element or elements. During processing these other elements are separated and frequently represent waste products of no commercial value thus constituting a materials loss to the system.

Writing C_g, C_c and C_s as the genuine, concentration and stoichiometric losses respectively, then for any system the overall materials conversion efficiency, C, will be:

$$C = C_g . C_c . C_s . \qquad (12.2)$$

Hence for an input run-of-mine ore containing 1% copper oxide, CuO, the system which extracts copper from this ore with a genuine overall loss of 5% will have:

$$C_g = 0.95,$$
$$C_c = 0.01,$$
$$C_s = 0.80,$$

and the overall conversion efficiency, C, will be:

$$C = 0.95 \times 0.01 \times 0.80$$
$$= 0.0076 \text{ or } 0.76\%.$$

For every kilogramme of copper produced, the system will require an input of 131.6 kg of ore. In this instance the dominant factors are C_c and C_s, with C_c being the most significant.

For the extraction operation alone, assuming that the concentrate is 80% copper oxide,

$$C_g = 0.95,$$
$$C_c = 0.80,$$
$$C_s = 0.80,$$

and so

$$C = 0.95 \times 0.80 \times 0.80$$
$$= 0.608 \text{ or } 60.8\%.$$

Here, although C_c and C_s are still the dominant factors, they are now of approximately equal significance. In contrast to both of these examples, the conversion of extracted metal into finished products will be dominated by C_g because both C_c and C_s are essentially unity.

As a metal-bearing material passes through the sequence of conversion operations from ore in the ground to some finished product, the overall materials conversion efficiency at each stage will tend to be dominated by a different component parameter. In particular, C_c and C_s tend to dominate in the earlier stages of processing and C_g dominates in the later stages.

Materials conversion efficiency strongly influences the energy requirement of sequential operations such as in metals production. The energy needed to drive any sub-system usually depends upon the mass throughput and the energy requirement of any sub-system, normalised with respect to unit output of useful product from that sub-system, therefore depends upon the proportion of the throughput that becomes useful product. This defines the materials conversion efficiency of the sub-system. In a sequence of operations such as that shown in Figure 12.2, the overall materials conversion efficiency for the whole system, C, is given by

$$C = C_1 . C_2 . C_3 . C_4 . C_5 , \tag{12.3}$$

where C_1, C_2 and so on are the conversion efficiencies of the component sub-systems. Because the efficiencies are multiplicative rather than additive, small changes in sub-system efficiencies rapidly introduce significant changes into the value of C and therefore into the energy requirement of the overall system.

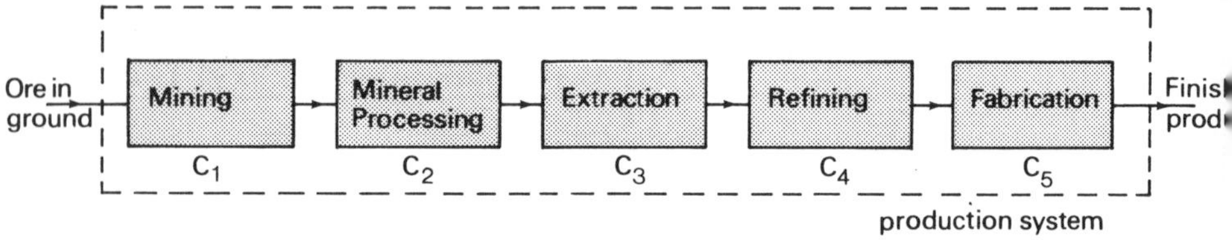

Figure 12.2 Sequence of operations from ore in the ground to finished artifact applicable to most metal processing operations.

Many processing systems where the genuine materials loss dominates have a conversion efficiency of less than 100%, but given unlimited time and money, they could probably be made to approach 100% materials efficient. However, the mining and mineral processing operations in the metals producing industries exhibit very low values of conversion efficiency because of the very low value of the concentration parameter, C_c. Similarly, it is the stoichiometric loss during extraction that results in conversion efficiencies well below 100%. The importance of this is twofold. First, very large quantities of materials must be processed in order to give a relatively small amount of saleable output. Secondly, materials losses from these two sources are inevitable and no amount of investment in plant can reduce them.

Consider a copper mine producing copper oxide, CuO. The run-of-mine ore contains 1% of the oxide and in order to win the ore an amount of overburden equal to the mass of ore extracted must also be removed. Figure 12.3 shows the materials flows for such a system assuming no genuine materials losses (that is $C_g = 1$ in Equation (12.2)). In this case, for every 1 kg of rock moved in the mining operation only 0.004 kg of copper results. Overall then, the mass of materials moved in the earth will be some 250 times greater than the mass of desired end product. If the next stage of processing involved a 1% loss, this would result in a 2.5% increase in input at the initial stage. It should be noted that in most cases the values will be much greater than this, as we shall see later in this Chapter.

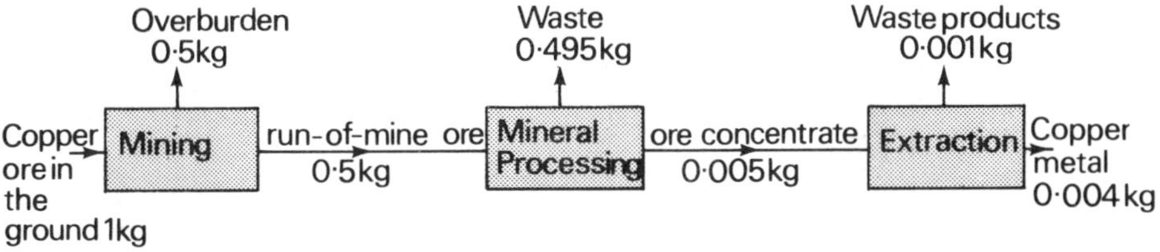

Figure 12.3 Simple sequence for producing copper metal from copper ore in the ground showing the mass flows between the different operations.

In general these low values of the overall conversion efficiency depend upon factors such as the location of the mine, the type of deposit, the type of mining operation employed, the concentration of the mineral in the deposit and so on. It is interesting to note that these are the same factors that would be taken into account initially in establishing the feasibility of a mining operation. No two mines are the same and these values will obviously vary from mine to mine. Moreover, even within a single mine the values cannot even be regarded as time-independent because factors such as ore concentration may well vary as the mining operation progresses. As a consequence there can be no single value for the energy required to produce a metal even when a precise production route is specified. Statements in the literature to the effect that, 'The energy to produce aluminium from ore in the ground is 100 MJ/kg', must therefore be treated with extreme caution even when the relevant system is closely defined, unless the ore grade is also specified. Furthermore, some care is also needed in interpreting the meaning of **ore grade**. Some authors refer to the elemental metal content whereas others refer to the mineral content. Part 2 of this book contains some widely used entries for operations where these problems exist.

A more detailed example demonstrates all of the problems associated with materials conversion efficiency that arise in the analysis of a metal production system, especially when the mining and mineral processing operations are included within the system. Consider the production of electrolytic-purity copper from a sulphide ore in the ground. Figure 12.4 shows a simplified sequence of operations whereby this might be achieved. If the materials conversion efficiency for each of the sub-systems 1, 2, ..., 5 is measured in terms of the copper present in the inputs and outputs, such efficiencies are of the genuine loss type C_g discussed above and typical values are shown in the Table immediately below the flow diagram in Figure 12.4. If the flows of materials between the individual sub-systems are measured in terms of the copper content, then they will be as shown and a copper content of 1.599 kg must be fed into the mining operation to yield an output of 1 kg of electro-refined copper.

Consider now the stoichiometric losses. The input to smelting, mineral processing and mining all contain copper chemically bonded to sulphur and so the stoichiometric conversion efficiency will not be unity if there is any change in the chemical form of the mineral. A typical sulphide ore is chalcopyrite, a

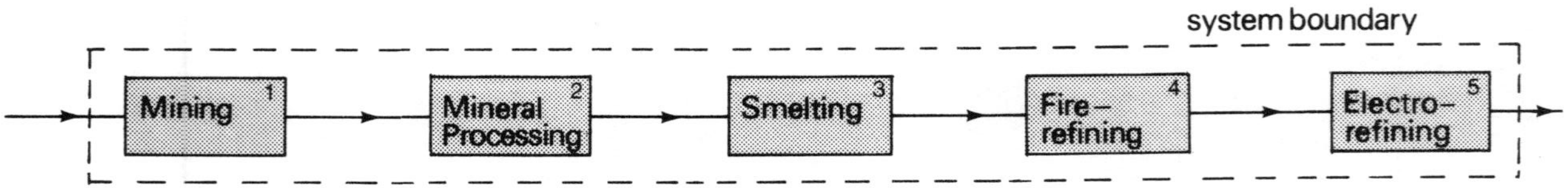

	Mining	Mineral Processing	Smelting	Fire-refining	Electro-refining
C_g	0·75	0·90	0·95	0·98	0·995
C_s	1	1	0·347	1	1
C_c	c/b	b/a	a	~1	~1
C	$0{\cdot}75\,c/b$	$0{\cdot}90\,b/a$	0·33a	0·98	0·995

Mass of copper/kg	1·599	1·1995	1·0795	1·0255	1·005	1
Mass of materials/kg	$4{\cdot}607/c$	$3{\cdot}455/b$	$3{\cdot}1077/a$	1·0255	1·005	1

Figure 12.4 Sequence of operations needed to produce electrolytically pure copper from ore in the ground. a is the mass fraction of mineral in the ore concentrate; b is the mass fraction of mineral in the ore as-mined and c is the mass fraction of mineral in the rock moved during mining. See the text for a more detailed explanation.

complex mixture of copper and iron sulphides with a chemical composition of $CuFeS_2$. During mining and mineral processing there is essentially no change of stoichiometric composition of the mineral and so stoichiometric conversion efficiencies for these processes are both unity.

Smelting however is a two stage process involving the formation of a matte (a liquid sulphide mixture) which is then blown with oxygen to remove iron as iron oxide into the slag, and sulphur as sulphur dioxide gas, leaving copper metal in the converter. In the matte smelting stage, the essential change is:

$$2CuFeS_2 \;=\; Cu_2S \;+\; \text{iron sulphides},$$

and the stoichiometry indicates that an input of 1 kg of chalcopyrite will only yield 0.434 kg of copper sulphide. During converting the reactions occurring can be simplified to:

$$Cu_2S \;=\; 2Cu \;+\; \text{sulphur compounds},$$

so that an input of 1 kg of copper sulphide will yield only 0.799 kg of copper. Since in Figure 12.4 we are considering the smelting stage as a single sub-system encompassing both matte smelting and converting, the overall stoichiometric conversion efficiency for the sub-system will be $0.434 \times 0.799 = 0.347$.

Consider now the losses arising from concentration. In both fire-refining and electro-refining there is removal of relatively small quantities of impurities and for both of these processes the concentration conversion efficiencies are sensibly unity.

For the smelting sub-system, the input is a copper concentrate containing the desired metal-bearing mineral together with a quantity of waste rock. During smelting, this waste rock is removed as part of the slag. If we define a as the mass fraction of *mineral* in the input ore concentrate, then the concentration conversion efficiency will also be a. Note that a is directly related to the *copper* content of the mineral. Using the stiochiometric conversion efficiency of 0.347 the concentration of copper in the input to the smelter will be $0.347a$.

During mineral processing, a significant proportion of the waste rock present in the ore as-mined is removed. If we define b as the mass fraction of *mineral* ($CuFeS_2$) in the input ore, then the concentration conversion efficiency which leads to a mass fraction a in the output will be b/a.

In mining, only a small fraction of the rock that must be moved is mineral. Not only is the mineral present in the deposit in low concentrations (usually of the order of 1% or less) but frequently large masses of rock as overburden must also be removed in order to expose the deposit. In open-pit mining for example the stripping ratio may be anything up to 2.5 to 3.0. That is, of the total rock moved, 1 kg of mineral bearing rock is recovered for every 2.5 to 3.0 kg of overburden removed. If we write c as the mass fraction of mineral ($CuFeS_2$) present in the total mass of rock moved during mining, the concentration efficiency that arises in order to give an output with a mass fraction of mineral b by disposing of the overburden will be c/b.

The Table in Figure 12.4 gives the component materials conversion efficiencies of each sub-system and the overall materials conversion efficiency arising from all sources may now be calculated to give the values shown using Equation (12.2). Using these overall materials conversion efficiencies, the actual masses of materials flowing between each of the sub-systems may be readily calculated and it can be seen that these require values for the three parameters a, b and c which depend upon factors such as ore grade, ease and efficiency of mining and efficiency of mineral processing. Since these factors vary from mine to mine it is necessary to obtain detailed information in each particular instance. However we can consider some of the more general aspects here in the absence of this detailed information.

Parameter b is directly related to the ore grade because it is a measure of the mineral present in the run-of-mine ore. However some care must be taken when using published information to ensure that the ore grade specified is appropriate to the definition of b. There are basically two problems. First, ore grade may be specified as either metal content of an ore or as mineral content. Thus a 1% copper ore specified in terms of metal content will contain 0.01 kg of copper per kg of ore, whereas a 1% ore specified in terms of mineral content ($CuFeS_2$) will only contain 0.0035 kg of copper per kg ore. The second problem is that ore grade may be specified as the average grade within a deposit or alternatively as the grade of recovered ore, known as the **head grade**. The head grade will differ from the actual grade being worked because it is seldom possible to win an ore from a deposit without incurring some dilution of the ore with waste non-metal-bearing rock. Average head grade, averaged over the lifetime of the working, will also differ from the average grade of the deposit because some of the mineral will be lost in the discarded overburden. In Figure 12.4, the conversion efficiency C_g for the mining operation will take account of the loss in the overburden but the dilution effect will form part of parameter b. From Figure 12.4, the head grade measured in terms of the copper content will be:

$$(1.1995b/3.455) \times 100 = 34.7b,$$

where grade is expressed as a percentage. Hence with a head grade of 1% copper, b will take the value 0.029. Such ores are currently being worked but many operations are using ores of a much lower grade. It is quite likely that in the not-too-distant future the average grade for most workings will be as low as 0.4% (that is $b = 0.011$).

From Figure 12.4, an apparent average ore grade measured as copper content can be written down for the input to the mining sub-system as:

$$(1.599c/4.607) \times 100 = 34.71c,$$

where ore grade is expressed as a percentage. This however is not a true ore grade in the sense that a mining engineer would use the term because it includes a contribution for the stripping ratio. For an ore grade $G(\%)$ and a stripping ratio S, then:

$$\frac{G}{S} = 34.71c,$$

and hence:

$$c = \frac{G}{34.71S}. \quad (12.5)$$

So for ore grades lying in the range 0.4 to 1.0% and stripping ratios in the range 2.5 to 3.0, c will take values in the range 0.004 to 0.012.

Evaluation of the energy requirement E_s of the overall production system shown in Figure 12.4 requires numerical values for the energy requirement of each of the component sub-systems and typical values are given in Table 12.1. Hence:

$$E_s = E_1\frac{4.607}{c} + E_2\frac{3.455}{b} + E_3\, 1.0255 + E_4\, 1.005 + E_5, \quad (12.6)$$

and using Table 12.1:

$$E_s = \frac{0.184}{c} + \frac{0.83}{b} + 36.19. \quad (12.7)$$

The presence of parameters b and c in Equation (12.7) immediately indicates that there can be no unique value for E_s because it will depend upon the grade of ore and the stripping ratio. Choosing a range of values for these variables and evaluating E_s, we can plot the results as shown in Figure 12.5. The dominant factor is ore grade, which causes the dramatic rise in energy requirement seen at low ore grades. Remembering that ore grades are currently 0.5% and greater, it is clear that future production of copper will demand significantly higher energy requirements than at present.

The energy required by the system to extract copper from the concentrate and subsequently refine the product remains sensibly constant, for the reasons discussed earlier, at 36 MJ/kg copper. The dramatic increase in energy requirement at lower ore grades shown in Figure 12.5 therefore arises in the mining and mineral processing areas. If the purpose of the analysis were to seek out those areas in which significant energy savings could be made, then at higher ore grades the mining/mineral processing and the extraction/refining areas are of equal importance. However, at the lower ore grades the most significant energy savings in absolute terms would result from improvements in the energy efficiency of mining and mineral processing.

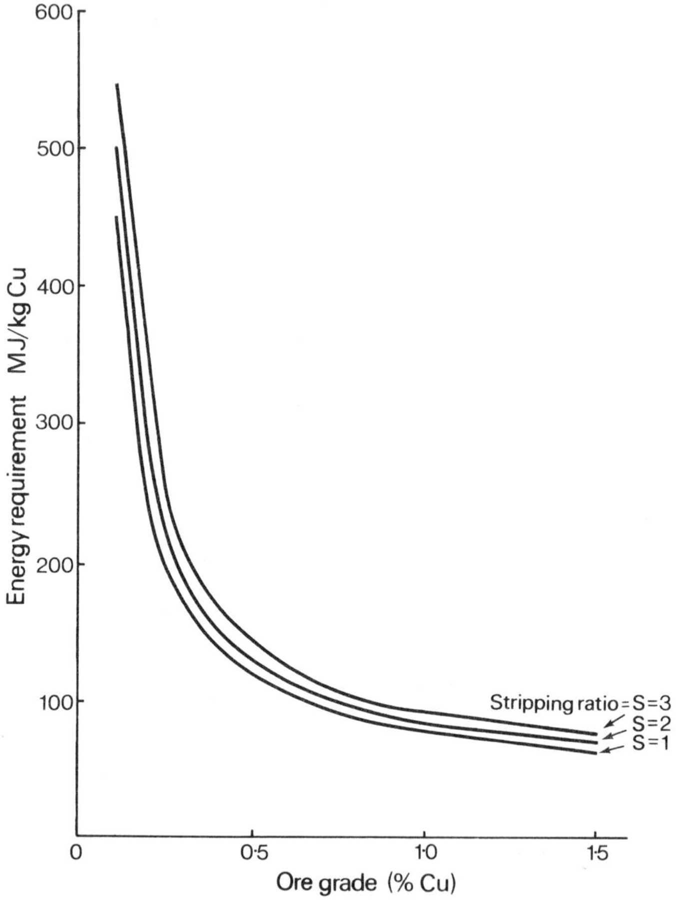

Figure 12.5 Energy required to produce electrolytically pure copper using the system of Figure 12.4. The curves show the dependence on ore grade and on stripping ratio.

It should be pointed out that throughout the calculations in this Section we have worked in terms of the total energy requirements of each of the sub-systems and have not considered the detailed energy breakdowns. Table 12.1 gives the data needed to perform these detailed calculations should they be needed and the procedure is simply to replace each of the energy parameters $E_1, E_2 \ldots, E_5$ in Equation (12.6) by the 16 element array discussed earlier.

Table 12.1
Typical energy requirements for the sub-systems shown in Figure 12.4

Operation	Electricity		Oil fuels		Other fuels		Total energy /MJ
	Production & delivery energy MJ	Energy content of fuel MJ	Production & delivery energy MJ	Energy content of fuel MJ	Production & delivery energy MJ	Energy content of fuel MJ	
Moving 1 kg of rock in mining (E_1)	0.03	0.01	nil	nil	nil	nil	0.04†
Treating 1 kg of run-of-mine ore to give ore concentrate (E_2)	0.18	0.06	nil	nil	nil	nil	0.24†
Smelting ore concentrate to produce 1 kg of blister copper (E_3)	1.59	0.50	nil	nil	3.28	22.99	28.36
Fire-refining to produce 1 kg of anode copper from blister copper (E_4)	nil	nil	nil	nil	0.52	3.62	4.14
Electro-refining to produce 1 kg of cathode copper from anode copper (E_5)	2.24	0.71	nil	nil	nil	nil	2.95

†Authors' estimates, all other values taken from Part 2.

This example highlights the difficulties inherent in specifying an energy requirement for a system producing metal products and illustrates why no unique value can be stated. A proper analysis must therefore state the starting grade of ore and the materials conversion efficiencies for each stage of processing. To avoid ambiguity, it is useful to break down the overall materials conversion efficiency for any sub-system into components C_g, C_s and C_c as described.

Part 2 gives numerical values for the energy requirements of a number of metals production systems. These data have been taken from published literature. Such variations as exist arise primarily from uncertainties in the definition of the system and from variables such as those discussed above. Before using any of the data in Part 2 it is important to refer to the relevant notes at the end of the Table in Part 2.

12.3 ALTERNATIVE ROUTES

There are frequently a number of routes by which materials can be converted from a given starting point to the same finished product. Thus not only will the different sub-systems within the overall system boundary exhibit different materials conversion efficiencies which influence the system energy, but the energy consumed by the different sub-systems will vary as well. This is particularly true in the metals producing industries. The effect may be readily illustrated in the general case by the two systems shown in Figure 12.6. Both systems are assumed to take in the same starting material and both produce 1 kg of the same unspecified product.

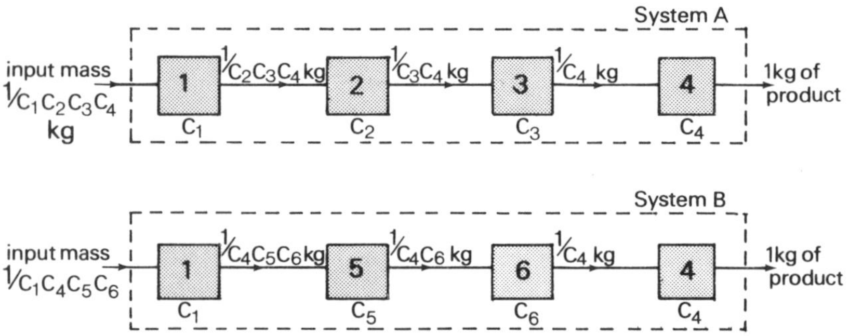

Figure 12.6 Two general systems *A* and *B* producing 1 kg of the same product by different routes. C_i denotes the materials conversion efficiency of the *i*'th sub-system.

The energy requirement for system *A* will be E_A per unit output, where

$$E_A = \frac{E_1}{C_2 C_3 C_4} + \frac{E_2}{C_3 C_4} + \frac{E_3}{C_4} + E_4, \tag{12.8}$$

and the energy requirement for system *B* will be E_B per unit output, where

$$E_B = \frac{E_1}{C_4 C_5 C_6} + \frac{E_5}{C_4 C_6} + \frac{E_6}{C_4} + E_4. \tag{12.9}$$

In both cases, E_i represents the energy requirement per unit output for the *i*'th sub-system. E_A will only be the same as E_B when:

$$E_2 = E_5, \tag{12.10}$$

$$E_3 = E_6, \qquad (12.11)$$
$$C_2 = C_5, \qquad (12.12)$$
$$C_3 = C_6. \qquad (12.13)$$

It is therefore clear that in general E_A will be different from E_B and the difference arises not only from the different materials conversion efficiencies but also from the different sub-system energies.

A typical example of this kind of problem is the production of 400 mm steel slab from scrap melted in an electric arc furnace. Figure 12.7 shown two possible routes by which this product could be made. The essential difference between the two routes arises from the nature of the cast product. In system *P*, continuous-casting is used to produce a bloom which can be rolled down to slab with only a single reheating stage. System *Q* however employs conventional casting to produce an ingot which then requires two reheating stages on the way to slab. Although the two systems described by Figure 12.7 are hypothetical, they are based on real practice. The materials conversion efficiency for each stage is given, together with the output mass from each sub-system. In this case, the materials conversion efficiency reflects genuine materials losses.

Figure 12.7 clearly indicates that the two systems will have different energy requirements due both to different materials conversion efficiencies as well as to different assemblies of sub-systems. The need for an extra heating stage in the case of ingot production arises from two effects. First, the continuous-casting machine generally forms part of an integrated steelworks so that the bloom does not lose all of its sensible heat before passing to the rolling stage. In contrast, ingot casting is usually remote from the rolling facility so that sensible heat is completely lost and soaking pits are required to reheat the ingot prior to rolling. This desire to utilise sensible heat is of course one of the driving forces for the construction of integrated metalworks. Secondly, the continuous-cast bloom is of smaller dimensions than an ingot so that less mechanical work is required to reduce it to the final product. This can result in fewer reheating stages in a given sequence. Note that system *P* requires a scrap input of 1.214 kg and system *Q* requires a scrap input of 1.35 kg. If therefore the horizontal extent of these systems were increased to include the supply of scrap, the energy requirements for the two extended systems would also differ because of the different materials input requirements, quite apart from the differences arising for the systems *P* and *Q*.

The energy requirements of the two processing systems illustrated in Figure 12.7 can be calculated using data from Part 2; these are shown in Table 12.2. By analogy with Equation (12.8), the energy requirement for system *P* is E_P, given by:

$$E_P = \frac{E_1}{C_2 C_3 C_4 C_5} + \frac{E_2}{C_3 C_4 C_5} + \frac{E_3}{C_4 C_5} + \frac{E_4}{C_5} + E_5 \qquad (12.14)$$

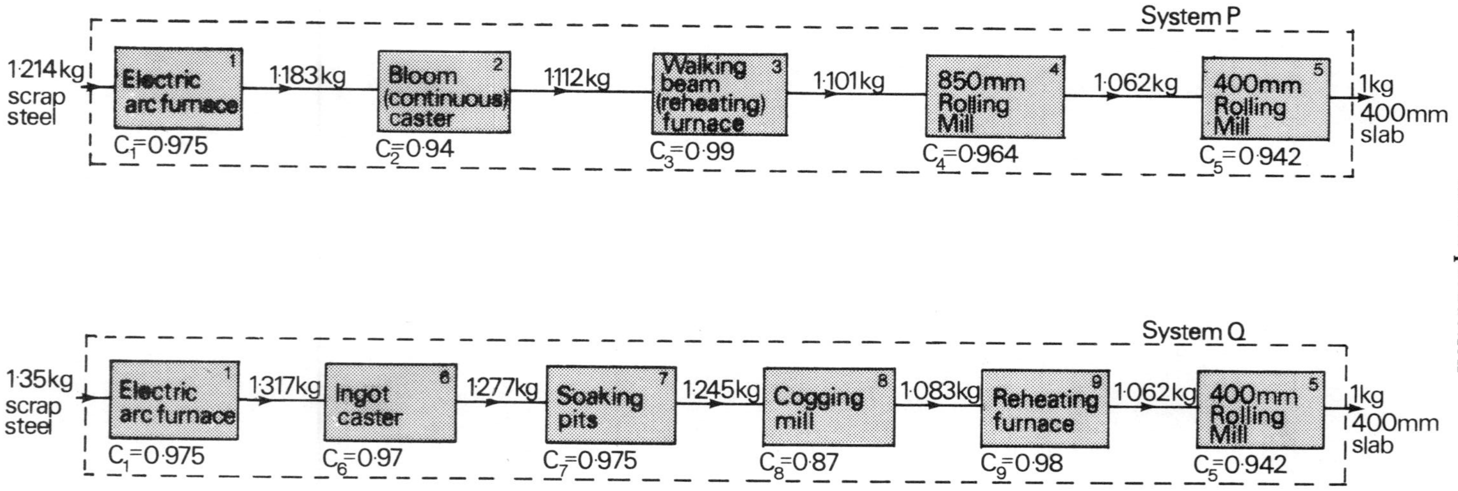

Figure 12.7 Two systems P and Q for producing 400 mm steel slab from steel scrap, C_i denotes the materials conversion efficiency of the i'th sub-system.

$$= 1.183E_1 + 1.112E_2 + 1.101E_3 + 1.062E_4 + E_5 \tag{12:15}$$

Using the multipliers shown in Equation (12.15) and the values of E_1 to E_5 from Table 12.2, the energy requirement for system P can be calculated to give Table 12.3.

Table 12.2

Typical values for the energy requirements of the sub-systems shown in Figure 12.7. All values are taken from Part 2 and are in units of MJ/kg output.

	Electricity		Oil fuels		Other fuels		
Parameter	Production & delivery energy	Energy content of fuel	Production & delivery energy	Energy content of fuel	Production & delivery energy	Energy content of fuel	Total energy
E_1	7.16	2.26	0.02	0.09	0.04	0.58	10.15
E_2	0.21	0.07	nil	nil	nil	nil	0.28
E_3	0.28	0.09	nil	nil	0.37	2.63	3.37
E_4	0.29	0.09	nil	nil	nil	nil	0.38
E_5	0.30	0.10	nil	nil	nil	nil	0.40
E_6	0.03	0.01	nil	nil	nil	nil	0.04
E_7	0.29	0.09	nil	nil	0.27	1.86	2.51
E_8	0.32	0.10	nil	nil	nil	nil	0.42
E_9	0.29	0.09	nil	nil	0.49	3.40	4.27

Table 12.3

Energy requirement per kg for the production of 400 mm slab from scrap by system P in Figure 12.7

Fuel type	Fuel production and delivery energy/MJ	Energy content of fuel /MJ	Feedstock energy /MJ	Total energy /MJ
Electricity	9.62	3.05	–	12.67
Oil fuels	0.02	0.11	–	0.13
Other fuels	0.46	3.59	–	4.05
Totals/MJ	10.10	6.75	–	16.85

Similarly for system Q in Figure 12.7, the system energy E_Q is given by:

$$E_Q = \frac{E_1}{C_6C_7C_8C_9C_5} + \frac{E_6}{C_7C_8C_9C_5} + \frac{E_7}{C_8C_9C_5} + \frac{E_8}{C_9C_5} + \frac{E_9}{C_5} + E_5 \tag{12.16}$$

$$= 1.317E_1 + 1.277E_6 + 1.245E_7 + 1.083E_8 + 1.062E_9 + E_5 \quad (12.17)$$

In a similar manner to that used for system *P*, the energy requirement for system *Q* may be calculated to give Table 12.4.

Table 12.4
Energy required per kg for the production of 400 mm slab from scrap by system *Q* shown in Figure 12.7

Fuel type	Fuel production and delivery energy/MJ	Energy content of fuel /MJ	Feedstock energy /MJ	Total energy /MJ
Electricity	10.79	3.41	–	14.20
Oil fuels	0.03	0.12	–	0.15
Other fuels	0.91	6.69	–	7.60
Totals/MJ	11.73	10.22	–	21.95

The two systems described are directly comparable in that they take into account the direct fuel supplies of each sub-system as well as energy contributions from materials such as electrode carbon and limestone supplies to the electric arc furnace. Materials supplies to other sub-systems as well as capital energy are excluded from the sub-systems (that is they are of type *A* as defined in Part 2). It can be seen therefore that at 6.75 MJ/kg slab, the direct energy requirement of system *P* is only 66% of that for system *Q*. Hence system *P* is the more energy efficient of the two. This difference in energy requirement clearly illustrates the care that must be taken to define the system properly. Many of the variations in the data given in Part 2 for particular production systems arise from unspecified variations in production routes even when the inputs and outputs are the same. To a lesser extent, variations in energy requirements also arise from the different performance characteristics of apparently identical machines within any one sub-system. This factor was highlighted in Chapter 10 where, for example, machine down-time was different from machine to machine even though machines were of the same type. Yet a further source of variation in the total energy requirements of a system is the different mixes of fuels that may be used.

It was this possible variation in fuel mixes that led to the comparison of systems *P* and *Q* on the basis of the direct energy consumption rather than the total energy requirement. Such a comparison gives a closer indication of the relative performance efficiencies of the plants, because it excludes the one variable that cannot be influenced by the metals manufacturer, namely the fuel production and delivery energy. However, for the more complete system which includes the fuel production and delivery energy, it is clear that the mix of fuels is of crucial importance in determining the total energy requirement of the system. The effect is not particularly marked in the above example, but examples can be found where this effect is extremely significant. Among these, aluminium metal may be melted using either a gas-fired furnace or an electric furnace. In both cases, the direct energy requirements are identical, but for gas firing the total energy requirement is only 4.51 MJ/kg of aluminium melted whereas for electric melting the total energy requirement is 16.46 MJ/kg. This difference arises from the much higher production and delivery energy of electricity. If the energy requirements for any metallurgical process are stated as a single number, as is often the case in the literature, the system must be carefully specified so that it is clear which type or types of fuels are being used. This problem does not arise if results are presented in tabular form as in Tables 12.3 and 12.4.

12.4 COMPLEX PRODUCTION PLANT - PARTITIONING ENERGY

Throughout this book we have repeatedly stressed that energy requirements can only be calculated for a *system* and not for a *product*. All of the systems considered so far in this Chapter produce only one product, that is electrolytically pure copper, 400 mm steel slab and molten aluminium. In such cases it is reasonable to equate the energy requirements of the system with the energy requirement of the product - more correctly, with the energy required to produce the product from the specified inputs using the defined system.

When two or more well-defined products are manufactured in the same plant, it is often possible to partition the energy requirement of the system amongst the products in such a way that the proportion attributed to any product reasonably represents the energy that would be needed to produce that product in a similar plant, if that were the only product being manufactured. The factory analysis of Chapter 10 is essentially an exercise in this technique. In many instances however, the partitioning of energy is extremely complex and subjective judgements are often required. It is then vital to have a very precise definition of the system analysed in order to make sense of numerical values. Nowhere is this complexity more immediately apparent than in some modern metal production facilities and the following simplified account of the energy requirements of such a plant illustrates the way in which apparently different energy requirements may arise from analysis of the same plant.

Figure 12.8 illustrates a system which produces copper from copper ore in the ground by mining and concentrating followed by flash smelting [82]. The inputs and outputs are as shown and for convenience the electrical inputs to the concentrating plant and the flash smelter are shown separately. Waste heat in the flue gases is dissipated and the slag formed during the smelting operation is discarded as waste; copper then is the only useful product. Figure 12.8 shows the requirements for the production of 1 kg of copper and the energy associated with this system may be readily calculated because the fuel inputs per kilogramme of copper output are (1.466 + 0.344) = 1.810 kWh of electricity and 0.152 kg of heavy fuel oil. The system energy requirement is therefore as shown in Table 12.5.

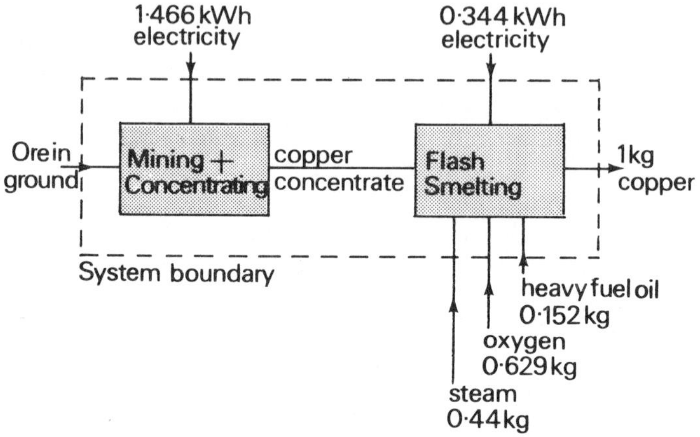

Figure 12.8 Production of copper from ore in the ground using flash-smelting. Electricity consumption in mining and mineral processing is based on a 2% ore producing a 25% copper concentrate.

Table 12.5
Energy requirement per kilogramme of copper for the system shown in Figure 12.8

Fuel type	Fuel production and delivery energy/MJ	Energy content of fuels /MJ	Feedstock energy /MJ	Total energy /MJ
Electricity	19.80	6.52	nil	26.32
Oil fuels	1.35	6.47	nil	7.82
Other fuels	nil	nil	nil	nil
Totals/MJ	21.15	12.99	nil	34.14

This process uses significant quantities of steam and oxygen and since the production of these services is excluded from the system of Figure 12.8, Table 12.5 does not truly reflect the energy required to operate the complete process. Table 12.6 summarises the energy required to produce both oxygen and steam using data from Part 2 and if these production facilities are included within the system boundary of Figure 12.8, the revised system energy requirement will be as shown in Table 12.7. Even in the revised form, capital energy and the energy

Table 12.6

Energy requirements for the production of steam and oxygen. Data from Part 2

(a) Steam from water in MJ/kg of steam delivered

Fuel type	Fuel production energy/MJ	Energy content of fuels/MJ	Feedstock energy /MJ	Total energy /MJ
Electricity	nil	nil	nil	nil
Oil fuels	0.62	2.98	nil	3.60
Other fuels	nil	nil	nil	nil
Totals/MJ	0.62	2.98	nil	3.60

(b) Oxygen from air in MJ/kg oxygen

Fuel type	Fuel production energy/MJ	Energy content of fuels/MJ	Feedstock energy /MJ	Total energy /MJ
Electricity	5.20	1.64	nil	6.84
Oil fuels	0.05	0.22	nil	0.27
Other fuels	0.25	1.77	nil	2.02
Totals/MJ	5.50	3.63	nil	9.13

Table 12.7

Energy required to produce 1 kg of copper from the flash-smelting system shown in Figure 12.8 but with the production of oxygen and steam included within the system boundary

Fuel type	Fuel production and delivery energy/MJ	Energy content of fuels /MJ	Feedstock energy /MJ	Total energy /MJ
Electricity	23.07	7.55	nil	30.62
Oil fuels	1.66	7.93	nil	9.59
Other fuels	0.16	1.11	nil	1.27
Totals/MJ	24.89	16.59	nil	41.48

associated with the provision of minor inputs to mining and mineral processing have been omitted. Although these are not expected to increase the system energy requirement significantly, they should be included in the most accurate form of analysis. For simplicity we shall omit them from the example considered here. The whole of the system energy may be reasonably attributed to the copper output because this is the only useful or saleable output. Hence the energy required for the production of 1 kg of copper from copper ore in the ground with the system as defined can be stated as 41.48 MJ. In other words, the energy associated with the product is the *same* as the system energy requirement.

In practice a flash smelting plant would make use of the heat present in the flue gases and Figure 12.9 shows how this can be done to generate steam. The steam in turn drives an electric generator and an oxygen plant, so satisfying the plant requirement for oxygen and electricity as well as steam. Surplus steam is produced and the inputs and outputs from the system are as shown. The energy required by the system of Figure 12.9 may be readily calculated from the input fuels (that is 1.466 kWh electricity [illegible]52 kg heavy fuel oil) and, if the energy requirement of the water supply is taken from Part 2, the system energy will be as shown in Table 12.8. This assumes that the excess steam is not used but is vented to air as waste.

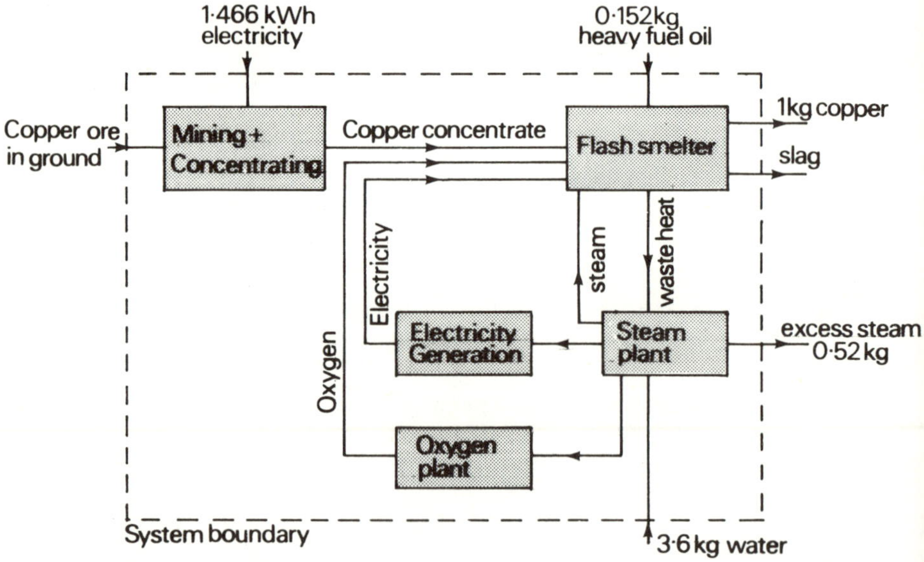

Figure 12.9 Production of copper using flash-smelting but with waste heat recovery. The recovered heat is converted to steam and used to provide the steam, oxygen and electricity needed by the flash-smelter. Surplus steam is exported from the system.

Table 12.8

Energy required to produce 1 kg of copper from a flash-smelter system employing waste heat recovery (Figure 12.9) and including the energy to supply water. It is assumed that excess steam is vented to air

Fuel type	Fuel production and delivery energy/MJ	Energy content of fuels /MJ	Feedstock energy /MJ	Total energy /MJ
Electricity	15.89	5.28	nil	21.17
Oil fuels	1.35	6.47	nil	7.82
Other fuels	nil	nil	nil	nil
Totals/MJ	17.24	11.75	nil	28.99

Again the energy associated with the product is equivalent to the system energy requirement because copper is still the only product, but the energy requirement for copper production using this system is only 70% of that needed when no heat recovery is practised. This underlines the importance of an adequate system definition because though in both instances copper is manufactured from concentrate using a flash smelter, it is the detail of the flash smelter circuit that gives rise to the energy difference.

Now suppose that the excess steam in Figure 12.9 is used elsewhere rather than being vented to air. The *system* energy remains at 28.99 MJ but this must now be apportioned between two products rather than being attributed to copper alone. The energy requirement attributed to the main product, copper, will therefore depend upon the method of apportionment used, but it will possess an upper limit of 28.99 MJ/kg.

The simplest procedure is partitioning energy in terms of mass and this gives values for the energy requirements for copper and steam as shown in Table 12.9. The energy requirement for copper is now only 19.07 MJ/kg, some 46% of that for the system with no waste heat recovery. However copper cannot be *produced* for an energy requirement of 19.07 MJ/kg using the system of Figure 12.9. On the contrary, the total energy content of the oil fuel (6.47 MJ) must always be supplied from fuel oil, or an equivalent fuel such as natural gas, if the system is to perform its function. In the same way, the 5.28 MJ due to direct electrical consumption must also be supplied. Therefore in the absence of fuel substitution, which will change the fuel production energy component of the energy requirement, and in the absence of changes in the energy associated with the materials supplies, the energy requirement of the *system* can never be less than 28.99 MJ/kg of copper output. The penalty for quoting an energy requirement of 19.07 MJ/kg for copper is that the plant must make use of the surplus steam.

Table 12.9

Energy required to produce 1 kg of copper or steam via a flash-smelter when waste heat recovery is practised and when excess steam is utilised. System energy has been partitioned on the basis of mass

Fuel type	Fuel production and delivery energy/MJ	Energy content of fuels /MJ	Feedstock energy /MJ	Total energy /MJ
Electricity	10.45	3.47	nil	13.92
Oil fuels	0.89	4.26	nil	5.15
Other fuels	nil	nil	nil	nil
Totals/MJ	11.34	7.73	nil	19.07

Partitioning of energy between co-products seeks to give some credit to the main product for the co-production of other useful materials and the mass apportioning method outlined above is one sensible means of achieving this. When the co-products are materials or artefacts, this method is usually satisfactory unless there are special reasons for preferring some other, but related, parameter such as volume. Frequently however co-products may be fuels such as blast furnace gas in iron-making or electricity generated from waste steam. In all cases except those where electricity is a co-product the system energy requirement can be partitioned on the basis of mass as used for steam in Table 12.9. This procedure has been adopted for many of the entries in Part 2†.

An alternative approach is to credit the system with the energy that would have been necessary to generate the electricity or manufacture the fuel with the same energy content as the co-product and to carry this as a negative entry in the energy table. This can be illustrated by the flash smelting system of Figure 12.9 where both copper and steam are co-products. Using the value for the production energy requirement for steam quoted in Table 12.8, we obtain an energy requirement as shown in Table 12.10 with an overall energy requirement for copper of 27.12 MJ/kg. Clearly the magnitude of this energy requirement for copper will depend on the assumed value for steam production and when making energy credits in this way, these assumed values must always be clearly stated.

†In Part 2, when electricity is a co-product, this technique attributes the whole of the system energy to the main product because all the mass resides there. However, the amount of electricity produced per unit output of product is quoted in the associated notes so that when the outputs from the system become the inputs to an adjacent system in a production sequence, the electrical output can be acknowledged and credited.

Table 12.10
Energy tables with credit for steam production. All values in MJ

	Electricity		Oil fuels			Other fuels			
	Fuel prod'n energy	Energy content of fuel	Fuel prod'n energy	Energy content of fuel	Feed-stock energy	Fuel prod'n energy	Energy content of fuel	Feed-stock energy	Total energy
Total system energy	15.89	5.28	1.35	6.47	–	–	–	–	28.99
Energy credit for steam	–	–	−0.32	−1.55	–	–	–	–	−1.87
Total energy attributable to copper	15.89	5.28	1.03	4.92	–	–	–	–	27.12

An alternative practice that might be employed by the operator of the flash smelter would be to use all of the steam to generate electricity. Any surplus to plant requirement could be fed to an external distribution network. The system in Figure 12.8 would then be modified to give that shown in Figure 12.10. Under these conditions, the inputs will be unchanged but the system outputs are now 1 kg of copper and 0.11 kWh of electricity. The energy requirement for copper is now 27.34 MJ/kg although again the system energy requirement is still 28.99 MJ/kg (Table 12.11).

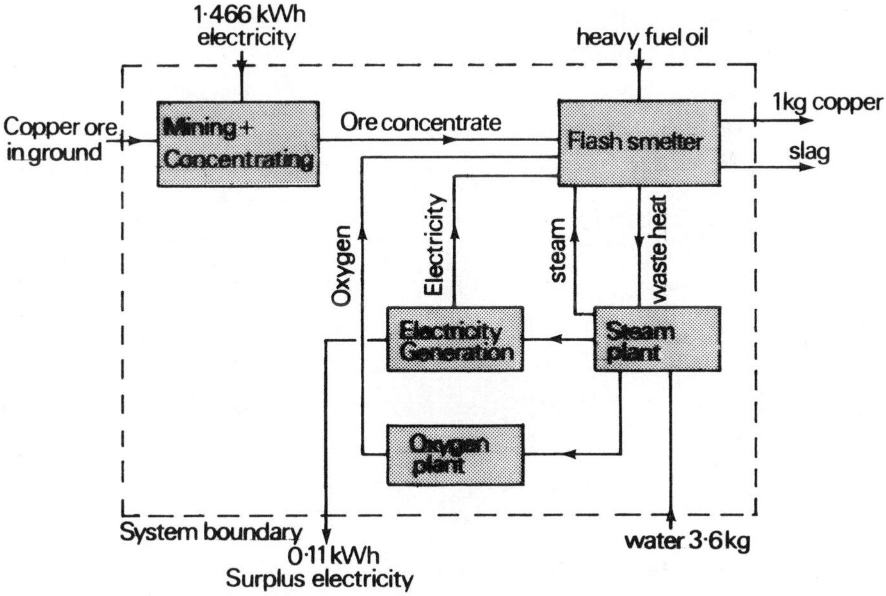

Figure 12.10 Production of copper by flash-smelting with waste heat recovery. Surplus steam is converted to electricity and then exported from the site.

Table 12.11
Energy required to produce 1 kg of copper when the outputs of the system are 1 kg of copper and 0.11 kWh of electricity (Figure 12.9)

	Electricity		Oil fuels			Other fuels			
	Fuel prod'n energy	Energy content of fuel	Fuel prod'n energy	Energy content of fuel	Feed-stock energy	Fuel prod'n energy	Energy content of fuel	Feed-stock energy	Total energy
Total system energy									
	15.89	5.28	1.35	6.47	–	–	–	–	28.99
credit due to 0.11 kWh electricity									
–	−1.25	−0.04	–	–	–	–	–	–	−1.65
Energy of system after allowing for credit									
	14.64	4.88	1.35	6.47	–	–	–	–	27.34

The values of 0.40 and 1.25 (in Table 12.11) can strictly only be subtracted from the fuel energy content and fuel production energy respectively of the electrical component of the system if the electricity generated as by-product of the flash smelter is fed back to the concentrator as shown in Figure 12.11. Under these conditions the electrical input to the concentrator is reduced from 1.466 kWh/kg copper to 1.356 kWh/kg copper because 0.11 kWh are produced within the system boundary. The energy requirements for this system are given in Table 12.12 and the overall requirement is 27.34 MJ/kg copper as before. This is now also the energy requirement for the production of copper, as might be expected by drawing in the system boundary to encompass all internal electrical generating facilities.

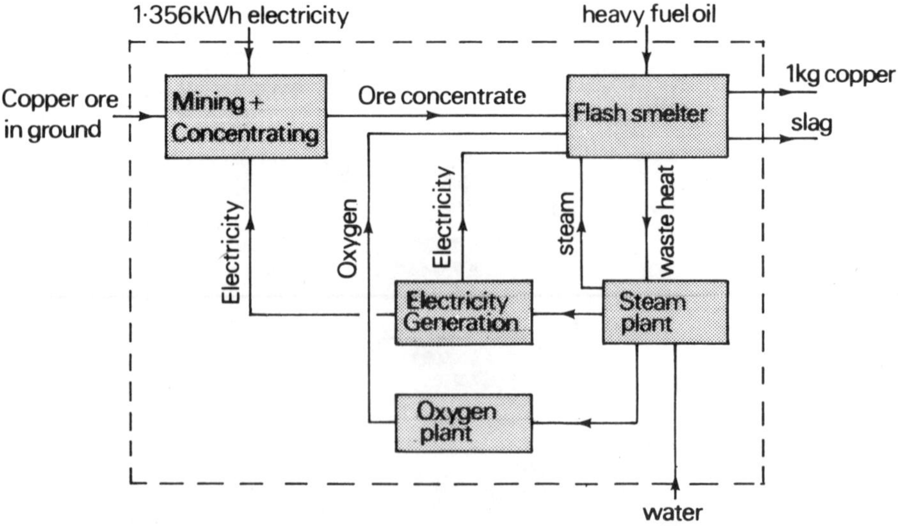

Figure 12.11 Flash-smelting in which all energy recovered from waste heat is used within the system.

Table 12.12
Energy requirement for the system described by Figure 12.10 for an output of 1 kg of copper

Fuel type	Fuel production energy/MJ	Energy content of fuel/MJ	Feedstock energy/MJ	Total energy/MJ
Electricity	14.64	4.88	nil	19.52
Oil fuels	1.35	6.47	nil	7.82
Other fuels	nil	nil	nil	nil
Totals/MJ	15.99	11.35	nil	27.34

It should be obvious from the above examples that there is frequently a clear distinction between system energy and the energy of a product derived from a given system. The two are only the same when there is a single useful or saleable product from the system. In all other circumstances the partitioning of energy between products or the introduction of energy credits will result in a lower value for the energy associated with the product compared with the system energy. This is yet a further reason for some of the disagreements between published values for the energy requirements of the same commodity as reported by different workers.

The need to define a system properly cannot be over-emphasised and it is essential to remember that it is the *system energy* that is of prime importance since this is the energy that must be supplied to effect a desired change. A product energy, no matter how carefully defined, will always be to some extent arbitrary and will depend both upon the method of apportionment and upon the method of accounting for energy credits. In all of the examples described above, copper is produced in a flash smelter but the energy required to produce 1 kg of copper ranges from 41.48 MJ/kg with no heat recovery to 27.34 MJ/kg with waste heat recovery and internal recycling of all available energy. Furthermore, if we were to choose to apportion all outputs on the basis of mass, the energy might fall as low as 19.07 MJ/kg. It is clear therefore that without additional information on the precise nature of the system, a quoted number may have very little meaning.

The recycling of energy within metalworks is common practice, as in many other industries. The above analysis highlights the care needed when seeking product energies to ensure that the basis on which energy credits have been made is understood. If the energy is recycled within the system boundary, it will reduce the primary fuel inputs. If energy appears as an output from the system it may be carried forward as a credit and so will reduce the primary energy inputs to the next processing stage.

12.5 COMPLEX PRODUCTION PLANTS - PRODUCTION NETWORKS

The complex production process discussed in the previous Section involves operations confined to a single plant and highlights the types of problems encountered in the detailed analysis of such systems. However, many production processes require systems that are even more complex than the operations carried out on a single site. Ideally it would be useful if operating data were available for all plants in a particular sector of the metals, and indeed other, industries, but this is not so. Moreover, as Figure 11.3 showed, not only are some commodities produced by quite different routes but the complete production process needs to take account of operations such as recycling that would not normally be regarded as part of the metals producing industries.

If therefore an average value were needed for the energy required to produce 1 kg of copper in the present state of the industry, only a part of that copper would be produced by the primary process. The remainder would be derived from recycled scrap. Moreover, even that produced by the primary route could be produced by at least two processes, the conventional smelter of Figure 12.4 and the flash smelter of Figure 12.9. An average value for the production of copper must therefore take account of these variables, even though at the present time the greatest proportion of primary copper is produced by the more conventional smelting route. However problems of sulphur disposal are gradually inducing changes within the industry so that eventually primary copper production may use a route quite different from those currently in use.

One industry in which production methods have already changed and are still changing is the iron and steel industry. Here new techniques have been developed less for environmental and pollution reasons than to accomodate technical improvements and to anticipate potential future shortages of good coking coal and rich iron ore. Nowhere are these changes more apparent than in the U.K., where dated capital plant centred on steel production by the open-hearth process is rapidly giving way to production by the basic oxygen and electric arc furnaces. The effect of these changes and the way in which they influence an 'average' steel production energy requirement are well worth examining, not only because they highlight the problems of complex inter-related plants but also because iron and steel are so widely used by the whole of industry.

An average energy requirement for steel production requires analysis of the complete steel production system and, as shown in Figure 11.3, even a very simplified representation leads to a complex network. In practice however the problem is considerably more complex than is suggested by Figure 11.3 because the sole output from such a system is unrecovered post-consumer scrap. Hence the iron ore input is merely that needed to replace this scrap loss in the steady state. Furthermore, if all post-consumer scrap is reclaimed the system requires no further input of ore, assuming that demand for steel products does not increase, and this is clearly an unrealistic situation. Despite efforts to conserve materials, that is iron, in the closed system, there will inevitably be some losses and any representation of the real steel production system must allow for such losses. Figure 12.12 shows a possible system. It includes both the blast furnace and the direct reduction furnace for the extraction of iron from ore and three different methods for producing steel from iron. The collection and distribution of scrap is also shown. In addition, each process is shown with a materials loss which can range from the very small iron loss in blast furnace flue gases to the more significant iron losses in slags.

If the network is analysed with respect to the iron content of the materials flowing between each of the operations, then for an input mass of 1 kg of iron, the parameters $a, b, c, \ldots, z, \alpha$ represent the quantities flowing in each of the links of the network. If now the operations are numbered 1, 2, ..., 10 as shown in Figure 12.12 and E_i represents the sub-system energy requirement for unit output from the i'th sub-system, then the systen energy E_s per unit *input* of iron will be:

$$\begin{aligned} E_s = {} & E_1(a + i + j + k) \\ & + E_2(b + l + m + s) \\ & + E_3(c + n) \\ & + E_4(d + p + q) \\ & + E_5(z + \alpha) \\ & + E_6(h + r) \\ & + E_7(e + x) \\ & + E_8(u + v + w) \\ & + E_9(g + y) \\ & + E_{10}(f + t). \end{aligned} \tag{12.18}$$

Unlike earlier examples, useful steel products do not form an output from the system but represent one of the internal flows, p, within the system. A value for the energy requirement to manufacture such products, E_f, *within the defined system* may be obtained by dividing through by the parameter p. Alternatively Equation (12.18) may be set equal to $p.E_f$.

To evaluate this system energy requires numerical values for 36 variables, that is the flows $a, b, \ldots, z, \alpha$ and the sub-system energies $E_1, E_2 \ldots, E_{10}$. In fact

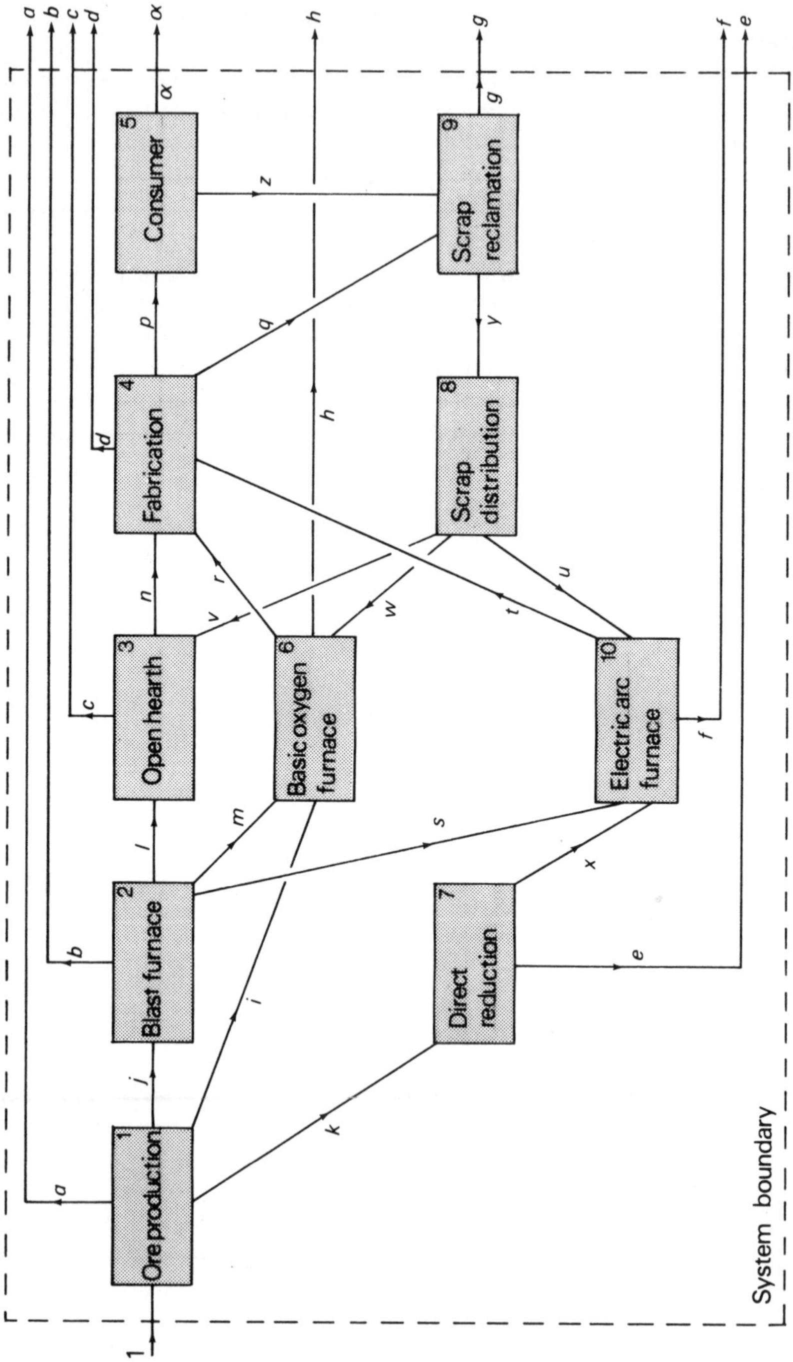

Figure 12.12 Simplified network of sub-systems required for the manufacture of steel products. The flows are those needed with an input of 1 kg of ore into the system.

not all of these variables are independent because for each sub-system the mass of the input materials must equal the mass of the output materials. Hence the following equations may be readily derived by carrying out a mass balance for each sub-system.

$$1 = a + i + j + k, \tag{12.19}$$

$$j = b + l + m + s, \tag{12.20}$$

$$l + v = c + n, \tag{12.21}$$

$$n + r + t = d + p + q, \tag{12.22}$$

$$p = z + \alpha, \tag{12.23}$$

$$i + m + w = h + r, \tag{12.24}$$

$$k = e + x, \tag{12.25}$$

$$y = u + v + w, \tag{12.26}$$

$$q + z = g + y, \tag{12.27}$$

$$s + u + x = f + t. \tag{12.28}$$

Using Equations (12.19) to (12.28), ten of the variables can be eliminated from expression (12.18) leaving a total of 26 parameters needing numerical values.

Given the system shown in Figure 12.12, expression (12.18) and Equations (12.19) to (12.28) represent a general equation for the energy requirements of the system and could be used to evaluate the energy performance of the network as flows of materials in the different branches vary. For example, it could be used to show changes in the system energy requirement as open-hearth steelmaking gives way to basic oxygen steelmaking, or to examine the influence of the introduction of the direct reduction of iron on the industry's energy requirements. Clearly such possibilities tend to move the analysis from the micro-level hitherto discussed into the macro-level of interest to policy makers. However, the value of such predictions will depend crucially upon the reliability of the results and this in turn depends upon the accuracy with which numerical values for the different variables can be determined.

Such a requirement brings us straight back to the need for accurate micro-analysis of the type described in Chapter 10 to provide the results on which to base this wider form of energy policy analysis. In general the numerical values for the parameters E_1 to E_{10} may be derived from data such as that presented in Part 2 and in principle a high degree of accuracy should be attainable because these refer to industrial techniques that are widely practised. However, it must be remembered that there will be no single numerical value that can be ascribed to each of these variables. Instead a range of values will be more appropriate. Nevertheless it should be possible to ascertain this range with some precision.

In contrast the flow parameters *a, b, c* and so on may be calculated or chosen to suit either the current state of the industry or to simulate some hypothetical future state. Numerical values for these parameters are less readily available than for energy requirements of specific operations and in general it will be more appropriate to choose a range of values rather than attempt to calculate specific values.

At first sight we would seem to have a method for calculating the dependence of the system energy requirement on all possible variables within the steel industry. If single values are chosen for each of the 26 independent variables, the solution is indeed relatively straightforward. However, if a range of possible values were chosen for each of these variables, the number of calculations and hence the number of solutions becomes unmanageable. For example, suppose that two possible values were assigned to each of the independent variables to indicate the range of values within which each variable could exist. In that case the number of possible solutions to the general equation is 2^{26}, or about 67 million, if all possible combinations are used. In practice, ranges containing more than two values are likely to be needed for many of the variables and if in addition the system is refined to include additional operations, it becomes clear that a complete numerical solution for all combinations of variables is impossible, even with the aid of a high-powered computer.

Some means must therefore be found to reduce the number of calculations needed whilst retaining the use that can be made of the general equation. To do this, we need guidance in choosing realistic values for the flow parameters *a, b, c,* and so on rather than simply choosing values at random.

Such guidance is available from published information concerning the component sub-systems of Figure 12.12 and an example of how such information might be used is given below. It must however be stressed that the following example is by way of illustration only and is not intended as a definitive analysis of the U.K. iron and steel industry. Such a study would represent a book in its own right and would also entail the use of a more complex network than that of Figure 12.12.

The system we have chosen to analyse is one in which there is:

(a) no direct reduction, so that:

$$e = 0, \tag{12.29}$$

$$k = 0, \tag{12.30}$$

$$x = 0, \tag{12.31}$$

(b) the electric arc furnace operates on a 100% scrap charge so that:

$$s = 0, \tag{12.32}$$

(c) all reclaimed scrap is used so that:

$$g = 0, \tag{12.33}$$

(d) no iron ore is fed directly to the basic oxygen furnace so that:

$$i = 0. \tag{12.34}$$

The remaining flow parameters can only be evaluated using data on the performance of either individual sub-systems or the industry as a whole.

Typically the basic oxygen furnace might take an input of 800 kg of liquid iron from the blast furnace together with 300 kg of scrap [131] and so the ratio w/m can be written as:

$$\frac{w}{m} = \frac{300}{800},$$

whence,

$$w = 0.375m. \tag{12.35}$$

Similarly an input of 800 kg of liquid iron to the basic oxygen furnace results in an iron loss of 47 kg of iron [6]:

$$\frac{h}{m} = \frac{47}{800},$$

or,

$$h = 0.059m. \tag{12.36}$$

Thornton and Williams [131] quote a materials efficiency for the electric arc furnace, measured in terms of iron contents of inputs and outputs as 97.5%, so that:

$$t = 0.975u. \tag{12.37}$$

Similarly Barnes [6] reports that typically an input of 757 kg of iron as ore to the blast furnace is accompanied by a loss of 4 kg of iron during smelting, so that:

$$j = 189b. \tag{12.38}$$

The performance of the open-hearth furnace has been reported by Berry, [12] who states that for an input of 0.761 kg of hot metal (pig iron) and 0.510 kg of scrap, a saleable output of 1.188 kg of ingot is obtained. Hence:

$$l = 1.49v \tag{12.39}$$

$$c = 0.11l \tag{12.40}$$

The information leading to Equations (12.35) to (12.40) is typical of that describing the performance of specific operations and relies on a detailed examination of such operations.

Often reasonable values for some parameters can be estimated using a mixture of operation-based information and industry-based information. For example, there is no satisfactory way of knowing how much scrap is actually fed to each of the different furnaces but this can be estimated knowing the relative proportions of scrap demanded by each type of furnace and the relative amounts of steel produced by each furnace. Heal [61] reports that in 1970, 47% of all steel made in the U.K. was by the open-hearth furnace and 33% was by the basic oxygen furnace. Almost the whole of the remainder was by the electric arc process. This information is readily converted into the flow parameters of Figure 12.12 as:

$$\frac{n}{n+r+t} = 0.47, \tag{12.41}$$

$$\frac{r}{n+r+t} = 0.33, \tag{12.42}$$

Note the third equation relating to the proportion of steel from the electric arc furnace should not be used because it is simply a rearrangement of equations (12.41) and (12.42). Hence the scrap feeds are given by the equations:

$$v = 0.39y, \tag{12.43}$$

$$w = 0.19y, \tag{12.44}$$

$$u = 0.42y. \tag{12.45}$$

In some instances, the actual values may vary with time and it may be necessary to assume a range of values to examine the effect of such changes. For example, the grade of ore available as an input will certainly change in such a way and this will cause a change in the value of *a*. For convenience we have chosen:

$$a = 0.2, \tag{12.46}$$

which implies that in total there is a 20% loss of iron ore between ore in the ground and the pellets entering the blast furnace.

Yet a further example of uncertainty arises in the area of post-consumer scrap reclamation. We have chosen to assume that 20% of all post-consumer scrap is reclaimed so that:

$$z = 0.2p. \tag{12.47}$$

However it is obvious that this parameter may be changed by the implementation of effective recycling schemes.

Equations (12.19) to (12.47) may now be solved for the flow parameters to give:

$$\begin{array}{ll}
a = 0.2, & b = 0.004, \\
c = 0.05, & d = 0.07, \\
e = 0, & f = 0.008, \\
g = 0, & h = 0.02, \\
i = 0, & j = 0.80, \\
k = 0, & l = 0.43, \\
m = 0.37, & n = 0.68, \\
p = 0.83, & q = 0.59, \\
r = 0.47, & s = 0, \\
t = 0.31, & u = 0.31, \\
v = 0.29, & w = 0.14, \\
x = 0, & y = 0.75, \\
z = 0.16, & \alpha = 0.66.
\end{array}$$

These numerical values arise from the use of a very limited set of data as described earlier and alternative sets can be obtained from other sources of information. It is difficult to obtain a totally consistent set of data for calculating these flow parameters and the values obtained for them will reflect this. It is however instructive to repeat the calculations for a number of different sets of information because the results give a clear indication of the sensitivity of each flow parameter to changes in input information. They also indicate the most suitable ranges of values that should be used in attempting more general numerical solutions than that given below.

When the above set of flow parameters are substituted into Equation (12.18), they give a system energy E_s as:

$$E_s = E_1 + 0.80E_2 + 0.73E_3 + 1.48E_4 + 0.82E_5 + 0.50E_6 + 0.75E_8 + 0.75E_9 + 0.32E_{10}. \quad (12.49)$$

Note that the term in E_7 has vanished because of the assumption that no direct reduction is carried out.

It is sensible to evaluate the flow parameters before the energy parameters because the energy parameters depend upon them when data are expressed in the form of Part 2. The reason for this is that energy requirements of sub-systems are usually normalised with respect to **useful output** from the sub-system, whereas the energy terms E_1 and so on, are dependent on materials throughput. For example, if the value of E_{10} taken from Part 2 were 8.99 MJ/kg of steel output, this essentially represents the sub-system energy normalised with respect to flow parameters $(t + f)$, hence:

$$E_{10} = 8.99 \times \frac{0.30}{(0.30 + 0.008)}$$

$$= 8.76 \text{ (MJ)}.$$

Other data must be treated in a similar manner. Note that the detailed breakdown of this total energy requirement into the form of Table 7.4 would be used in more detailed analysis, in which case each of the 16 entries would be treated in the same way.

In the present exercise we are interested primarily in the technique rather than the accuracy of the final result and so we have chosen, for simplicity, the following approximate values for the energy requirements of the sub-systems of Figure 12.12 in MJ normalised with respect to throughput:

$E_1 = 3$, $E_2 = 25$,

$E_3 = 5$, $E_4 = 10$,

$E_5 = 0$, $E_6 = 1$,

$$E_8 + E_9 = 2, \qquad E_{10} = 8.$$

Substituting these values into Equation (12.49) gives:

$$E_s = 46.65 \text{ MJ}.$$

This energy requirement represents the energy consumed by the system defined by Figure 12.12 and Equations (12.29) to (12.47) to maintain the flows of materials resulting from an input of 1 kg of iron as ore. It is clear that this system energy requirement can be expressed in a number of ways. For example, the mass flow from the fabrication sub-system (sub-system 4 in Figure 12.12) is $p = 0.83$ kg and the overall system energy may be normalised with respect to this parameter to give $46.65/0.83 = 56.20$ MJ per kg of fabricated steel. Alternatively the system energy could be normalised with respect to the output of liquid steel from the three furnaces, that is 1.46 kg, to give 31.95 MJ per kg of liquid steel. Obviously then the numerical value will vary with the choice of normalising parameter and it is impossible to identify the total system energy with a single flow within the system.

Inspection of the sources of some of the data for iron and steel products given in Part 2 soon shows that the majority of analysts tend to select only a part of the network of Figure 12.12 for analysis. As a consequence their results are expected to differ from the results that would be obtained from a more detailed network analysis because of the inter-dependence of the sub-system behaviour. Thus not only do results differ because of different sets of operating data but because different systems are chosen as well.

Part 2

TABLES OF ENERGY REQUIREMENTS OF SOME INDUSTRIAL PROCESSES

Energy requirements of some industrial processes

For most purposes a satisfactory energy analysis of an industrial system can be carried out using primary data for the first order analysis and secondary or published data for the higher orders of the analysis. Published data relating to the performance of industrial systems is unfortunately scattered throughout the literature and a search for specific values can be very time-consuming. Part 2 of this book therefore presents a collection of data from a wide variety of published sources and will, it is hoped, supply values for many of the industrial systems likely to be encountered.

For most of the entries the source of the data is quoted in the notes at the end of the Table and the basis of the calculation is also given. The systems are arranged in alphabetical order of output material and the energy requirements are stated in terms of the energy per unit output. The system detail may be identified from the nature of the inputs and outputs and from the column labelled 'System Type', which attempts to identify the 'vertical completeness' of the system (see Chapter 3). In addition the supplementary notes at the end of the Table give as much detail as is available from the Reference to identify the precise processing route described.

The major difficulty encountered in compiling the Table has been the imprecise way in which most published values define the system being described. Indeed some published values have had to be omitted because it proved impossible to identify the system in any way. Most authors identify the output. Some give only a vague notion of the input materials but very few specify the precise processing route or indicate whether the energy associated with transport, services or capital plant have been included. Wherever possible we have attempted to identify the system on the basis of the published descriptions, but even so there must inevitably be uncertainties remaining in some of the values.

Again, wherever possible we have included values which indicate the spread of published energy requirements for processes. This range of possible values can arise from a number of causes of which it is well to be aware.

(1) Apparently similar processing routes may use different mixes of fuels even though the energy directly consumed by the processes is the same.

For example, two of the total energy requirements for the production of oxygen are 9.13 MJ/kg and 15.00 MJ/kg. If however the detailed fuel breakdown is examined, it is found that the corresponding fuel energies directly consumed by the two plants are 3.63 and 3.60 MJ/kg respectively. The very different total energy requirements arise because of the latter value is consumed as electricity alone, with a conversion efficiency of 25%, whereas the former consumes only 1.64 MJ/kg as electricity, the remainder being oil and natural gas. Where this effect occurs it can usually be readily detected in the fuel energy breakdown.

(2) Different companies operating apparently identical systems frequently consume different process energies, indeed this effect frequently occurs in different factories within the same company. Such variations arise from different ages of plant, different levels of maintenance and different practices. Variations due to this cause are often extremely difficult to identify because the company or factory to which the data relates is seldom identified, but variations of 5% in the energy content of the fuels consumed by the plant are common and even greater variations are not unknown.

(3) Where the Table shows apparently similar systems in terms of input and output materials together with the same system type, this description may be inadequate to identify the system completely and hence the total system energy requirements will be different. In such cases the more detailed description must be obtained by reference to the supplementary notes at the end of the Table. An example of this is the production of 1,2.dichloroethane from ethylene, where the system energy varies from 30.30 MJ/kg to 488.88 MJ/kg but the detailed routes are very different.

(4) Differences may arise because of the inclusion of some products as useable by-products in some instances and the exclusion of these same products in others. For example the electrolysis of brine produces chlorine, sodium hydroxide and hydrogen, yet not all operators recover the hydrogen. In such a case, the energy of the process will be apportioned over only two products rather than over three and will be consequently higher. Similarly waste heat in the form of steam may be used elsewhere on a large production site but where no immediate use is available would be sent to cooling towers. In consequence, steam sent to cooling towers would be charged to the production process whereas steam used elsewhere would represent an energy credit to the process. Where such differences are indicated in the source material, this is noticed in the supplementary notes to the Table.

System Type	Electricity		Oil Fuels			Other Fuels			Total System Energy Requirement	Notes (see page 376 *et seq.*)
	Fuel production energy	Energy content of fuel	Fuel production energy	Energy content of fuel	Feedstock	Fuel production energy	Energy content of fuel	Feedstock		
ABS RESIN from ABS scrap (MJ/kg)										
A	0.94	0.30	–	–	–	–	–	–	1.24	(1), (16), (18)
ABS RESIN from crude oil (MJ/kg)										
A	3.60	1.14	5.77	28.22	74.73	0.09	0.65	–	114.20	(1), (519)
ABS RESIN from monomers (MJ/kg)										
A	5.17	1.63	–	–	–	0.79	0.65	–	13.13	(1), (16), (17)
ACETYLENE from crude oil (MJ/kg)										
D	8.68	2.89	8.58	41.81	24.03	–	–	–	85.99	(1), (19), (120)
H	3.04	0.96	11.74	57.33	24.00	–	–	–	97.07	(2), (20), (123)
H	*	*	*	*	*	*	*	*	*198.55*	(2), (15), (108)
D	9.77	3.26	134.37	654.27	11.61	–	–	–	813.28	(1), (19), (122)
ACETYLENE from naphtha (MJ/kg)										
D	8.68	2.89	7.74	37.73	–	–	–	–	57.04	(1), (19), (119)
D	99.77	3.26	133.97	652.30	–	–	–	–	799.30	(1), (19), (121)
ACETYLENE from natural gas (MJ/kg)										
D	7.45	2.48	33.25	161.79	–	0.30	2.10	30.00	237.37	(1), (19), (117)
ACRYLIC FIBRE from crude oil (MJ/kg)										
A	73.69	23.27	29.07	141.93	81.00	–	–	–	348.96	(2), (15), (352)
ACRYLONITRILE from crude oil (MJ/kg)										
H	2.19	0.69	4.55	22.20	62.80	–	–	–	92.43	(1), (463)
H	53.20	16.80	2.04	9.95	70.64	–	–	–	152.63	(2), (15), (124)

System Type	Electricity		Oil Fuels		Other Fuels				Total System Energy Requirement	Notes (see page 376 *et seq.*)
	Fuel production energy	Energy content of fuel	Fuel production energy	Energy content of fuel	Feedstock	Fuel production energy	Energy content of fuel	Feedstock		
ACRYLONITRILE from propylene and ammonia (MJ/kg)										
A	1.75	0.55	–	–	–	–	–	–	2.30	(1), (16), (21)
AGGREGATE from raw materials in the ground (MJ/kg)										
B	–	–	*(0.02)*	*(0.10)*	–	–	–	–	*(0.12)*	(2), (24), (125)
H	–	–	0.82	4.00	–	–	–	–	4.82	(6), (22), (118)
AIR intake to chemical processes (MJ/kg)										
A	0.64	0.20	–	–	–	–	–	–	0.84	(2), (15), (514)
ALLOY STEEL – see under STEEL (ALLOY)										
ALUMINA from aluminium hydroxide (MJ/kg)										
A	–	–	–	–	–	*(0.38)*	*(2.69)*	–	*(3.07)*	(1), (28), (30)
A	–	–	–	–	–	*(0.66)*	*(4.64)*	–	*(5.30)*	(1), (28), (29)
A	–	–	0.96	4.66	–	–	–	–	5.62	(2), (38), (520)
ALUMINA from clay (MJ/kg)										
A	8.19	2.73	–	–	–	5.93	44.37	–	53.24	(1), (31), (33)
A	1.14	0.36	–	–	–	4.64	47.16	–	53.30	(1), (31), (33)
ALUMINA from bauxite at the plant (MJ/kg)										
A	*	*	*	*	*	*	*	*	*6.80*	(1), (36), (37)
A	4.78	1.51	0.35	1.73	–	0.61	5.70	–	14.68	(1), (26), (27)
A	*	*	*	*	*	*	*	*	*16.20*	(7), (25), (126)
A	2.82	0.89	3.41	16.57	–	–	–	–	23.69	(14), (547), (548)
A	2.70	0.90	3.85	18.65	–	–	–	–	26.10	(2), (38), (39)
A	8.78	2.77	0.66	3.17	–	1.09	10.18	–	26.65	(1), (16), (35)

System Type	Electricity: Fuel production energy	Electricity: Energy content of fuel	Oil Fuels: Fuel production energy	Oil Fuels: Energy content of fuel	Oil Fuels: Feedstock	Other Fuels: Fuel production energy	Other Fuels: Energy content of fuel	Other Fuels: Feedstock	Total System Energy Requirement	Notes (see page 376 *et seq.*)
ALUMINA from bauxite in the ground (MJ/kg)										
A	3.45	1.09	–	–	–	3.25	22.73	–	30.52	(1), (31), (34)
A	2.97	0.93	4.97	25.89	–	–	–	–	34.76	(2), (15)
ALUMINA from feldspar ore (MJ/kg)										
A	8.19	2.73	–	–	–	5.93	44.37	–	61.22	(1), (31), (32)
ALUMINIUM from alumina (MK/kg)										
A	163.24	51.55	–	–	–	–	–	–	214.79	(1), (28), (137)
A	171.00	54.00	–	–	–	–	–	–	225.00	(1), (49), (133)
A	176.70	55.80	–	–	–	–	–	–	232.50	(7), (25), (137)
A	179.36	56.64	–	–	–	–	–	–	236.00	(1), (704), (705)
D	149.44	47.19	0.06	0.28	–	6.36	44.49	–	247.82	(1), (26), (142)
D	193.80	61.20	–	–	–	0.48	2.52	–	258.00	(1), (4), (54), (137), (143)
A	200.86	63.43	–	–	–	–	–	–	264.29	(1), (28), (137)
A	160.51	50.51	0.07	0.32	–	6.63	46.39	–	264.61	(1), (16), (132)
A	213.59	67.45	–	–	–	0.10	1.97	–	283.11	(1), (31), (134)
F	*	*	*	*	*	*	*	*	*241.74*	(1), (36)
ALUMINIUM from aluminium chloride (MJ/kg)										
A	112.99	35.68	–	–	–	–	–	–	148.67	(1), (28), (139)
ALUMINIUM from bauxite (MJ/kg)										
H	*	*	*	*	*	*	*	*	95.58	(2), (50), (127)
D	*	*	*	*	*	*	*	*	120.00	(6), (22)
J	*	*	*	*	*	*	*	*	226.56	(1), (40), (125)
F	*	*	*	*	*	*	*	*	233.64	(1), (44), (125)
H	*	*	*	*	*	*	*	*	245.03	(2), (48)
J	*	*	*	*	*	*	*	*	259.11	(1), (42)

System Type	Electricity: Fuel production energy	Electricity: Energy content of fuel	Oil Fuels: Fuel production energy	Oil Fuels: Energy content of fuel	Oil Fuels: Feedstock	Other Fuels: Fuel production energy	Other Fuels: Energy content of fuel	Other Fuels: Feedstock	Total System Energy Requirement	Notes (see page 376 *et seq.*)
J	*	*	*	*	*	*	*	*	288.00	(2), (51)
D	220.53	69.64	–	–	–	5.57	40.24	–	335.98	(1), (31), (135)
J	*	*	*	*	*	*	*	*	342.00	(2), (52), (127)
D	231.01	72.95	–	–	–	8.34	59.67	–	371.97	(1), (31), (136)
ALUMINIUM RING-PULL CAN ENDS from aluminium sheet (MJ/kg)										
A	1.06	0.34	–	–	–	0.48	2.52	–	2.73	(1), (16), (144)
ALUMINIUM CASTINGS/INGOTS from ore in the ground (MJ/kg)										
D	*	*	*	*	*	*	*	*	*222.20*	(1), (43), (689)
F	*	*	*	*	*	*	*	*	*253.36*	(1), (36), (690)
D	*	*	*	*	*	*	*	*	*290.19*	(1), (46), (129)
ALUMINIUM CASTINGS/INGOTS from aluminium scrap or solid aluminium (MJ/kg)										
F	*	*	*	*	*	*	*	*	*10.99*	(1), (43), (692)
F	4.51	1.43	0.39	1.86	–	0.77	5.40	–	14.36	(1), (16), (128)
A	*	*	*	*	*	*	*	*	*23.24*	(1), (36), (691)
ALUMINIUM CASTINGS/INGOTS from aluminium liquid (MJ/kg)										
A	–	–	–	–	–	*(0.86)*	*(6.00)*	–	*(6.86)*	(7), (25), (693)
F	–	–	–	–	–	*(0.88)*	*(6.19)*	–	*(7.07)*	(1), (44), (694)
ALUMINIUM FORGINGS from ore in the ground (MJ/kg)										
F	*	*	*	*	*	*	*	*	*259.64*	(1), (43), (695)
ALUMINIUM LIQUID from aluminium solid or scrap (MJ/kg)										
A	–	–	–	–	–	0.35	2.45	–	2.80	(1), (47), (130)
A	–	–	–	–	–	0.56	3.95	–	4.51	(1), (47), (131)

System Type	Electricity		Oil Fuels			Other Fuels			Total System Energy Requirement	Notes (see page 376 *et seq.*)
	Fuel production energy	Energy content of fuel	Fuel production energy	Energy content of fuel	Feedstock	Fuel production energy	Energy content of fuel	Feedstock		
A	–	–	–	–	–	0.92	4.88	–	5.80	(1), (47), (138)
A	–	–	–	–	–	(*0.76*)	(*5.31*)	–	(*6.07*)	(1), (44), (696)
H	*	*	*	*	*	*	*	*	10.80	(2), (50)
J	*	*	*	*	*	*	*	*	12.45	(1), (42)
J	*	*	*	*	*	*	*	*	*13.00*	(2), (52), (127)
A	10.26	3.24	–	–	–	–	–	–	13.50	(7), (53), (140)
A	14.69	4.64	–	–	–	–	–	–	19.33	(1), (31), (140)
H	*	*	*	*	*	*	*	*	*23.40*	(2), (50), (127)
ALUMINIUM (REFINED) from aluminium from the Hall cell (MJ/kg)										
A	201.78	63.72	–	–	–	–	–	–	265.50	(1), (545), (546)
ALUMINIUM SCRAP from delacquering scrap aluminium cans (MJ/kg)										
A	–	–	(*0.58*)	(*2.80*)	–	–	–	–	(*3.38*)	(1), (44), (697)
ALUMINIUM SHEET from aluminium ingot (MJ/kg)										
A	4.91	1.55	–	–	–	1.16	8.14	–	15.76	(1), (16), (147)
A	*	*	*	*	*	*	*	*	*18.76*	(1), (44), (698)
A	*	*	*	*	*	*	*	*	*30.34*	(1), (36)
F	10.17	3.21	–	–	–	2.51	17.60	–	33.49	(1), (26), (149)
A	*	*	*	*	*	*	*	*	*42.30*	(1), (36)
ALUMINIUM SHEET from aluminium scrap (MJ/kg)										
F	2.53	0.80	0.21	1.02	–	0.44	3.11	–	8.11	(1), (26), (148)
ALUMINIUM SHEET from ore in the ground (mj/kg)										
A	*	*	*	*	*	*	*	*	*259.97*	(1), (43), (699)
A	*	*	*	*	*	*	*	*	*266.43*	(1), (55), (127), (700)
A	*	*	*	*	*	*	*	*	*266.61*	(1), (36)

System Type	Electricity: Fuel production energy	Electricity: Energy content of fuel	Oil Fuels: Fuel production energy	Oil Fuels: Energy content of fuel	Oil Fuels: Feedstock	Other Fuels: Fuel production energy	Other Fuels: Energy content of fuel	Other Fuels: Feedstock	Total System Energy Requirement	Notes (see page 376 *et seq.*)
A	*	*	*	*	*	*	*	*	*295.20*	(1), (36)
A	189.48	59.40	11.98	61.01	–	1.79	26.18	–	349.84	(2), (15)
H	220.84	69.74	–	–	–	14.62	77.53	–	382.73	(4), (45)
ALUMINIUM SHEET (HOT ROLLED) from aluminium ingot from reheating furnace (MJ/kg)										
A	1.71	0.54	–	–	–	–	–	–	2.25	(7), (53), (145)
A	3.04	0.96	–	–	–	–	–	–	4.00	(7), (25), (146)
ALUMINIUM STRIP (CONTINUOUS-CAST) from aluminium liquid (MJ/kg)										
A	*(0.35)*	*(0.11)*	–	–	–	–	–	–	*(0.46)*	(7), (53), (150)
ALUMINIUM WIRE from aluminium ingot (MJ/kg)										
A	*	*	*	*	*	*	*	*	*30.34*	(1), (36)
ALUMINIUM WIRE from ore in the ground (MJ/kg)										
A	*	*	*	*	*	*	*	*	*295.20*	(1), (36)
ALUMINIUM OXIDE – see under ALUMINA.										
AMMONIA GAS from ammonium chloride (MJ/kg)										
A	–	–	–	–	–	0.40	2.14	–	2.54	(2), (56), (155)
AMMONIA GAS from annonium chloride (MJ/kg)										
D	0.74	0.23	3.58	17.48	–	–	–	–	22.03	(1), (19), (211)
AMMONIA GAS from crude oil (MJ/kg)										
H	0.03	0.01	3.40	16.58	24.00	–	–	–	44.02	(2), (20), (154)
H	*	*	*	*	*	*	*	*	*49.75*	(2), (15), (151)

System Type	Electricity: Fuel production energy	Electricity: Energy content of fuel	Oil Fuels: Fuel production energy	Oil Fuels: Energy content of fuel	Oil Fuels: Feedstock	Other Fuels: Fuel production energy	Other Fuels: Energy content of fuel	Other Fuels: Feedstock	Total System Energy Requirement	Notes (see page 376 *et seq.*)
AMMONIA GAS from natural gas (MJ/kg)										
A	0.61	0.18	1.59	7.72	–	0.09	0.60	8.61	19.40	(1), (16), (152)
A	0.78	0.26	0.08	0.38	–	(0.20)	(1.39)	(19.90)	22.99	(1), (31), (153)
A	2.05	0.67	0.10	0.51	–	0.27	1.89	26.95	32.44	(1), (106), (513)
A	1.00	0.33	0.09	0.44	–	2.59	18.15	23.13	45.73	(1), (704), (725)
D	*(7.54)*	*(2.47)*	*(0.37)*	*(1.88)*	–	*(0.99)*	*(6.95)*	*(28.00)*	*48.20*	(2), (56)
A	*	*	*	*	*	*	*	*	*58.04*	(1), (57)
AMMONIA LIQUID from ammonia gas (MJ/kg)										
A	–	–	–	–	–	3.26	22.80	–	26.06	(1), (106), (512)
AMMONIA SOLUTION from natural gas (MJ/kg)										
A	0.71	0.24	0.09	0.45	–	2.53	16.05	23.68	43.75	(2), (56), (158)
A	3.45	1.10	0.09	0.46	–	2.30	16.08	24.19	47.67	(2), (56), (157)
A	7.79	2.47	0.09	0.46	–	1.57	11.01	24.00	47.79	(2), (56), (156)
AMMONIUM NITRATE from ammonia gas and nitric acid (MJ/kg)										
D	0.36	0.11	1.09	5.27	–	–	–	–	6.83	(2), (56), (540)
D	0.57	0.18	–	–	–	0.49	9.35	–	10.59	(1), (31), (536)
D	0.65	0.21	2.13	10.33	–	–	–	–	13.32	(1), (31), (535)
D	0.78	0.25	2.13	10.33	–	–	–	–	13.49	(2), (56), (542)
AMMONIUM NITRATE from natural gas/crude oil (MJ/kg)										
D	2.65	0.83	1.36	6.61	5.62	0.52	9.57	5.50	32.66	(1), (31), (538)
D	3.37	1.05	2.57	12.43	6.09	0.41	2.87	5.96	34.75	(2), (56), (539)
D	2.73	0.86	3.49	16.94	5.62	0.03	0.22	5.50	35.39	(1), (31), (537)
D	3.56	1.12	3.49	16.93	5.62	0.38	2.64	4.40	39.24	(2), (56), (541)
AMMONIUM SULPHATE from ammonia solution (MJ/kg)										
D	0.34	0.11	0.31	1.53	–	–	–	–	2.29	(2), (56), (544)

System Type	Electricity: Fuel production energy	Electricity: Energy content of fuel	Oil Fuels: Fuel production energy	Oil Fuels: Energy content of fuel	Oil Fuels: Feedstock	Other Fuels: Fuel production energy	Other Fuels: Energy content of fuel	Other Fuels: Feedstock	Total System Energy Requirement	Notes (see page 376 *et seq.*)
AMMONIUM SULPHATE from crude oil/natural gas (MJ/kg)										
D	3.20	1.02	0.33	1.65	–	0.44	3.12	6.50	16.26	(2), (56), (543)
AUTOMOBILES (see Chapter 9)										
BATTERY (LEAD-ACID) for road vehicle, see Chapter 9 (9.4.6)										
BAUXITE from ore in the ground (MJ/kg)										
A	*	*	*	*	*	*	*	*	*0.63*	(1), (36), (161)
A	0.09	0.03	0.01	0.07	–	0.07	0.46	–	0.73	(1), (16), (160)
A	0.16	0.05	0.02	0.10	–	0.11	0.79	–	1.23	(1), (26)
BENTONITE (MJ/kg)										
A	*	*	*	*	*	*	*	*	*1.39*	(1), (704)
BENZENE from crude oil (MJ/kg)										
A	1.23	0.30	–	–	44.50	1.99	13.93	–	62.04	(1), (16), (162)
H	2.50	0.79	4.19	20.47	51.83	–	–	–	79.78	(2), (20), (163)
BLOW MOULDING – see under BOTTLES										
BOARD from timber (MJ/kg)										
F	4.78	1.51	1.67	8.16	–	–	–	12.50	28.62	(4), (45), (159)
F	*	*	*	*	*	*	*	*	*34.87*	(1), (44), (159), (701)
H	*	*	*	*	*	*	*	*	*61.26*	(2), (48), (159), (702)
BOARD from waste paper (MJ/kg)										
F	*	*	*	*	*	*	*	*	*3.43*	(1), (44), (703)

System Type	Electricity: Fuel production energy	Electricity: Energy content of fuel	Oil Fuels: Fuel production energy	Oil Fuels: Energy content of fuel	Oil Fuels: Feedstock	Other Fuels: Fuel production energy	Other Fuels: Energy content of fuel	Other Fuels: Feedstock	Total System Energy Requirement	Notes (see page 376 *et seq.*)
BOARD (CORRUGATED) from card and paper (MJ/kg)										
D	0.06	0.02	0.01	0.07	–	–	–	–	0.16	(4), (45)
BOTTLES (GLASS) from raw materials (MJ/kg)										
A	*(4.10)*	*(1.30)*	*(1.87)*	*(9.15)*	–	–	–	–	*(16.42)*	(4), (45), (618)
A	2.80	0.88	0.32	1.60	–	1.40	9.79	–	16.79	(1), (704), (718)
A	2.85	0.90	0.20	0.10	–	1.58	11.06	–	17.59	(1), (26), (166)
F	*	*	*	*	*	*	*	*	*17.99*	(1), (58), (127), (616)
A	3.09	1.03	2.67	13.06	–	–	–	–	19.85	(1), (19), (165)
A	3.30	1.04	0.24	1.15	–	1.77	12.41	–	19.91	(1), (16), (169)
A	*	*	*	*	*	*	*	*	*21.02*	(2), (48), (617)
H	*	*	*	*	*	*	*	*	*31.72*	(1), (55), (127)
BOTTLES (PLASTIC) from polymer resin (MJ/kg)										
A	10.55	3.33	–	–	–	–	–	–	13.88	(1), (44), (533)
BOTTLES (POLYETHYLENE) from polymer resin (MJ/kg)										
D	31.95	10.09	–	–	–	–	–	–	42.04	(2), (15), (530)
BOTTLES (PVC) from polymer resin (MJ/kg)										
A	2.98	0.94	–	–	–	–	–	–	3.92	(4), (45), (534)
BOTTLE FILLING (MJ/1000 bottles)										
A	5.00	1.58	–	–	–	–	–	–	6.58	(4), (45)
A	*(2.51)*	*(0.79)*	–	–	–	–	–	–	*(3.30)*	(1), (44), (619)
BOTTLE FILLING (MJ/gallon)										
A	1.54	0.49	0.39	1.86	–	0.30	2.10	–	6.68	(1), (58), (620)

System Type	Electricity: Fuel production energy	Electricity: Energy content of fuel	Oil Fuels: Fuel production energy	Oil Fuels: Energy content of fuel	Oil Fuels: Feedstock	Other Fuels: Fuel production energy	Other Fuels: Energy content of fuel	Other Fuels: Feedstock	Total System Energy Requirement	Notes (see page 376 *et seq.*)
BOX (CARDBOARD) from cardboard (MJ/box)										
A	*(0.19)*	*(0.06)*	*(0.03)*	*(0.13)*	–	–	–	–	*(0.41)*	(4), (45), (621)
BOX (CORRUGATED CARDBOARD) from wood fibre (MJ/kg)										
F	*	*	*	*	*	*	*	*	*36.46*	(1), (44), (171)
A	4.01	1.27	0.42	2.12	–	4.34	30.39	–	42.55	(1), (16), (170)
BOX (PAPERBOARD) from wood fibre (MJ/kg)										
A	4.01	1.27	0.44	2.23	–	4.84	33.87	–	46.66	(1), (16), (172)
BRICKS (UNSPECIFIED) from raw materials (MJ/brick)										
A	*(0.06)*	*(0.02)*	–	–	–	*(0.10)*	*(0.70)*	–	*(0.88)*	(2), (41)
BRICKS (BAKED CLAY) from raw materials (MJ/kg)										
H	*	*	*	*	*	*	*	*	*1.80*	(2), (50)
J	*	*	*	*	*	*	*	*	*2.00*	(2), (52)
J	0.19	0.07	0.08	1.00	–	0.01	1.15	–	2.50	(2), (59), (127), (173)
D	0.40	0.13	0.04	0.20	–	0.41	2.98	–	4.16	(1), (704), (713)
J	*(0.33)*	*(0.10)*	–	–	–	*(0.60)*	*(4.20)*	–	*(5.23)*	(6), (22), (615)
BRICKS – REFRACTORY (CHROME-MAGNESITE) from raw materials (MJ/kg)										
A	0.83	0.26	–	–	–	0.79	5.40	–	7.28	(1), (704), (729)
BRICKS – REFRACTORY (FIRECLAY) from raw materials (MJ/kg)										
A	0.39	0.12	0.04	0.20	–	0.59	4.10	–	5.44	(1), (704), (730)
BRICKS (SAND-LIME) from raw materials (MJ/kg)										
J	*(0.06)*	*(0.02)*	–	–	–	*(0.11)*	*(0.82)*	–	*(1.01)*	(6), (22), (615)

System Type	Electricity: Fuel production energy	Electricity: Energy content of fuel	Oil Fuels: Fuel production energy	Oil Fuels: Energy content of fuel	Oil Fuels: Feedstock	Other Fuels: Fuel production energy	Other Fuels: Energy content of fuel	Other Fuels: Feedstock	Total System Energy Requirement	Notes (see page 376 *et seq.*)
BRINE – see under SODIUM CHLORIDE										
BUILDING (APARTMENT) from raw materials (MJ/m²)										
	*	*	*	*	*	*	*	*	*1700–2680*	(2), (59), (623)
	*	*	*	*	*	*	*	*	*3185–4083*	(2), (59), (624)
BUILDING (2-STOREY HOUSE) from raw materials (MJ/m²)										
	*	*	*	*	*	*	*	*	*1235–2266*	(2), (59), (625)
	*	*	*	*	*	*	*	*	*1207–1831*	(2), (59), (626)
BUILDING (INDUSTRIAL) from raw materials (MJ/m²)										
	*	*	*	*	*	*	*	*	*36,000*	(2), (38), (627)
1.3 – BUTADIENE from naphtha (MJ/kg)										
A	1.71	0.54	–	–	–	0.68	4.73	–	7.66	(1), (16), (175)
1.3 – BUTADIENE from n-butanes (MJ/kg)										
A	8.21	2.59	–	–	–	2.87	20.07	–	33.74	(1), (16), (174)
1.3 – BUTADIENE from n-butenes (MJ/kg)										
A	5.47	1.73	–	–	–	0.89	6.22	–	14.31	(1), (16), (173)
1.3 – BUTADIENE from crude oil (MJ/kg)										
H	2.98	0.94	7.37	35.99	53.50	–	–	–	100.78	(2), (15), (20), (176)
BUTANE – L.P.G. (fuel content and production energy) see Chapter 6 and Table 6.13										
BUTENES (MIXED) from crude oil (MJ/kg)										
A	0.56	0.18	2.79	13.55	44.50	–	–	–	61.58	(1), (19), (265)

System Type	Electricity: Fuel production energy	Electricity: Energy content of fuel	Oil Fuels: Fuel production energy	Oil Fuels: Energy content of fuel	Oil Fuels: Feedstock	Other Fuels: Fuel production energy	Other Fuels: Energy content of fuel	Other Fuels: Feedstock	Total System Energy Requirement	Notes (see page 376 *et seq.*)
H	0.41	0.13	2.96	14.46	50.16	–	–	–	68.12	(2), (15), (20), (261)
A	8.68	2.89	8.58	41.81	24.03	–	–	–	85.99	(1), (19), (120)
BUTENES (MIXED) from ethane (MJ/kg)										
A	0.78	0.24	5.42	26.39	–	–	–	–	32.83	(1), (19), (264)
BUTENES (MIXED) from naphtha (MJ/kg)										
A	8.68	2.89	7.74	37.73	–	–	–	–	57.04	(1), (19), (119)
CALCIUM FLUORIDE (MJ/kg)										
J	*	*	*	*	*	*	*	*	*1.59*	(1), (704), (706)
CALCIUM OXIDE from limestone (MJ/kg)										
A	–	–	–	–	–	0.45	3.18	–	3.63	(2), (60), (497)
F	0.35	0.11	0.18	0.89	–	0.41	2.84	–	4.78	(1), (26), (181)
A	0.35	0.11	0.03	0.17	–	0.40	4.46	–	5.52	(1), (16), (179)
A	–	–	–	–	–	0.70	4.93	–	5.63	(1), (31), (180)
F	*	*	*	*	–	*	*	–	*5.81*	(1), (57), (628)
F	*	*	*	*	*	*	*	–	*6.30*	(6), (22), (629)
CALCIUM OXIDE from limestone in the ground (MJ/kg)										
H	*	*	*	*	*	*	*	*	*9.71*	(2), (38)
A	0.64	0.20	0.13	0.64	–	0.69	7.65	–	9.95	(1), (704), (709)
CALCIUM SULPHATE (CRUSHED) from natural gypsum in the ground (MJ/kg)										
D	0.48	0.15	neg	0.02	–	–	–	–	0.65	(2), (15), (182)
D	0.67	0.21	neg	0.01	–	–	–	–	0.89	(2), (15), (182)

System Type	Electricity: Fuel production energy	Electricity: Energy content of fuel	Oil Fuels: Fuel production energy	Oil Fuels: Energy content of fuel	Oil Fuels: Feedstock	Other Fuels: Fuel production energy	Other Fuels: Energy content of fuel	Other Fuels: Feedstock	Total System Energy Requirement	Notes (see page 376 *et seq.*)
CALUMITE BRAND SLAG from selected blast furnace slag (MJ/kg)										
D	0.44	0.14	0.16	0.77	–	–	–	–	1.51	(2), (15), (183)
CAPITAL ENERGY OF RAIL TRANSPORT see under RAIL TRANSPORT and Chapter 9										
CAPITAL ENERGY OF SEA TRANSPORT see under SEA TRANSPORT and Chapter 9										
CANS see under METAL CANS										
CARBON BLACK from crude oil (MJ/kg)										
H	22.83	7.16	16.00	51.21	32.79	–	–	–	129.99	(2), (15), (498)
CARBON DISULPHIDE from natural gas (MJ/kg)										
D	–	–	0.29	1.41	–	2.10	14.69	–	18.49	(1), (106), (511)
CARBON ELECTRODES from raw materials in the ground (MJ/kg)										
H	*(114.00)*	*(36.00)*	–	–	–	–	–	*(50.40)*	*(200.40)*	(2), (38), (185)
CARBON ELECTRODES from raw materials at the plant (MJ/kg)										
A	*	*	*	*	*	*	*	*	*22.90*	(2), (61), (187)
A	*	*	*	*	*	*	*	*	*25.20*	(2), (49), (186)
J	*	*	*	*	*	*	*	*	*46.49*	(1), (57), (189)
A	75.37	23.80	–	–	–	0.20	3.95	–	103.32	(1), (31), (188)
A	78.81	24.88	–	–	–	3.79	26.54	–	134.02	(1), (704), (738)
CARBON ELECTRODES from petroleum coke at the plant (MJ/kg)										
A	*	*	*	*	*	*	*	*	*5.89*	(1), (36), (190)
A	–	–	–	–	–	0.36	2.55	–	2.91	(2), (15), (630)

System Type	Electricity: Fuel production energy	Electricity: Energy content of fuel	Oil Fuels: Fuel production energy	Oil Fuels: Energy content of fuel	Oil Fuels: Feedstock	Other Fuels: Fuel production energy	Other Fuels: Energy content of fuel	Other Fuels: Feedstock	Total System Energy Requirement	Notes (see page 376 *et seq.*)
CARBON DIOXIDE GAS from ammonium chloride (MJ/kg)										
D	0.74	0.23	3.58	17.48	–	–	–	–	22.03	(1), (19), (211)
CARBON DIOXIDE GAS from coke (MJ/kg)										
A	0.09	0.03	0.35	1.70	–	0.17	1.11	–	3.45	(1), (106), (505)
CARBON DIOXIDE LIQUID from by-product gases (MJ/kg)										
A	0.53	0.17	–	–	–	–	–	–	0.70	(2), (56), (191), (192)
A	2.25	0.71	–	–	–	–	–	–	2.96	(2), (15), (191)
CARDBOARD – see under BOARD										
CARDBOARD BOXES – see under BOXES										
CARTONS – see under BOXES										
CASTING – see under the relevant metal										
CAST IRON – see under IRON										
CELLULOSE FILM from raw materials (MJ/kg)										
H	*	*	*	*	*	*	*	*	*192.53*	(2), (48), (193)
CEMENT from raw materials (MJ/kg)										
J	*	*	*	*	*	*	*	*	*4.60*	(1), (40)
A	1.12	0.35	0.04	0.20	–	0.52	3.60	–	5.83	(6), (22), (199)
F	1.14	0.36	–	–	–	0.78	5.40	–	7.68	(1), (4), (54), (127), (164)
H	*	*	*	*	*	*	*	*	*7.79*	(2), (50), (195)

System Type	Electricity		Oil Fuels			Other Fuels			Total System Energy Requirement	Notes (see page 376 *et seq.*)
	Fuel production energy	Energy content of fuel	Fuel production energy	Energy content of fuel	Feedstock	Fuel production energy	Energy content of fuel	Feedstock		
J	*	*	*	*	*	*	*	*	*8.00*	(2), (52), (127)
A	1.12	0.35	0.04	0.18	–	0.85	5.93	–	8.47	(6), (22), (200)
D	1.33	0.42	0.20	0.98	–	0.58	5.62	–	9.13	(1), (704), (712)
F	1.14	0.36	–	–	–	0.99	6.84	–	9.33	(1), (4), (54), (127), (614)
H	*(1.14)*	*(0.36)*	*(0.04)*	*(0.20)*	–	*(0.99)*	*(6.84)*	–	*(9.57)*	(1), (55), (127), (196)
J	11.62	3.67	0.07	0.34	–	0.36	5.38	–	21.44	(2), (59), (194)
CEMENT (ASBESTOS HEAT RESISTANT) from raw materials (MJ/kg)										
J	*	*	*	*	*	*	*	*	*5.1*	(6), (22)
A	*	*	*	*	*	*	*	*	*7.7*	(2), (41)
CEMENT (H.C.) from raw materials										
A	0.77	0.24	0.04	0.21	–	0.20	1.41	–	2.87	(6), (22), (197)
A	0.77	0.24	0.04	0.21	–	0.33	2.32	–	3.91	(6), (22), (198)
CHLORINE from ammonium chloride (MJ/kg)										
D	0.74	0.23	3.58	17.48	–	–	–	–	22.03	(1), (19), (211)
CHLORINE from brine (MJ/kg)										
A	9.12	2.88	–	–	–	–	–	–	12.00	(4), (45), (215)
H	21.19	6.69	0.34	1.66	–	–	–	–	29.88	(2), (20), (202)
H	20.62	6.19	0.66	3.45	–	–	–	–	30.92	(2), (15)
F	*	*	*	*	*	*	*	*	*33.70*	(1), (57), (212)
CHLORINE from hydrogen chloride (MJ/kg)										
D	0.11	0.19	1.00	4.85	–	–	–	–	6.15	(1), (19), (205)
D	2.16	0.72	3.02	14.68	–	0.06	0.44	–	21.08	(1), (19), (209)
A	19.73	6.23	–	–	–	–	–	–	25.96	(1), (19), (201)

System Type	Electricity: Fuel production energy	Electricity: Energy content of fuel	Oil Fuels: Fuel production energy	Oil Fuels: Energy content of fuel	Oil Fuels: Feedstock	Other Fuels: Fuel production energy	Other Fuels: Energy content of fuel	Other Fuels: Feedstock	Total System Energy Requirement	Notes (see page 376 *et seq.*)
CHLORINE from sea water (MJ/kg)										
A	0.25	0.08	1.56	7.56	–	–	–	–	9.45	(1), (19), (210)
CHLORINE from solid sodium chloride (MJ/kg)										
D	9.50	3.00	0.39	1.86	–	–	–	–	14.75	(1), (19), (214)
D	8.48	2.68	2.19	10.64	–	–	–	–	23.99	(1), (19), (213)
CHROME ORE from ore in ground (MJ/kg)										
A	*	*	*	*	*	*	*	*	*0.33*	(1), (704)
CLAY from raw material in ground (MJ/kg)										
A	*	*	*	*	*	*	*	*	*0.07*	(1), (704), (710)
CLAY PIPES from raw materials in the ground (MJ/kg)										
H	0.55	0.17	0.43	2.12	–	–	–	–	3.27	(2), (15), (415)
CLOSURES (ALUMINIUM SCREW) from aluminium sheet (MJ/1000 closures)										
A	22.80	7.20	–	–	–	–	–	–	30.00	(4), (45)
CLOSURES (STEEL CROWN) from sheet tinplate (MJ/1000 closures)										
A	5.13	1.62	0.71	3.44	–	1.22	8.56	–	20.68	(4), (45), (631)
H	*	*	*	*	*	*	*	*	*25.00*	(1), (58)
A	12.86	4.03	1.61	8.25	–	0.06	0.15	–	26.96	(2), (15)
CLOSURES (STEEL CROWN) from raw materials in the ground (MJ/1000 closures)										
H	34.77	10.88	4.23	26.92	–	13.76	86.21	–	176.77	(2), (15), (632)
CLOSURES (STEEL CROWN) from raw materials in the ground (MJ/kg)										
A	7.85	2.48	0.64	3.12	–	3.79	52.72	0.58	71.18	(1), (26)

COAL (fuel content and production energy) see Chapter 6 and Table 6.13

System Type	Electricity: Fuel production energy	Electricity: Energy content of fuel	Oil Fuels: Fuel production energy	Oil Fuels: Energy content of fuel	Oil Fuels: Feedstock	Other Fuels: Fuel production energy	Other Fuels: Energy content of fuel	Other Fuels: Feedstock	Total System Energy Requirement	Notes (see page 376 *et seq.*)
COKE (fuel content and production energy) see Chapter 6 and Table 6.13										
COMPRESSED AIR from natural air (MJ/m³)										
A	1.27	0.40	–	–	–	–	–	–	1.67	(2), (62), (216)
CONCRETE from raw materials (MJ/kg)										
J	*	*	*	*	*	*	*	*	*0.75–1.00*	(2), (59), (634)
J	*	*	*	*	*	*	*	*	*1.00–1.30*	(2), (59), (635)
J	*	*	*	*	*	*	*	*	*1.77*	(2), (41), (127), (633)
CONCRETE (FABRICATED) from raw materials (MJ/kg)										
H	*	*	*	*	*	*	*	*	*2.72*	(2), (20), (636)
CONCRETE BLOCKS from raw materials (MJ/kg)										
J	*	*	*	*	*	*	*	*	*1.42*	(2), (41), (127), (633)
CONCRETE (LIGHTWEIGHT) (MJ/kg)										
J	*	*	*	*	*	*	*	*	*2.3*	(6), (22), (638)
J	*	*	*	*	*	*	*	*	*2.5–3.5*	(2), (59), (637)
CONCRETE (LIGHTWEIGHT REINFORCED) from raw materials (MJ/kg)										
J	*	*	*	*	*	*	*	*	*3.8*	(6), (22), (639)
CONCRETE PIPE from raw materials (MJ/kg)										
J	*	*	*	*	*	*	*	*	*1.42*	(2), (41), (633)
CONCRETE (READY MIXED) from raw materials (MJ/kg)										
B	0.01	neg	0.01	0.05	–	neg	neg	–	0.07	(6), (640), (641)
J	*	*	*	*	*	*	*	*	*0.80*	(6), (22), (638)

System Type	Electricity		Oil Fuels			Other Fuels			Total System Energy Requirement	Notes (see page 376 *et seq.*)
	Fuel production energy	Energy content of fuel	Fuel production energy	Energy content of fuel	Feedstock	Fuel production energy	Energy content of fuel	Feedstock		
CONCRETE (REINFORCED) from raw materials (MJ/kg)										
J	*	*	*	*	*	*	*	*	*2.5*	(6), (22), (638)
CONSTRUCTION OF ROAD VEHICLES see Chapter 9 (9.4.1)										
COPPER from scrap (MJ/kg)										
A	*	*	*	*	*	*	*	*	*15.24*	(1), (42), (219)
F	*	*	*	*	*	*	*	*	*20.53*	(1), (43), (555)
COPPER ANODE from blister copper (MJ/kg)										
A	–	–	–	–	–	0.03	0.21	–	0.24	(1), (675), (677)
A	–	–	0.22	1.04	–	–	–	–	1.62	(2), (38), (556)
A	–	–	–	–	–	0.52	3.62	–	4.14	(1), (31), (223)
COPPER ANODE from copper concentrate (MJ/kg)										
A	1.80	0.57	0.41	1.98	–	0.04	0.28	–	5.08	(5), (64), (613)
COPPER (BLISTER) from calcined concentrate (MJ/kg)										
A	19.38	6.12	–	–	–	–	–	–	25.50	(1), (63), (220)
A	1.59	0.50	–	–	–	3.28	22.99	–	28.36	(1), (63), (221)
COPPER (BLISTER) from green concentrate (MJ/kg)										
A	1.75	0.55	–	–	–	3.90	27.28	–	33.48	(1), (63), (222)
COPPER (BLISTER) from unspecified concentrate (MJ/kg)										
A	15.29	4.83	–	–	–	–	–	–	20.12	(1), (46), (244)
A	–	–	–	–	–	3.61	25.31	–	28.92	(1), (31), (232)

System Type	Electricity: Fuel production energy	Electricity: Energy content of fuel	Oil Fuels: Fuel production energy	Oil Fuels: Energy content of fuel	Oil Fuels: Feedstock	Other Fuels: Fuel production energy	Other Fuels: Energy content of fuel	Other Fuels: Feedstock	Total System Energy Requirement	Notes (see page 376 _et seq._)
COPPER (BLISTER) from matte (MJ/kg)										
_A_	0.76	0.24	–	–	–	–	–	–	1.00	(5), (64), (557)
COPPER CASTINGS from ore in the ground (MJ/kg)										
_H_	*	*	*	*	*	*	*	*	_52.51_	(2), (48), (553)
_F_	*	*	*	*	*	*	*	*	_129.16_	(1), (43), (554)
COPPER (ELECTROLYTIC) from anode copper (MJ/kg)										
_A_	2.24	0.71	–	–	–	–	–	–	2.95	(1), (31), (562)
_H_	(_3.17_)	(_1.00_)	–	–	–	(_0.01_)	(_0.06_)	–	(_4.24_)	(2), (38), (226)
_A_	3.42	1.08	–	–	–	–	–	–	4.50	(1), (675), (678)
_A_	5.60	1.77	0.01	0.04	–	–	–	–	7.42	(2), (65), (224)
COPPER (ELECTROLYTIC) from copper concentrate (MJ/kg)										
_A_	3.87	1.22	–	–	–	3.88	27.21	–	36.18	(2), (15), (650)
_F_	*	*	*	*	*	*	*	*	_46.49_	(1), (57), (249)
_A_	(_5.79_)	(_1.83_)	–	–	–	(_5.82_)	(_40.81_)	–	(_54.25_)	(1), (36), (217)
COPPER (ELECTROLYTIC) from impure scrap (MJ/kg)										
_A_	2.24	0.71	–	–	–	0.53	3.73	–	7.21	(1), (31), (227)
_D_	10.51	3.32	2.40	11.53	–	2.83	18.30	–	48.89	(2), (563), (564)
COPPER (ELECTROLYTIC) from ore (MJ/kg)										
_A_	11.21	3.54	2.63	12.57	–	–	–	–	29.95	(2), (66), (231)
_A_	14.63	4.62	1.38	6.97	–	3.99	27.93	–	59.52	(1), (46), (225)
_A_	56.05	17.70	–	–	–	–	–	–	73.75	(2), (66), (230)
_A_	40.81	13.52	–	–	–	3.88	27.21	–	85.42	(2), (15), (644)
_A_	57.44	18.14	–	–	–	4.84	33.91	–	114.33	(1), (31), (229)
_A_	*	*	*	*	*	*	*	*	_120.00_	(2), (52), (127)

System Type	Electricity: Fuel production energy	Electricity: Energy content of fuel	Oil Fuels: Fuel production energy	Oil Fuels: Energy content of fuel	Oil Fuels: Feedstock	Other Fuels: Fuel production energy	Other Fuels: Energy content of fuel	Other Fuels: Feedstock	Total System Energy Requirement	Notes (see page 376 *et.seq.*)
F	*	*	*	*	*	*	*	*	*124.79*	(1), (57), (248)
D	54.88	17.33	3.80	19.21	–	7.97	56.01	–	159.20	(1), (704), (715)
A	183.73	58.02	–	–	–	5.49	38.43	–	285.67	(1), (31), (228)
COPPER (FINISHED GENERAL PRODUCTS) from ore in the ground (MJ/kg)										
H	*	*	*	*	*	*	*	*	*45.13*	(2), (50), (127)
F	*	*	*	*	*	*	*	*	*107.06*	(1), (42), (218)
COPPER LIQUID from cathode copper (MJ/kg)										
A	0.07	0.02	–	–	–	0.19	1.36	–	1.64	(1), (57), (551)
A	–	–	–	–	–	0.24	1.67	–	1.91	(1), (675), (679)
A	2.50	0.79	–	–	–	–	–	–	3.29	(1), (57), (552)
A	0.05	0.02	–	–	–	0.44	3.08	–	3.59	(1), (57), (550)
A	3.42	1.08	–	–	–	–	–	–	4.50	(1), (67), (237)
A	–	–	0.94	4.47	–	–	–	–	5.41	(1), (67), (236)
COPPER LIQUID from pure copper scrap (MJ/kg)										
A	2.85	0.90	–	–	–	–	–	–	3.75	(1), (31), (238)
COPPER MATTE from copper concentrate (MJ/kg)										
A	*	*	*	*	*	*	*	*	*13.82*	(2), (38), (233)
H	3.15	0.99	–	–	–	*(2.18)*	*(15.25)*	–	*(21.57)*	(2), (38), (234), (642)
H	21.91	6.92	–	–	–	*(0.12)*	*(0.88)*	–	*(29.83)*	(2), (38), (235), (317)
COPPER (ORE CONCENTRATE) from run-of-mine ore (MJ/kg)										
D	*	*	*	*	*	*	*	*	*(0.24)*	(1), (57), (253)
H	0.17	0.05	*(0.01)*	*(0.07)*	–	–	–	–	*(0.30)*	(2), (38), (251)
F	*(0.26)*	*(0.08)*	*(0.02)*	*(0.11)*	–	–	–	–	*(0.47)*	(1), (36), (250)

System Type	Electricity: Fuel production energy	Electricity: Energy content of fuel	Oil Fuels: Fuel production energy	Oil Fuels: Energy content of fuel	Oil Fuels: Feedstock	Other Fuels: Fuel production energy	Other Fuels: Energy content of fuel	Other Fuels: Feedstock	Total System Energy Requirement	Notes (see page 376 *et.seq.*)
COPPER (ORE CONCENTRATE) from ore in the ground (MJ/kg)										
A	0.28	0.09	–	–	–	–	–	–	0.37	(1), (675), (680)
D	0.47	0.15	–	–	–	–	–	–	0.62	(1), (31), (252)
COPPER ROD from copper cathode (MJ/kg)										
A	1.35	0.42	0.02	0.08	–	0.34	2.35	–	4.56	(1), (565), (566)
COPPER ROD from copper wire bar (MJ/kg)										
A	0.80	0.25	0.24	1.15	–	–	–	–	2.44	(2), (67), (239)
COPPER ROD (PICKLED) from copper rod (oxidised) (MJ/kg)										
D	0.07	0.02	–	–	–	0.01	0.06	–	0.16	(2), (67), (240)
COPPER ROD (SHAVED) from copper rod (MJ/kg)										
D	0.63	0.20	–	–	–	–	–	–	0.83	(2), (67), (241)
COPPER SHEET from copper ingot (MJ/kg)										
F	*	*	*	*	*	*	*	*	*40.44*	(1), (36), (643)
COPPER SHEET from ore in the ground (MJ/kg)										
H	*	*	*	*	*	*	*	*	*83.26*	(1), (55), (127), (242)
H	*	*	*	*	*	*	*	*	*132.75*	(1), (43), (558)
F	*	*	*	*	*	*	*	*	*152.71*	(1), (36), (559)
COPPER STRIP from blister copper and scrap (MJ/kg)										
A	0.26	0.08	–	–	–	0.33	1.82	–	2.49	(2), (567), (568)
COPPER SPONGE from copper ore (MJ/kg)										
A	43.77	13.82	–	–	–	–	–	–	57.59	(1), (704), (714)

System Type	Electricity: Fuel production energy	Electricity: Energy content of fuel	Oil Fuels: Fuel production energy	Oil Fuels: Energy content of fuel	Oil Fuels: Feedstock	Other Fuels: Fuel production energy	Other Fuels: Energy content of fuel	Other Fuels: Feedstock	Total System Energy Require-ment	Notes (see page 376 *et.seq.*)
A	79.80	25.20	–	–	–	–	–	–	105.00	(1), (46), (243)
COPPER WIRE from copper ore (MJ/kg)										
H	*	*	*	*	*	*	*	*	*110.20*	(1), (43), (560)
COPPER WIRE (0.25mm) from copper wire (8mm) (MJ/kg)										
A	4.32	1.36	–	–	–	–	–	–	5.68	(2), (67), (246)
COPPER WIRE (0.25mm ANNEALED) from wire bar (MJ/kg)										
D	6.00	1.88	0.24	1.15	–	1.05	1.46	–	11.78	(2), (67), (561)
COPPER WIRE (ANNEALED) from wire as drawn (MJ/kg)										
A	0.64	0.20	–	–	–	–	–	–	0.84	(2), (67), (245)
CORRUGATED CARDBOARD – see under BOARD										
CRATES (POLYETHYLENE) from polymer resin (MJ/kg)										
A	6.05	1.91	–	–	–	–	–	–	7.96	(4), (45), (254)
CROWN CORKS – see under CLOSURES										
CRYOLITE from raw materials (MJ/kg)										
D	*	*	*	*	*	*	*	*	*0.25*	(2), (49), (645)
CYCLOHEXANE from crude oil (MJ/kg)										
H	–	–	4.49	21.94	52.25	–	–	–	78.68	(2), (15), (274)
1.2 – DICHLOROETHANE from acetylene and chlorine (MJ/kg)										
A	24.90	7.86	3.93	19.09	–	0.07	0.47	–	56.32	(1), (19), (160)

System Type	Electricity: Fuel production energy	Electricity: Energy content of fuel	Oil Fuels: Fuel production energy	Oil Fuels: Energy content of fuel	Oil Fuels: Feedstock	Other Fuels: Fuel production energy	Other Fuels: Energy content of fuel	Other Fuels: Feedstock	Total System Energy Requirement	Notes (see page 376 *et.seq.*)
1.2 – DICHLORETHANE from ethylene (MJ/kg)										
A	3.37	1.07	4.42	21.44	–	–	–	–	30.30	(1), (19), (255)
A	36.92	11.66	0.38	1.83	–	–	–	–	50.79	(1), (19), (259)
A	11.33	3.58	7.17	34.81	–	–	–	–	56.89	(1), (19), (258)
A	32.05	10.13	12.52	60.81	–	–	–	–	115.51	(1), (19), (257)
A	59.77	18.92	70.05	340.14	–	–	–	–	488.88	(1), (19), (256)
DIESEL OIL (fuel content and production energy) see Chapter 6 and Table 6.13										
DOLOMITE (DEAD BURNED) from DOLOMITE ROCK (MJ/kg)										
A	*	*	*	*	*	*	*	*	*9.23*	(1), (704)
DRILLING (α-BRASS) (MJ/kg metal removed)										
A	0.25	0.08	–	–	–	–	–	–	0.33	(2), (68)
DRILLING MILD STEEL (MJ/kg metal removed)										
A	2.79	0.88	–	–	–	–	–	–	3.67	(2), (68)
DRUMS (FIBRE) from raw materials (MJ/kg)										
J	*	*	*	*	*	*	*	*	*43.19*	(1), (44), (646)
DUCTILE IRON – see under IRON										
ELECTRICITY (fuel content and production energy) see Chapter 6 and Table 6.13										
ENAMEL from input chemicals (MJ/kg)										
A	*	*	*	*	*	*	*	*	*65.63*	(4), (45), (647)

System Type	Electricity: Fuel production energy	Electricity: Energy content of fuel	Oil Fuels: Fuel production energy	Oil Fuels: Energy content of fuel	Oil Fuels: Feedstock	Other Fuels: Fuel production energy	Other Fuels: Energy content of fuel	Other Fuels: Feedstock	Total System Energy Requirement	Notes (see page 376 *et.seq.*)
ETHANE from crude oil (MJ/kg)										
A	0.56	0.18	2.79	13.55	44.50	–	–	–	61.58	(1), (19), (265)
ETHANOL from ethylene by hydrolysis (MJ/kg)										
D	2.97	0.93	2.51	12.21	–	–	–	–	18.62	(1), (106), (507)
ETHANOL from crude oil (MJ/kg)										
D	3.81	1.19	4.32	21.03	30.50	–	–	–	60.85	(1), (106), (508)
ETHANOLAMINE (MJ/kg)										
J	*	*	*	*	*	*	*	*	*46.49*	(1), (704), (726)
ETHYLENE from crude oil (MJ/kg)										
A	0.56	0.18	2.79	13.55	44.50	–	–	–	61.58	(1), (19), (265)
H	0.41	0.13	2.96	14.46	50.16	–	–	–	68.12	(2), (15), (20), (261)
A	0.67	0.21	3.65	17.80	50.00	–	–	–	72.33	(2), (15)
D	9.77	3.26	134.37	654.27	11.61	–	–	–	813.28	(1), (19), (122)
ETHYLENE from ethane (MJ/kg)										
A	0.78	0.24	5.42	26.39	–	–	–	–	32.83	(1), (19), (264)
ETHYLENE from L.P.G. (MJ/kg)										
A	1.50	0.48	–	–	–	2.14	14.96	–	19.08	(1), (16), (262)
ETHYLENE from naphtha (MJ/kg)										
A	9.02	2.85	6.41	31.30	–	–	–	–	49.58	(4), (45), (266)
D	8.68	2.89	7.74	37.73	–	–	–	–	57.04	(1), (19), (119)
H	*	*	*	*	*50.16*	*	*	*	*80.67*	(2), (15), (263)

System Type	Electricity: Fuel production energy	Electricity: Energy content of fuel	Oil Fuels: Fuel production energy	Oil Fuels: Energy content of fuel	Oil Fuels: Feedstock	Other Fuels: Fuel production energy	Other Fuels: Energy content of fuel	Other Fuels: Feedstock	Total System Energy Requirement	Notes (see page 376 *et.seq.*)
D	8.68	2.89	8.58	41.81	24.03	–	–	–	85.99	(1), (19), (120)
D	9.77	3.26	133.97	652.30	–	–	–	–	799.30	(1), (19), (121)
ETHYLENE DICHLORIDE – see under DICHLOROETHANE										
ETHYLENE GLYCOL from crude oil (MJ/kg)										
A	2.78	0.87	2.38	11.62	40.31	–	–	–	57.96	(1), (106), (515)
A	9.14	2.89	6.92	33.72	41.38	–	–	–	94.05	(2), (15)
ETHYLENE OXIDE from crude oil (MJ/kg)										
D	21.16	6.68	2.98	14.53	49.26	–	–	–	94.61	(1), (106), (510)
D	43.05	13.60	10.90	53.05	35.81	0.57	6.37	–	163.35	(1), (106), (509)
EXPLOSIVES from raw materials (MJ/kg)										
H	*	*	*	*	*	*	*	*	*9.87*	(2), (38), (267)
F	*	*	*	*	*	*	*	*	*23.23*	(2), (56), (268)
F	*	*	*	*	*	*	*	*	*34.86*	(1), (57)
J	*	*	*	*	*	*	*	*	*69.73*	(1), (704), (708)
FELDSPAR from raw materials in the ground (MJ/kg)										
F	*	*	*	*	*	*	*	*	*0.94*	(1), (58), (269)
A	*	*	*	*	*	*	*	*	*0.95*	(1), (44), (648)
F	0.63	0.20	0.08	0.40	–	0.01	0.04	–	1.36	(1), (26), (270)
A	0.70	0.22	0.14	0.69	–	0.01	0.07	–	1.83	(1), (16), (271)
F	1.23	0.39	0.16	0.78	–	–	–	–	2.56	(5), (15), (272)

FIBRES – see under POLYESTER FIBRE
NYLON 66 FIBRE
ACRYLIC FIBRE

System Type	Electricity: Fuel production energy	Electricity: Energy content of fuel	Oil Fuels: Fuel production energy	Oil Fuels: Energy content of fuel	Oil Fuels: Feedstock	Other Fuels: Fuel production energy	Other Fuels: Energy content of fuel	Other Fuels: Feedstock	Total System Energy Requirement	Notes (see page 376 *et.seq.*)
FILLING CONTAINERS – see under BOTTLES and METAL CANS										
FILMS (PLASTIC) from polymer resin (MJ/kg)										
A	30.94	9.77	–	–	–	–	–	–	40.71	(1), (44), (532)
FLOTATION REAGENTS from raw materials (MJ/kg)										
F	*	*	*	*	*	*	*	*	*46.48*	(1), (57), (173)

FROTH FLOTATION – see under the relevant metal

FUEL OIL – see Chapter 6 and Table 6.13

FUEL requirement of Rail Transport – see Chapter 9 (9.6.1)

FUEL requirement of Road Vehicles – see Chapter 9 (9.3)

FUEL requirements of Sea Transport – see Chapter 9 (9.7.1)

GALVANISED STEEL – see under STEEL

GARAGING energy of Road Vehicles – see Chapter 9 (9.4.5)

GAS – MANUFACTURED (fuel content and production energy) see Chapter 6 and Table 6.13

GAS – NATURAL (fuel content and production energy) see Chapter 6 and Table 6.13

GAS OIL (fuel content and production energy) see Chapter 6 and Table 6.13

GASOLINE (fuel content and production energy) see under PETROL in Chapter 6 and Table 6.13

System Type	Electricity		Oil Fuels		Other Fuels				Total System Energy Require-ment	Notes (see page 376 *et.seq.*)
	Fuel production energy	Energy content of fuel	Fuel production energy	Energy content of fuel	Feedstock	Fuel production energy	Energy content of fuel	Feedstock		
GLASS from raw materials in the ground (MJ/kg)										
J	*	*	*	*	*	*	*	*	*12.39*	(1), (40)
J	*	*	*	*	*	*	*	*	*18.24*	(1), (42)
F	*	*	*	*	*	*	*	*	*21.00*	(6), (22), (651)
H	*(2.82)*	*(0.89)*	*(1.42)*	*(7.27)*	–	*(2.84)*	*(7.29)*	–	*(22.53)*	(2), (50), (127), (275)
J	*	*	*	*	*	*	*	*	*25.00*	(2), (52), (127)
F	3.16	0.99	1.58	8.09	–	3.16	8.11	–	25.09	(2), (15), (649)
GLASS (FOR AUTOMOBILE WINDOWS) from raw materials in the ground (MJ/kg)										
F	*	*	*	*	*	*	*	*	*30.22*	(1), (36), (653)
GLASS BOTTLES – see under BOTTLES										
GLASS CONTAINERS – see under BOTTLES										
GLASS FORMING (MJ/kg)										
A	3.64	1.15	0.01	0.04	–	–	–	–	4.84	(4), (45)
GLASS MELTING (MJ/kg)										
A	0.44	0.14	0.19	0.92	–	–	–	–	1.69	(4), (45)
GLASS (PLATE) from raw materials in the ground (MJ/kg)										
D	*	*	*	*	*	*	*	*	*21.77*	(1), (704), (716)
H	*(4.59)*	*(1.45)*	*(1.91)*	*(9.80)*	–	*(3.83)*	*(9.84)*	–	*(31.42)*	(1), (55), (652)
GLASS (CULLET) from refuse (MJ/kg)										
A	5.02	1.58	neg	0.01	–	neg	0.01	–	6.62	(1), (16), (276)
F	*	*	*	*	*	*	*	*	*2.56*	(1), (36), (277)

System Type	Electricity: Fuel production energy	Electricity: Energy content of fuel	Oil Fuels: Fuel production energy	Oil Fuels: Energy content of fuel	Oil Fuels: Feedstock	Other Fuels: Fuel production energy	Other Fuels: Energy content of fuel	Other Fuels: Feedstock	Total System Energy Requirement	Notes (see page 376 *et.seq.*)
GLASS (WINDOW) from raw materials in the ground (MJ/kg)										
F	*	*	*	*	*	*	*	*	*17.43*	(1), (36), (654)
H	*(2.76)*	*(0.87)*	*(1.39)*	*(7.11)*	–	*(2.77)*	*(7.13)*	–	*(22.03)*	(2), (50), (127), (275)
F	*	*	*	*	*	*	*	*	*28.59*	(1), (36), (655)
GLUE from raw materials (MJ/kg)										
	27.91	8.75	6.68	35.00	–	–	–	–	78.34	(2), (15)
GOLD from copper anode slime (MJ/kg)										
A	750.21	231.91	796.73	3817.59	–	–	–	–	5601.44	(1), (31), (278)
GRAVEL from raw materials in the ground (MJ/kg)										
H	*	*	*	*	*	*	*	*	*0.08*	(1), (55), (127), (279)
D	*	*	*	*	*	*	*	*	*0.10*	(6), (22), (656)
GREY IRON – see under IRON										
GRINDING WASTE PLASTIC (MJ/kg plastic output)										
A	1.44	0.36	–	–	–	–	–	–	1.50	(2), (167)
GYPSUM – see CALCIUM SULPHATE										
GYPSUM (PLASTER) from raw materials in the ground (MJ/kg)										
H	*	*	*	*	*	*	*	*	*3.24*	(2), (50), (127), (275)
J	*	*	*	*	*	*	*	*	*3.60*	(2), (52), (127)
GYPSUM PLASTERBOARD from raw materials (MJ/sq.yd)										
H	*	*	*	*	*	*	*	*	*54.00*	(2), (50), (127), (275)

System Type	Electricity: Fuel production energy	Electricity: Energy content of fuel	Oil Fuels: Fuel production energy	Oil Fuels: Energy content of fuel	Oil Fuels: Feedstock	Other Fuels: Fuel production energy	Other Fuels: Energy content of fuel	Other Fuels: Feedstock	Total System Energy Requirement	Notes (see page 376 *et seq.*)
HEAVY FUEL OIL (fuel content and production energy) see Chapter 6 and Table 6.13										
HYDROCHLORIC ACID (MJ/kg)										
J	*	*	*	*	*	*	*	*	*9.59*	(1), (704)
HYDROGEN as a by-product from various reactions (MJ/kg)										
D	0.25	0.08	1.56	7.56	–	–	–	–	9.45	(1), (19), (210)
A	9.12	2.88	–	–	–	–	–	–	12.00	(4), (45), (215)
D	9.50	3.00	0.39	1.86	–	–	–	–	14.75	(1), (19), (214)
D	8.48	2.68	2.19	10.64	–	–	–	–	23.99	(1), (19), (213)
A	19.73	6.23	–	–	–	–	–	–	25.96	(1), (19), (201)
HYDROGEN from coke (MJ/kg)										
A	0.09	0.03	0.35	1.70	–	0.17	1.11	–	3.45	(1), (106), (505)
A	2.12	0.68	8.06	39.15	–	3.93	25.45	–	79.39	(1), (106), (504)
HYDROGEN from natural gas (MJ/kg)										
A	10.93	3.47	6.58	31.99	–	2.91	20.34	13.83	90.05	(1), (106), (506)
INCINERATION – see under WASTE DISPOSAL										
IRON (CAST) from ore in the ground (MJ/kg)										
J	*	*	*	*	*	*	*	*	*16.39*	(1), (42)
IRON (CAST, GREY) from pig iron and scrap (MJ/kg)										
A	–	–	–	–	–	(2.56)	(17.94)	–	(20.50)	(1), (36), (280)
F	*	*	*	*	*	*	*	*	*29.06*	(1), (36), (281)

System Type	Electricity: Fuel production energy	Electricity: Energy content of fuel	Oil Fuels: Fuel production energy	Oil Fuels: Energy content of fuel	Oil Fuels: Feedstock	Other Fuels: Fuel production energy	Other Fuels: Energy content of fuel	Other Fuels: Feedstock	Total System Energy Requirement	Notes (see page 376 *et seq.*)
IRON (CAST, GREY, DUCTILE) from scrap and ferro-alloy (MJ/kg)										
F	*	*	*	*	*	*	*	*	*23.94*	(1), (36), (283)
F	–	–	–	–	–	(3.07)	(21.15)	–	(24.22)	(1), (36), (282)
IRON (CAST, MALLEABLE) from pig iron, scrap and ferro-alloy (MJ/kg)										
F	*	*	*	*	*	*	*	*	*36.03*	(1), (36), (284)
IRON (CAST, NODULAR) from pig iron, scrap and ferro-alloy (MJ/kg)										
F	*	*	*	*	*	*	*	*	*32.66*	(1), (36), (285)
IRON (PIG) from ore concentrate (MJ/kg)										
A	–	–	–	–	–	2.16	13.98	–	16.14	(1), (72), (574)
A	1.04	0.33	–	–	–	2.07	13.38	–	16.82	(2), (74), (297)
A	1.12	0.35	–	–	–	2.16	13.98	–	17.61	(1), (69), (571)
A	–	–	0.57	2.74	–	1.93	12.48	–	17.72	(1), (73), (572)
A	–	–	0.17	0.80	–	2.41	15.63	–	18.04	(1), (72), (573)
A	–	–	–	–	–	2.58	16.70	–	19.28	(1), (31), (295)
A	–	–	0.14	0.70	–	2.95	19.06	–	22.85	(1), (71), (287)
F	*	*	*	*	*	*	*	*	*24.28*	(1), (43)
F	–	–	–	–	–	(3.31)	(21.45)	–	(24.76)	(1), (57), (569)
A	–	–	–	–	–	1.19	24.00	–	25.19	(1), (60), (596)
F	(0.95)	(0.30)	–	–	–	(3.31)	(23.17)	–	(27.73)	(1), (36), (288)
A	–	–	0.70	3.40	–	2.19	23.35	–	29.64	(2), (60), (286)
A	22.42	7.08	–	–	–	1.23	7.94	–	38.67	(1), (73), (570)
A	24.66	7.79	–	–	–	1.19	7.72	–	41.36	(1), (31), (289)
IRON (PIG) from ore in the ground (MJ/kg)										
A	0.25	0.08	neg	0.02	–	2.30	14.89	–	17.54	(1), (31), (290)
A	0.95	0.30	0.13	0.85	–	2.30	14.89	–	19.42	(1), (31), (291)

System Type	Electricity: Fuel production energy	Electricity: Energy content of fuel	Oil Fuels: Fuel production energy	Oil Fuels: Energy content of fuel	Oil Fuels: Feedstock	Other Fuels: Fuel production energy	Other Fuels: Energy content of fuel	Other Fuels: Feedstock	Total System Energy Requirement	Notes (see page 376 *et seq.*)
A	0.72	0.23	0.42	2.71	–	2.30	14.89	–	21.27	(1), (31), (292)
A	1.43	0.45	0.38	2.65	–	2.30	14.89	–	22.10	(1), (31), (293)
A	0.77	0.24	0.95	6.67	–	2.30	14.89	–	25.82	(1), (31), (294)
F	*	*	*	*	*	*	*	*	*27.89*	(1), (36)
IRON (SPONGE) from ore (MJ/kg)										
A	–	–	–	–	–	neg	13.47	–	13.47	(2), (75), (579)
A	–	–	–	–	–	1.78	12.44	–	14.22	(1), (70), (584)
A	–	–	–	–	–	0.71	14.40	–	15.11	(3), (79), (580)
A	–	–	–	–	–	1.98	13.85	–	15.83	(2), (60), (301)
A	–	–	–	–	–	1.98	13.89	–	15.87	(2), (75), (577)
A	1.68	0.53	–	–	–	1.88	13.18	–	17.27	(2), (74), (298)
A	–	–	–	–	–	2.17	15.22	–	17.39	(8), (77), (578)
A	1.37	0.43	–	–	–	1.97	13.82	–	17.59	(2), (76), (300)
A	–	–	–	–	–	2.22	15.55	–	17.77	(1), (69), (585)
A	–	–	–	–	–	2.32	16.27	–	18.59	(1), (78), (576)
A	0.77	0.24	–	–	–	0.89	17.92	–	19.82	(2), (76), (299)
A	–	–	–	–	–	2.52	17.62	–	20.14	(10), (81), (582)
A	0.78	0.24	–	–	–	1.57	17.64	–	20.23	(9), (80), (581)
A	–	–	–	–	–	2.59	18.10	–	20.69	(2), (77), (575)
A	–	–	–	–	–	3.01	21.05	–	24.06	(10), (82), (583)
IRON ORE from ore in the ground (MJ/kg)										
A	0.28	0.09	0.01	0.03	–	neg	neg	–	0.41	(1), (704), (735)
F	*	*	*	*	*	*	*	*	*0.91*	(1), (36)
F	0.35	0.11	0.02	0.11	–	0.06	0.44	–	1.09	(1), (26)
A	0.35	0.11	0.03	0.14	–	0.06	0.42	–	1.11	(1), (16), (302)
A	3.42	1.08	0.22	1.08	–	–	–	–	5.80	(4), (45)

System Type	Electricity: Fuel production energy	Electricity: Energy content of fuel	Oil Fuels: Fuel production energy	Oil Fuels: Energy content of fuel	Oil Fuels: Feedstock	Other Fuels: Fuel production energy	Other Fuels: Energy content of fuel	Other Fuels: Feedstock	Total System Energy Requirement	Notes (see page 376 *et seq.*)
IRON ORE BLENDING (MJ/kg)										
A	0.09	0.03	–	–	–	–	–	–	0.12	(2), (74), (586)
IRON ORE PELLETS from ore and flux (MJ/kg)										
A	–	–	–	–	–	0.08	0.58	–	0.66	(1), (31), (587)
A	*	*	*	*	*	*	*	*	*1.01*	(1), (69), (304)
A	–	–	–	–	–	0.25	1.74	–	1.99	(1), (31), (588)
A	0.51	0.16	–	–	–	0.21	1.47	–	2.35	(2), (74), (303)
A	1.34	0.42	0.07	0.32	–	0.04	0.29	–	2.48	(2), (704), (736)
IRON ORE SINTER from ore (MJ/kg)										
A	0.34	0.11	–	–	–	0.02	1.77	–	2.24	(2), (704), (737)
KEROSINE (fuel content and production energy) see Chapter 6 and Table 6.13										
LABELS (BOTTLE) from paper (MJ/1000 labels)										
A	2.42	0.76	0.28	1.47	–	–	–	–	4.93	(2), (15)
A	*	*	*	*	*	*	*	*	*16.20*	(4), (45), (657)
LABELS (BOTTLE) from raw materials in the ground (MJ/1000 labels)										
F	15.50	4.86	2.30	12.00	–	0.23	7.58	8.55	51.02	(2), (15)
LEAD from ore in the ground (MJ/kg)										
A	9.89	3.12	0.18	0.89	–	1.81	12.76	–	28.65	(1), (704), (720)
J	*	*	*	*	*	*	*	–	*29.00*	(2), (52), (127)
J	*	*	*	*	*	*	*	–	*31.12*	(1), (42)
J	*	*	*	*	*	*	*	–	*31.09*	(1), (46), (309)
F	*	*	*	*	*	*	*	–	*51.14*	(1), (36)
H	*	*	*	*	*	*	*	–	*51.14*	(1), (55), (127), (307)

System Type	Electricity: Fuel production energy	Electricity: Energy content of fuel	Oil Fuels: Fuel production energy	Oil Fuels: Energy content of fuel	Oil Fuels: Feedstock	Other Fuels: Fuel production energy	Other Fuels: Energy content of fuel	Other Fuels: Feedstock	Total System Energy Requirement	Notes (see page 376 *et seq.*)
H	*	*	*	*	*	*	*	–	*54.00*	(2), (50), (127), (275)
A	6.40	2.00	1.37	3.53	–	4.80	35.90	–	54.00	(2), (15)
LEAD from ore concentrate (MJ/kg)										
A	0.60	0.19	–	–	–	2.15	14.37	–	17.31	(1), (63), (308)
LEAD from scrap (MJ/kg)										
H	*	*	*	*	*	*	*	*	*7.20*	(2), (50), (127), (275)
J	–	–	–	–	–	(1.18)	(8.30)	–	(9.48)	(1), (42), (658)

LIGHT FUEL OIL (fuel content and production energy) see Chapter 6 and Table 6.13

LIME – see under CALCIUM OXIDE

System Type	Electricity: Fuel production energy	Electricity: Energy content of fuel	Oil Fuels: Fuel production energy	Oil Fuels: Energy content of fuel	Oil Fuels: Feedstock	Other Fuels: Fuel production energy	Other Fuels: Energy content of fuel	Other Fuels: Feedstock	Total System Energy Requirement	Notes (see page 376 *et seq.*)
LIMESTONE from rock (MJ/kg)										
A	neg	neg	neg	0.02	–	neg	0.01	–	0.03	(1), (704), (707)
F	*	*	*	*	*	*	*	*	*0.07*	(1), (58), (659)
A	*	*	*	*	*	*	*	*	*0.07*	(4), (45), (311)
F	*	*	*	*	*	*	*	*	*0.08*	(1), (36)
A	*	*	*	*	*	*	*	*	*0.08*	(1), (44)
A	0.02	0.01	0.01	0.04	–	neg	0.01	–	0.09	(1), (16), (310)
F	0.03	0.01	0.01	0.03	–	neg	0.02	–	0.10	(1), (26), (311)
A	*	*	*	*	*	*	*	*	*0.12*	(1), (57)
A	0.13	0.04	0.04	0.20	–	–	–	–	0.41	(2), (15), (312)
A	0.16	0.05	0.07	0.35	–	–	–	–	0.63	(2), (15), (313)
A	0.49	0.16	–	–	–	–	–	–	0.65	(1), (31)

LPG–BUTANE (fuel content and production energy) see Chapter 6 and Table 6.13

System Type	Electricity: Fuel production energy	Electricity: Energy content of fuel	Oil Fuels: Fuel production energy	Oil Fuels: Energy content of fuel	Oil Fuels: Feedstock	Other Fuels: Fuel production energy	Other Fuels: Energy content of fuel	Other Fuels: Feedstock	Total System Energy Requirement	Notes (see page 376 *et seq.*)
LPG–PROPANE (fuel content and production energy) see Chapter 6 and Table 6.13										
LORRY TRANSPORT – see ROAD TRANSPORT and Chapter 9										
LUBRICATION ENERGY OF ROAD VEHICLES – see Chapter 9 (9.4.3)										
LUMBER – see under WOOD										
MACHINING α-BRASS (MJ/kg metal removed)										
A	0.57	0.18	–	–	–	–	–	–	0.75	(2), (68)
MACHINING MILD STEEL (MJ/kg metal removed)										
A	0.86	0.27	–	–	–	–	–	–	1.13	(2), (68)
MACHINING NICKEL CHROME STEEL (MJ/kg metal removed)										
A	1.52	0.48	–	–	–	–	–	–	2.00	(2), (68)
MAGNESIA (DEAD BURNED) from MAGNESITE ROCK (MJ/kg)										
A	*	*	*	*	*	*	*	*	*14.72*	(1), (704)
MAGNESIA (SYNTHETIC) from raw materials (MJ/kg)										
A	*	*	*	*	*	*	*	*	*30.48*	(1), (704)
MAGNESIUM from ore (MJ/kg)										
A	250.12	79.30	–	–	–	–	–	–	329.42	(1), (84), (315)
A	190.35	60.11	–	–	–	29.52	206.91	–	486.89	(1), (31), (314)
MAINTENANCE ENERGY OF RAIL TRANSPORT – see Chapter 9 (9.6.2)										

System Type	Electricity: Fuel production energy	Electricity: Energy content of fuel	Oil Fuels: Fuel production energy	Oil Fuels: Energy content of fuel	Oil Fuels: Feedstock	Other Fuels: Fuel production energy	Other Fuels: Energy content of fuel	Other Fuels: Feedstock	Total System Energy Requirement	Notes (see page 376 *et seq.*)
MAINTENANCE ENERGY OF ROAD VEHICLES – see Chapter 9 (9.4.2)										
MAINTENANCE ENERGY OF SEA TRANSPORT – see Chapter 9 (9.7.2)										
MANGANESE from leach liquor (MJ/kg)										
A	112.99	35.68	–	–	–	–	–	–	148.67	(1), (84), (316)
MANGANESE DIOXIDE (MJ/kg)										
J	*	*	*	*	*	*	*	*	*11.62*	(1), (704)
MANUFACTURED GAS (fuel content and production energy) see Chapter 6 and Table 6.13										
MALLEABLE IRON – see under IRON										
MEDIUM FUEL OIL (fuel content and production energy) see Chapter 6 and Table 6.13										
METAL CAN (ALUMINIUM) from raw materials (MJ/can)										
F	*	*	*	*	*	*	*	*	*6.88*	(1), (44), (318)
METAL CAN (ALUMINIUM) from raw materials (MJ/kg)										
H	*	*	*	*	*	*	*	*	*328.28*	(1), (55), (127)
METAL CAN (ALUMINIUM) from aluminium sheet (MJ/kg)										
F	7.41	2.34	–	–	–	1.15	8.07	–	18.97	(1), (26)
A	23.60	7.45	–	–	–	3.58	25.05	–	59.68	(1), (16), (319)
METAL CAN FILLING (MJ/can)										
A	0.02	0.01	–	–	–	neg	0.02	–	0.05	(1), (16), (321)
A	*	*	*	*	*	*	*	*	0.08	(4), (45)

System Type	Electricity: Fuel production energy	Electricity: Energy content of fuel	Oil Fuels: Fuel production energy	Oil Fuels: Energy content of fuel	Oil Fuels: Feedstock	Other Fuels: Fuel production energy	Other Fuels: Energy content of fuel	Other Fuels: Feedstock	Total System Energy Requirement	Notes (see page 376 *et seq.*)
A	0.02	0.01	neg	0.02	–	0.01	0.06	–	0.09	(1), (16), (322)
A	0.21	0.07	0.09	0.47	–	–	–	–	0.84	(2), (15), (660)
METAL CAN (TINPLATE) from raw materials in the ground (MJ/can)										
F	*	*	*	*	*	*	*	*	*4.21*	(1), (44)
F	1.55	0.48	0.09	0.46	0.02	0.27	1.65	neg	4.52	(2), (15), (660)
METAL CAN (TINPLATE) from raw materials in the ground (MJ/kg)										
H	*	*	*	*	*	*	*	*	*79.29*	(1), (55), (127)
F	32.27	10.12	1.87	9.55	0.52	5.62	34.45	0.04	94.44	(2), (15), (661)
METAL CAN (TINPLATE) from sheet tinplate (MJ/can)										
A	0.09	0.03	0.01	0.05	–	0.03	0.10	–	0.31	(2), (15), (660)
A	*	*	*	*	*	*	*	*	*0.32*	(1), (44)
METAL CAN (TINPLATE) from sheet tinplate (MJ/kg)										
A	2.23	0.70	0.42	2.01	–	–	–	–	5.36	(4), (45), (662)
A	3.42	1.08	–	–	–	0.36	2.55	–	7.41	(1), (16), (320)
A	3.45	1.09	–	–	–	0.38	2.64	–	7.56	(1), (26)
METHANOL from crude oil (MJ/kg)										
H	0.66	0.22	2.40	11.70	28.01	*	*	*	42.99	(2), (15), (324)
H	4.50	1.42	3.62	17.68	28.00	–	–	–	55.22	(2), (20), (323)
MILLING ALUMINIUM (MJ/kg metal removed)										
A	1.96	0.62	–	–	–	–	–	–	2.58	(2), (68)
MILLING MILD STEEL (MJ/kg of metal removed)										
A	2.57	0.81	–	–	–	–	–	–	3.38	(2), (68)

System Type	Electricity: Fuel production energy	Electricity: Energy content of fuel	Oil Fuels: Fuel production energy	Oil Fuels: Energy content of fuel	Oil Fuels: Feedstock	Other Fuels: Fuel production energy	Other Fuels: Energy content of fuel	Other Fuels: Feedstock	Total System Energy Requirement	Notes (see page 376 *et seq.*)
MINING – DRILLING AND BLASTING(MJ/ton of rock removed)										
A	6.11	1.93	–	–	–	–	–	–	8.04	(2), (38), (664)
MINING – GENERAL (MJ/m³)										
	11.40	3.60	–	–	–	–	–	–	15.00	(2), (56), (663)
MINING – ORE CRUSHING (MJ/kg ore)										
A	0.04	0.01	–	–	–	–	–	–	0.05	(1), (31), (326)
MINING – OPEN PIT (MJ/kg rock mined)										
H	*	*	*	*	*	*	*	*	*0.04*	(1), (57), (327)
MINING – ORE GRINDING (MJ/kg rock milled)										
A	0.21	0.07	–	–	–	–	–	–	0.28	(1), (31), (328)
MINING – SHOVEL LOADING (MJ/kg rock loaded)										
H	*	*	*	*	*	*	*	*	*0.04*	(2), (38), (325)
NAPHTHA from crude oil (MJ/kg)										
H	–	–	0.93	4.54	44.50	–	–	–	49.97	(2), (20), (329)
A	–	–	1.03	5.04	44.50	–	–	–	50.57	(4), (45), (330)
A	0.56	0.18	2.79	13.55	44.50	–	–	–	61.58	(1), (19), (265)
NATURAL GAS (fuel content and production energy) see Chapter 6 and Table 6.13										
NICKEL from ore concentrate (MJ/kg)										
J	*	*	*	*	*	*	*	*	*232.23*	(1), (46), (333)
A	176.36	56.64	–	–	–	–	–	–	236.00	(3), (66), (331)
J	*	*	*	*	*	*	*	*	*696.71*	(1), (46), (332)

System Type	Electricity: Fuel production energy	Electricity: Energy content of fuel	Oil Fuels: Fuel production energy	Oil Fuels: Energy content of fuel	Oil Fuels: Feedstock	Other Fuels: Fuel production energy	Other Fuels: Energy content of fuel	Other Fuels: Feedstock	Total System Energy Requirement	Notes (see page 376 *et seq.*)
NITRIC ACID from ammonia (MJ/kg)										
D	0.10	0.03	–	–	–	0.23	1.63	–	1.99	(1), (31), (517)
H	2.47	0.78	–	–	–	–	–	–	3.25	(2), (20), (334)
NITRIC ACID from crude oil (MJ/kg)										
H	2.47	0.78	0.40	1.93	2.80	–	–	–	8.38	(2), (20), (503)
A	2.12	0.66	1.65	7.98	6.89	–	–	–	19.30	(2), (56), (516)
NITROGEN (GASEOUS) from air (MJ/kg)										
A	1.27	0.40	–	–	–	–	–	–	1.67	(1), (704), (723)
NITROGEN (LIQUID) from air (MJ/kg)										
A	8.69	2.74	–	–	–	–	–	–	11.43	(1), (704), (724)
NODULAR IRON – see under IRON										
NYLON 66 from crude oil (MJ/kg)										
H	43.44	14.48	22.63	110.50	62.28	–	–	–	253.33	(2), (15), (305)
NYLON 66 fibre from nylon resin (MJ/kg)										
A	23.85	7.95	–	–	–	–	–	–	31.80	(2), (15), (306)
ORE CONCENTRATION – see under the relevant metal										
ORES – see under the relevant metal										
OXYGEN (LIQUID) from air (MJ/kg)										
A	*	*	*	*	*	*	*	*	*3.72*	(1), (57), (206)
A	2.98	0.94	–	–	–	–	–	–	3.92	(1), (85), (206)

System Type	Electricity		Oil Fuels			Other Fuels			Total System Energy Requirement	Notes (see page 376 *et seq.*)
	Fuel production energy	Energy content of fuel	Fuel production energy	Energy content of fuel	Feedstock	Fuel production energy	Energy content of fuel	Feedstock		
A	4.71	1.49	–	–	–	–	–	–	6.20	(1), (704), (719)
A	5.38	1.70	–	–	–	–	–	–	7.08	(2), (88), (206)
A	5.73	1.81	–	–	–	–	–	–	7.54	(1), (31), (206)
A	6.43	2.03	–	–	–	–	–	–	8.46	(2), (74), (206)
A	5.20	1.64	0.05	0.22	–	0.25	1.77	–	9.13	(1), (16), (206), (208)
F	5.35	1.69	0.04	0.19	–	0.26	1.80	–	9.33	(1), (26), (206), (207)
A	10.45	3.30	–	–	–	–	–	–	13.75	(2), (61), (206)
A	10.45	3.30	–	–	–	–	–	–	13.75	(2), (86), (206)
A	10.86	3.43	–	–	–	–	–	–	14.29	(2), (60), (206)
A	11.40	3.60	–	–	–	–	–	–	15.00	(2), (87), (206)
A	12.54	3.96	–	–	–	–	–	–	16.50	(1), (63), (206)
A	13.33	4.21	–	–	–	–	–	–	17.54	(2), (87), (206)
PAPER from wood pulp (MJ/kg)										
A	3.38	1.07	0.38	1.85	–	–	–	–	6.68	(2), (159), (521), (526)
PAPER (AVERAGE VALUES) from standing timber (MJ/kg)										
A	4.78	1.51	1.67	8.16	–	–	–	–	16.12	(4), (45), (341)
A	*	*	*	*	*	*	*	*	*17.63*	(1), (90), (159)
J	*	*	*	*	*	*	*	*	*17.70*	(1), (40), (159)
F	*	*	*	*	*	*	*	*	*25.37*	(1),(55),(127),(159),(336)
J	*	*	*	*	*	*	*	*	*25.57*	(1), (36), (159)
H	*	*	*	*	*	*	*	*	*26.55*	(1),(50),(127),(159),(275)
A	4.85	1.53	–	–	–	1.15	21.83	–	29.36	(5), (524), (528)
J	*	*	*	*	*	*	*	*	*30.00*	(2), (52), (127), (159)
J	*	*	*	*	*	*	*	*	*35.17*	(1), (58), (159)
F	*	*	*	*	*	*	*	*	*40.86*	(4),(54),(127),(159),(337)
F	*	*	*	*	*	*	*	*	*51.30*	(1),(54),(127),(159),(338)

System Type	Electricity: Fuel production energy	Electricity: Energy content of fuel	Oil Fuels: Fuel production energy	Oil Fuels: Energy content of fuel	Oil Fuels: Feedstock	Other Fuels: Fuel production energy	Other Fuels: Energy content of fuel	Other Fuels: Feedstock	Total System Energy Requirement	Notes (see page 376 *et seq.*)
A	20.17	6.37	–	–	–	1.35	25.60	–	53.49	(5), (524), (527)
H	*	*	*	*	*	*	*	*	*61.26*	(2), (48), (159), (340)
H	20.46	6.42	2.59	13.04	–	0.52	16.64	18.76	78.34	(2), (15), (159), (502)
A	4.40	1.39	0.44	2.15	–	2.61	18.24	115.23	144.46	(1), (26), (159), (335)
PAPER from waste paper (MJ/kg)										
A	*	*	*	*	*	*	*	*	*5.18*	(1), (90)
PAPER (AEROCOTE) from wood pulp (MJ/kg)										
A	38.98	12.31	–	–	–	1.30	24.66	–	77.25	(2), (521), (529)
PAPER (BASE) from wood pulp (MJ/kg)										
H	10.49	3.29	2.45	12.76	–	–	–	–	28.99	(2), (15), (501)
PAPER (MACHINE-GLAZED BASKET WEAVE MANILLA) from wood pulp (MJ/kg)										
A	24.83	7.84	–	–	–	0.83	15.69	–	49.19	(2), (521), (529)
PAPER (MACHINE-GLAZED BLEACHED KRAFT) from wood pulp (MJ/kg)										
A	21.56	6.81	–	–	–	0.72	13.64	–	42.73	(2), (521), (529)
PAPER (MACHINE-GLAZED LIGHTWEIGHT) from wood pulp (MJ/kg)										
A	28.94	9.14	–	–	–	0.96	18.31	–	57.35	(2), (521), (529)
PAPER (MACHINE-GLAZED SULPHITE) from wood pulp (MJ/kg)										
A	27.77	8.77	–	–	–	0.92	17.56	–	55.02	(2), (521), (529)
PAPER (MACHINE-GLAZED KRAFT) from wood pulp (MJ/kg)										
A	29.54	9.33	–	–	–	0.98	18.68	–	58.53	(2), (521), (529)

System Type	Electricity		Oil Fuels			Other Fuels			Total System Energy Requirement	Notes (see page 376 *et seq.*)
	Fuel production energy	Energy content of fuel	Fuel production energy	Energy content of fuel	Feedstock	Fuel production energy	Energy content of fuel	Feedstock		
PAPER (MECHANICAL PRINTING) from wood pulp (MJ/kg)										
A	20.08	6.34	–	–	–	0.67	12.71	–	39.80	(2), (521), (529)
PAPER (NEWSPRINT) from wood pulp (MJ/kg)										
A	16.97	5.36	–	–	–	0.60	11.45	–	34.38	(2), (521), (529)
PAPER (POLLOCK KRAFT) from wood pulp (MJ/kg)										
A	54.34	17.16	–	–	–	1.81	34.38	–	103.35	(2), (521), (529)
PAPER (SACK KRAFT UNGLAZED) from wood pulp (MJ/kg)										
A	16.97	5.36	–	–	–	0.60	11.45	–	34.38	(2), (521), (529)
PAPER (UNBLEACHED KRAFT) from wood pulp (MJ/kg)										
F	*	*	*	*	*	*	*	*	*41.77*	(1), (44), (339)
PAPER (UNGLAZED MACHINE CREPE) from wood pulp (MJ/kg)										
A	25.40	8.02	–	–	–	0.85	16.07	–	50.34	(2), (521), (529)
PAPER (UNGLAZED PURE KRAFT) from wood pulp (MJ/kg)										
A	38.09	12.03	–	–	–	1.27	24.10	–	75.49	(2), (521), (529)
PAPER (UNGLAZED IMITATION GREASEPROOF) from wood pulp (MJ/kg)										
A	56.11	17.72	–	–	–	1.87	35.50	–	111.20	(2), (521), (529)
PAPER (WET STRENGTH CREPE) from wood pulp (MJ/kg)										
A	29.54	9.33	–	–	–	0.98	18.68	–	58.53	(2), (521), (529)

PETROL (fuel content and production energy) see Chapter 6 and Table 6.13

System Type	Electricity: Fuel production energy	Electricity: Energy content of fuel	Oil Fuels: Fuel production energy	Oil Fuels: Energy content of fuel	Oil Fuels: Feedstock	Other Fuels: Fuel production energy	Other Fuels: Energy content of fuel	Other Fuels: Feedstock	Total System Energy Requirement	Notes (see page 376 *et seq.*)
PETROLEUM COKE from crude oil (MJ/kg)										
A	0.56	0.18	2.79	13.55	44.50	–	–	–	61.58	(1), (19), (265)
PHOSPHATE CONCENTRATE from ore in the ground (MJ/kg)										
J	*	*	*	*	*	*	*	*	*0.28*	(2), (56), (665)
H	*	*	*	*	*	*	*	*	*3.60*	(2), (20), (342)
o-PHOSPHORIC ACID from phosphate concentrate (MJ/kg)										
A	1.20	0.38	0.38	1.84	–	–	–	–	3.80	(2), (56), (666)
A	1.26	0.40	0.48	2.31	–	–	–	–	4.45	(2), (56), (668)
A	3.57	1.13	0.79	3.86	–	0.10	0.73	–	10.18	(1), (704), (728)
A	2.83	0.89	2.91	14.15	–	–	–	–	20.78	(2), (56), (667)
PHOSPHORUS from phosphate rock (MJ/kg)										
A	141.27	44.61	6.11	0.55	–	0.24	33.14	–	219.92	(1), (704), (727)
H	171.48	54.15	5.00	24.39	–	–	–	–	255.02	(2), (20), (343)
PHOSPHORUS PENTOXIDE from phosphate rock concentrate (MJ/kg)										
A	0.28	0.09	0.02	0.09	–	–	–	–	0.48	(2), (56), (669)
PIG IRON – See under IRON										
PIPES (CLAY) from raw material (MJ/kg)										
H	0.55	0.17	0.43	2.12	–	–	–	–	3.27	(2), (15), (415)
PIPES (H.D. POLYETHYLENE) from polymer granules (MJ/kg)										
A	11.40	3.60	–	–	–	–	–	–	15.00	(2), (15), (428)

System Type	Electricity: Fuel production energy	Electricity: Energy content of fuel	Oil Fuels: Fuel production energy	Oil Fuels: Energy content of fuel	Oil Fuels: Feedstock	Other Fuels: Fuel production energy	Other Fuels: Energy content of fuel	Other Fuels: Feedstock	Total System Energy Requirement	Notes (see page 376 *et seq.*)
PIPES (PVC) from polymer resin (MJ/kg)										
A	8.99	2.84	–	–	–	–	–	–	11.83	(2), (15), (549)
PLASTER – see under GYPSUM										
POLYBUTADIENE from 1.3 – butadiene (MJ/kg)										
A	–	–	–	–	–	0.77	5.41	–	6.18	(1), (16), (344)
POLYBUTADIENE from crude oil (MJ/kg)										
A	3.02	0.95	7.48	36.53	54.30	0.77	5.41	–	108.46	(1), (518)
POLYESTER FIBRE (POLY-DIMETHYL TEREPHTHALATE) from crude oil (MJ/kg)										
A	50.45	15.93	11.38	55.55	36.78	–	–	–	170.09	(2), (15), (347)
H	57.28	18.12	17.01	83.03	16.60	–	–	–	222.14	(2), (15), (20), (177)
POLYETHYLENE (HIGH DENSITY) from ethylene (MJ/kg)										
A	7.85	2.48	0.50	2.43	–	–	–	–	13.26	(4), (45), (345)
A	7.54	2.39	7.26	35.26	–	–	–	–	52.45	(1), (19), (346)
A	*	*	*	*	*	*	*	*	*87.22*	(1), (44)
POLYETHYLENE (LOW DENSITY) from crude oil (MJ/kg)										
H	16.59	5.24	4.33	21.16	45.00	–	–	–	92.32	(2), (20), (350)
POLYETHYLENE (LOW DENSITY) from ethylene (MJ/kg)										
A	9.43	2.99	16.93	82.18	–	–	–	–	111.53	(1), (19), (351)
POLYPROPYLENE from crude oil (MJ/kg)										
H	8.84	2.77	7.65	39.70	51.20	–	–	–	110.16	(2), (15), (499)
F	*	*	*	*	*53.81*	*	*	*	*149.95*	(1), (44)

System Type	Electricity: Fuel production energy	Electricity: Energy content of fuel	Oil Fuels: Fuel production energy	Oil Fuels: Energy content of fuel	Oil Fuels: Feedstock	Other Fuels: Fuel production energy	Other Fuels: Energy content of fuel	Other Fuels: Feedstock	Total System Energy Requirement	Notes (see page 376 *et seq.*)
POLYSTYRENE from crude oil (MJ/kg)										
F	*	*	*	*	*	*	*	*	*62.49*	(1), (44)
A	10.65	3.37	6.82	33.26	54.36	–	–	–	108.46	(2), (15)
POLYSTYRENE FOAM from crude oil (MJ/kg)										
A	24.60	7.77	–	–	(50.00)	6.99	48.94	–	138.30	(1), (16), (353)
H	*	*	*	*	*(56.88)*	*	*	*	*139.14*	(2), (48), (355)
A	25.07	7.93	0.06	0.30	42.99	7.15	50.05	15.80	149.35	(1), (16), (354)
POLYVINYLCHLORIDE (PVC) from crude oil (MJ/kg)										
H	*	*	*	*	*24.07*	*	*	*	*85.32*	(2), (48), (356)
H	22.17	7.00	7.73	37.73	24.20	–	–	–	98.83	(2), (15), (20), (360)
H	*	*	*	*	*	*	*	*	*106.62*	(1), (44)
POLYVINYLCHLORIDE (PVC) from vinyl chloride monomer (MJ/kg)										
A	2.18	0.69	0.11	0.52	–	–	–	–	3.50	(1), (19), (357)
A	3.71	1.17	0.98	4.78	–	–	–	–	10.64	(1), (19). (358)
A	*	*	*	*	*	*	*	*	*14.58*	(1), (44)
A	4.89	1.54	0.26	1.24	–	0.92	6.45	–	15.30	(1), (19), (359)
A	7.28	2.30	1.56	7.63	–	–	–	–	18.77	(4), (45), (361)
POTASSIUM CHLORIDE from sylvinite ore (MJ/kg)										
D	0.60	0.19	0.86	4.19	–	–	–	–	5.84	(2), (56), (362)
PRINTING (MJ/m^2)										
A	1.90	0.06	–	–	–	–	–	–	1.96	(4), (45), (363)
PROPANE from crude oil (MJ/kg)										
A	0.56	0.18	2.79	13.55	44.50	–	–	–	61.58	(1), (19), (265)

System Type	Electricity: Fuel production energy	Electricity: Energy content of fuel	Oil Fuels: Fuel production energy	Oil Fuels: Energy content of fuel	Oil Fuels: Feedstock	Other Fuels: Fuel production energy	Other Fuels: Energy content of fuel	Other Fuels: Feedstock	Total System Energy Requirement	Notes (see page 376 *et seq.*)
PROPANE–LPG (fuel content and production energy) see Chapter 6 and Table 6.13										
PROPYLENE from crude oil (MJ/kg)										
A	0.56	0.18	2.79	13.55	44.50	–	–	–	61.58	(1), (19), (265)
H	0.41	0.13	2.96	14.46	50.16	–	–	–	68.12	(2), (15), (20), (261)
A	8.68	2.89	8.58	41.81	24.03	–	–	–	85.99	(1), (19), (120)
PROPYLENE from ethane (MJ/kg)										
A	0.78	0.24	5.42	26.39	–	–	–	–	32.83	(1), (19), (264)
PROPYLENE from naphtha (MJ/kg)										
A	8.86	2.89	7.74	37.73	–	–	–	–	57.04	(1), (19), (119)

PULP (WOOD) see under WOOD PULP

RAILROAD TRANSPORT see under RAIL TRANSPORT ENERGY and Chapter 9

RAIL TRANSPORT ENERGY see Chapter 9

Capital energy of rail transport	9.6.2
Fuel requirements of rail transport	9.6.1
Maintenance energy	9.6.2

REFRACTORY BRICKS – see under BRICKS

ROAD TRANSPORT ENERGY – see Chapter 9

Battery energy	9.4.6
Construction of road vehicles	9.4.1
Fuel requirements of road vehicles	9.3
Garaging energy	9.4.5

System Type	Electricity: Fuel production energy	Electricity: Energy content of fuel	Oil Fuels: Fuel production energy	Oil Fuels: Energy content of fuel	Oil Fuels: Feedstock	Other Fuels: Fuel production energy	Other Fuels: Energy content of fuel	Other Fuels: Feedstock	Total System Energy Requirement	Notes (see page 376 *et seq.*)
Lubrication energy					9.4.3					
Maintenance energy					9.4.2					
Route facilities					9.5					
Spares					9.4.6					
Tyres					9.4.4					
ROAD VEHICLES – see under ROAD TRANSPORT and Chapter 9										
ROCK SALT – see under SODIUM CHLORIDE										
ROCKWOOL from raw materials (MJ/kg)										
F	*	*	*	*	*	*	*	*	*14.00*	(6), (22)
ROLLING – BLOOMING MILL (MJ/kg output)										
A	0.19	0.06	–	–	–	–	–	–	0.25	(1), (31)
ROLLING – 400mm MILL (MJ/kg output)										
A	0.30	0.10	–	–	–	–	–	–	0.40	(2), (91), (364)
ROLLING – 850mm MILL (MJ/kg output)										
A	0.29	0.09	–	–	–	–	–	–	0.38	(2), (91), (365)
ROLLING – STEEL PLATE MILL (MJ/kg output)										
A	0.44	0.14	–	–	–	–	–	–	0.58	(1), (31)
ROLLING – 5 STAND REDUCTION MILL (MJ/kg output)										
A	1.14	0.36	–	–	–	–	–	–	1.50	(1), (31), (366)

System Type	Electricity: Fuel production energy	Electricity: Energy content of fuel	Oil Fuels: Fuel production energy	Oil Fuels: Energy content of fuel	Oil Fuels: Feedstock	Other Fuels: Fuel production energy	Other Fuels: Energy content of fuel	Other Fuels: Feedstock	Total System Energy Requirement	Notes (see page 376 *et seq.*)
ROLLING – SLABBING MILL (MJ/kg output)										
A	0.16	0.05	–	–	–	–	–	–	0.21	(1), (31)
ROLLING – STRIP MILL (MJ/kg output)										
A	0.76	0.24	–	–	–	–	–	–	1.00	(1), (31)
ROLLING – WIDE STRIP MILL (MJ/kg output)										
A	1.52	0.48	–	–	–	–	–	–	2.00	(1), (31)
ROUTE FACILITIES FOR ROAD VEHICLES – see under ROAD TRANSPORT and Chapter 9										
SAND (GENERAL) from sand in the ground (MJ/kg)										
D	*	*	*	*	*	*	*	*	*0.02*	(2), (24)
B	–	–	0.01	0.07	–	–	–	–	0.08	(4), (45)
H	*	*	*	*	*	*	*	*	*0.08*	(1), (55), (127), (368)
F	*	*	*	*	*	*	*	*	*0.08*	(1), (44)
F	*	*	*	*	*	*	*	*	*0.10*	(6), (22)
A	0.10	0.03	0.01	0.04	–	0.05	0.39	–	0.62	(1), (26), (367)
A	*	*	*	*	*	*	*	*	*0.63*	(1), (704), (711)
SAND (DRY) from sand in the ground (MJ/kg)										
A	0.29	0.09	0.03	0.17	–	–	–	–	0.58	(2), (15), (367)
A	0.16	0.05	0.07	0.32	–	–	–	–	0.60	(2), (15), (367)
A	1.14	0.36	0.09	0.44	–	–	–	–	2.03	(2), (15), (367)
SAND (GLASSMAKING) from raw materials in the ground (MJ/kg)										
A	–	–	0.01	0.07	–	–	–	–	0.08	(4), (45), (367)
H	*	*	*	*	*	*	*	*	*0.60*	(1), (58), (367)
A	0.17	0.05	0.02	0.08	–	0.08	0.66	–	1.06	(1), (16), (369)

System Type	Electricity: Fuel production energy	Electricity: Energy content of fuel	Oil Fuels: Fuel production energy	Oil Fuels: Energy content of fuel	Oil Fuels: Feedstock	Other Fuels: Fuel production energy	Other Fuels: Energy content of fuel	Other Fuels: Feedstock	Total System Energy Requirement	Notes (see page 376 *et seq.*)
SAND (WET) from raw materials in the ground (MJ/kg)										
A	0.22	0.07	0.03	0.13	–	–	–	–	0.45	(2), (15), (367)
A	0.13	0.04	0.05	0.05	–	–	–	–	0.45	(2), (15), (367)
A	0.86	0.27	0.07	0.33	–	–	–	–	1.53	(2), (15), (367)
S.B.R. (RUBBER) from crude oil (MJ/kg)										
H	8.68	2.74	10.82	52.83	55.00	–	–	–	130.07	(2), (15), (20), (370)
H	*	*	*	*	(55.00)	*	*	*	*145.14*	(2), (50), (127)
SEA TRANSPORT ENERGY see Chapter 9										
Capital energy of sea transport					9.7.2					
Fuel requirements of sea transport					9.7.1					
Maintenance energy					9.7.2					
SEALING COMPOUND from raw materials (MJ/kg)										
A	*	*	*	*	*	*	*	*	*89.79*	(4), (45), (371)
A	3.02	0.95	7.48	36.53	54.30	0.77	5.41	–	108.46	(2), (15), (670)
SHALE from raw materials in the ground (MJ/kg)										
A	*	*	*	*	*	*	*	*	*0.07*	(1), (704), (710)
SHEET – see under the relevant metal										
SHRINKWRAP FILM from polymer resin (MJ/kg)										
A	0.29	0.09	–	–	–	–	–	–	0.38	(4), (45), (372)
see also FILMS (PLASTIC)										
SELENIUM from anode slimes in copper refining (MJ/kg)										
D	55.29	17.46	49.82	238.73	–	–	–	–	361.30	(1), (31), (373)

System Type	Electricity: Fuel production energy	Electricity: Energy content of fuel	Oil Fuels: Fuel production energy	Oil Fuels: Energy content of fuel	Oil Fuels: Feedstock	Other Fuels: Fuel production energy	Other Fuels: Energy content of fuel	Other Fuels: Feedstock	Total System Energy Requirement	Notes (see page 376 *et seq.*)
SHIP TRANSPORT see under SEA TRANSPORT and Chapter 9										
SILICA (CRUSHED) from silicious rock (MJ/kg)										
A	0.60	0.19	–	–	–	–	–	–	0.79	(1), (31)
SILICON from silica (MJ/kg)										
A	156.65	49.50	–	–	–	3.38	21.86	–	231.49	(5), (15), (671)
H	*	*	*	*	*	*	*	*	*234.71*	(1), (55), (127)
SILVER from anode slimes in copper refining (MJ/kg)										
D	17.67	5.58	18.12	86.81	–	–	–	–	128.18	(1), (31), (374)
SODA ASH – see under SODIUM CARBONATE										
SODIUM from sodium chloride										
A	125.56	39.65	–	–	–	–	–	–	165.21	(1), (84), (375)
SODIUM CARBONATE from sodium chloride (MJ/kg)										
A	–	–	0.60	2.85	–	0.20	1.40	–	5.05	(1), (16), (377)
H	0.57	0.18	–	–	–	0.57	10.97	–	12.29	(2), (20), (376)
A	–	–	1.59	7.75	–	0.20	3.86	–	13.40	(2), (15), (376)
J	*	*	*	*	*	*	*	*	*18.97*	(1), (704)
F	*	*	*	*	*	*	*	*	*44.16*	(1), (57)
SODIUM CARBONATE from naturally occurring sodium sesquicarbonate (MJ/kg)										
D	0.22	0.07	0.09	0.44	–	0.76	5.34	–	6.92	(1), (26)
F	*	*	*	*	*	*	*	*	*7.44*	(1), (58)
F	*	*	*	*	*	*	*	*	*8.40*	(1), (44)

System Type	Electricity: Fuel production energy	Electricity: Energy content of fuel	Oil Fuels: Fuel production energy	Oil Fuels: Energy content of fuel	Oil Fuels: Feedstock	Other Fuels: Fuel production energy	Other Fuels: Energy content of fuel	Other Fuels: Feedstock	Total System Energy Requirement	Notes (see page 376 *et seq.*)
J	*	*	*	*	*	*	*	*	*8.50*	(1), (704), (717)
A	–	–	–	–	–	0.69	14.01	–	14.70	(1), (31), (672)
SODIUM CHLORIDE (BRINE) from mineral in the ground (MJ/kg)										
A	0.01	neg	neg	0.01	–	0.03	0.20	–	0.26	(1), (16), (379)
SODIUM CHLORIDE (ROCK SALT) from mineral in the ground (MJ/kg)										
A	1.07	0.34	neg	0.02	–	0.03	0.19	–	1.65	(1), (16), (378)
SODIUM CHLORIDE (VACUUM RECRYSTALLISED) from brine (MJ/kg)										
A	–	–	0.88	4.29	–	–	–	–	5.17	(4), (45), (380)
SODIUM HYDROXIDE from sodium chloride (MJ/kg)										
A	10.26	3.24	–	–	–	–	–	–	13.50	(4), (45), (215)
D	9.50	3.00	0.39	1.86	–	–	–	–	14.75	(1), (19), (214)
J	*	*	*	*	*	*	*	*	*14.83*	(7), (25)
A	18.13	5.72	–	–	–	–	–	–	23.85	(1), (16), (381)
D	8.48	2.68	2.19	10.64	–	–	–	–	23.99	(1), (19), (213)
H	21.19	6.69	0.34	1.66	–	–	–	–	29.88	(2), (20), (202)
J	*	*	*	*	*	*	*	*	*33.69*	(1), (57)
A	*	*	*	*	*	*	*	*	*47.70*	(1), (704), (721)
J	*	*	*	*	*	*	*	*	*70.00*	(2), (49)
SODIUM NITRATE from material in the ground (MJ/kg)										
A	0.10	0.03	neg	0.02	–	–	–	–	0.15	(2), (15)
SODIUM NITRATE (synthetic)										
A	*	*	*	*	*	*	*	*	*43.80*	(1), (704), (722)

System Type	Electricity: Fuel production energy	Electricity: Energy content of fuel	Oil Fuels: Fuel production energy	Oil Fuels: Energy content of fuel	Oil Fuels: Feedstock	Other Fuels: Fuel production energy	Other Fuels: Energy content of fuel	Other Fuels: Feedstock	Total System Energy Requirement	Notes (see page 376 *et seq.*)
SODIUM SULPHATE from sodium chloride and sulphuric acid (MJ/kg)										
A	0.86	0.27	0.57	2.99	–	–	–	–	4.69	(2), (15)
SPARES FOR ROAD VEHICLES see under ROAD TRANSPORT and Chapter 9										
STAINLESS STEEL – see under STEEL										
STEAM from water (MJ/kg)										
A	–	–	–	–	–	0.11	2.32	–	2.43	(1), (31), (110)
A	–	–	0.51	2.45	–	–	–	–	2.96	(4), (45), (114)
A	–	–	0.56	2.75	–	–	–	–	3.31	(2), (56), (116)
A	–	–	–	–	–	0.16	3.34	–	3.50	(1), (31), (109)
A	–	–	0.62	2.98	–	–	–	–	3.60	(1), (5), (673)
A	–	–	0.64	3.11	–	–	–	–	3.75	(1), (92), (111)
A	–	–	0.73	3.58	–	–	–	–	4.31	(2), (49), (112)
H	0.03	0.01	0.76	3.69	–	–	–	–	4.49	(2), (20), (115)
A	–	–	0.77	3.74	–	–	–	–	4.51	(1), (93), (113)
STEEL ALLOY (HOT-ROLLED) from ore (MJ/kg)										
F	*	*	*	*	*	*	*	*	*51.52*	(1), (43)
STEEL ALLOY (COLD ROLLED) from ore (MJ/kg)										
D	*	*	*	*	*	*	*	*	*51.83*	(1), (36), (392)
F	*	*	*	*	*	*	*	*	*57.44*	(1), (43)
STEEL BILLETS from ore in the ground (MJ/kg)										
H	*	*	*	*	*	*	*	*	*43.75*	(2), (48), (674)

System Type	Electricity: Fuel production energy	Electricity: Energy content of fuel	Oil Fuels: Fuel production energy	Oil Fuels: Energy content of fuel	Oil Fuels: Feedstock	Other Fuels: Fuel production energy	Other Fuels: Energy content of fuel	Other Fuels: Feedstock	Total System Energy Requirement	Notes (see page 376 *et seq.*)
STEEL (CONTINUOUS CASTING) from liquid metal (MJ/kg)										
A	0.06	0.02	–	–	–	–	–	–	0.08	(2), (83), (394)
A	0.21	0.07	–	–	–	–	–	–	0.28	(2), (97), (395)
STEEL COGGING MILL (MJ/kg)										
A	0.32	0.10	–	–	–	–	–	–	0.42	(2), (97), (393)
STEEL FORGINGS from ore (MJ/kg)										
F	*	*	*	*	*	*	*	*	*78.85*	(1), (43)
STEEL (FINISHED GENERAL PRODUCTS) from ore in the ground (MJ/kg)										
D	*	*	*	*	*	*	*	*	*22.06*	(1), (99), (400)
H	*	*	*	*	*	*	*	*	*24.21*	(1), (90)
D	–	–	0.43	2.09	–	1.59	21.83	–	25.94	(1), (71), (401)
A	*	*	*	*	*	*	*	*	*27.76*	(2), (41), (590)
J	*	*	*	*	*	*	*	*	*30.00*	(6), (22)
A	1.87	0.59	–	–	–	4.49	29.61	–	35.97	(1), (46), (591)
H	*	*	*	*	*	*	*	*	*37.31*	(2), (50), (127)
D	*	*	*	*	*	*	*	*	*40.98*	(1), (99), (399)
J	*	*	*	*	*	*	*	*	*43.36*	(1), (78), (595)
D	*	*	*	*	*	*	*	*	*44.81*	(1), (99), (398)
H	*	*	*	*	*	*	*	*	*47.52*	(2), (50), (397)
J	*	*	*	*	*	*	*	*	*49.56*	(1), (40)
D	*	*	*	*	*	*	*	*	*52.00*	(2), (52), (127), (589)
STEEL (FINISHED GENERAL PRODUCTS) from ore (MJ/kg)										
H	*	*	*	*	*	*	*	*	*23.40*	(2), (50), (127), (275)
F	*	*	*	*	*	*	*	*	*40.68*	(1), (36), (406)

System Type	Electricity: Fuel production energy	Electricity: Energy content of fuel	Oil Fuels: Fuel production energy	Oil Fuels: Energy content of fuel	Oil Fuels: Feedstock	Other Fuels: Fuel production energy	Other Fuels: Energy content of fuel	Other Fuels: Feedstock	Total System Energy Require-ment	Notes (see page 376 *et seq.*)
F	*	*	*	*	*	*	*	*	*42.96*	(1), (99), (396)
F	*	*	*	*	*	*	*	*	*43.23*	(1), (36), (404)
F	*	*	*	*	*	*	*	*	*43.70*	(1), (36), (406)
F	*	*	*	*	*	*	*	*	*45.79*	(1), (36), (403)
F	*	*	*	*	*	*	*	*	*48.35*	(1), (36), (405)
F	*	*	*	*	*	*	*	*	*48.81*	(1), (36), (403)
H	*	*	*	*	*	*	*	*	*49.97*	(1), (36)
H	*	*	*	*	*	*	*	*	*60.25*	(2), (20), (407)
STEEL (FINISHED) from scrap (MJ/kg)										
A	7.66	2.42	–	–	–	0.22	1.53	–	11.83	(2), (62), (450)
J	*	*	*	*	*	*	*	*	*16.17*	(1), (42), (593)
H	*	*	*	*	*	*	*	*	*23.49*	(1), (44), (594)
STEEL (GALVANIZED) from ore in the ground (MJ/kg)										
D	*	*	*	*	*	*	*	*	*49.97*	(1), (36), (596)
STEEL (GALVANIZED) from steel casting (MJ/kg)										
J	*	*	*	*	*	*	*	*	*25.30*	(11), (98)
STEEL (GALVANIZED) from steel sheet (MJ/kg)										
A	0.38	0.12	–	–	–	–	–	–	0.50	(1), (31), (402)
STEEL HEAT TREATMENT FURNACE (MJ/kg steel treated)										
A	0.32	0.10	–	–	–	0.23	1.64	–	2.29	(2), (97), (408)
STEEL (HEAVY PLATE) from steel casting (MJ/kg)										
A	*	*	*	*	*	*	*	*	*21.80*	(11), (98), (597)

System Type	Electricity: Fuel production energy	Electricity: Energy content of fuel	Oil Fuels: Fuel production energy	Oil Fuels: Energy content of fuel	Oil Fuels: Feedstock	Other Fuels: Fuel production energy	Other Fuels: Energy content of fuel	Other Fuels: Feedstock	Total System Energy Requirement	Notes (see page 376 *et seq.*)
STEEL (HOT-ROLLED) from ore (MJ/kg)										
F	*	*	*	*	*	*	*	*	*49.63*	(1), (43)
STEEL (HOT-ROLLED) from ingot (MJ/kg)										
D	*	*	*	*	*	*	*	*	*18.78*	(1), (36), (409)
A	*	*	*	*	*	*	*	*	*19.60*	(11), (98)
STEEL (COLD-ROLLED STRIP) from ingot (MJ/kg)										
D	*	*	*	*	*	*	*	*	*23.30*	(11), (98)
STEEL (COLD-ROLLED) from ore (MJ/kg)										
F	*	*	*	*	*	(*1.24*)	(*24.93*)	*	*54.71*	(1), (43), (498)
STEEL INGOT from liquid steel (MJ/kg)										
A	0.03	0.01	–	–	–	–	–	–	0.04	(2), (83), (410)
STEEL INGOT from ore (MJ/kg)										
D	*	*	*	*	*	*	*	*	*27.99*	(1), (46), (412)
A	*	*	*	*	*	*	*	*	*49.75*	(1), (78), (411)
STEEL (LIQUID) from ore (MJ/kg)										
D	1.90	0.60	–	–	–	0.83	5.38	–	8.71	(2), (74), (422)
D	2.18	0.69	–	–	–	1.44	9.34	–	13.65	(2), (74), (423)
A	2.22	0.70	–	–	–	1.49	9.61	–	14.02	(1), (98), (430)
D	2.22	0.70	–	–	–	1.66	11.46	–	15.34	(2), (74), (424)
D	4.84	1.53	–	–	–	1.87	13.10	–	21.34	(2), (60), (414)
D	0.73	0.23	–	–	–	0.98	19.40	–	21.34	(2), (60), (604)
D	0.66	0.21	–	–	–	1.12	22.21	–	24.20	(2), (60), (603)
D	0.70	0.22	–	–	–	1.12	22.27	–	24.31	(2), (60), (605)

System Type	Electricity		Oil Fuels			Other Fuels			Total System Energy Requirement	Notes (see page 376 *et seq.*)
	Fuel production energy	Energy content of fuel	Fuel production energy	Energy content of fuel	Feedstock	Fuel production energy	Energy content of fuel	Feedstock		
D	8.33	2.63	–	–	–	1.70	11.92	–	24.58	(2), (74), (426)
A	8.33	2.63	–	–	–	1.79	12.50	–	25.25	(11), (98), (599)
D	1.36	0.43	0.56	2.72	–	1.75	18.68	–	25.50	(2), (60), (418)
D	0.76	0.24	–	–	–	2.38	24.55	–	26.93	(2), (60), (600)
D	0.82	0.26	–	–	–	1.62	24.85	–	27.55	(2), (60), (601)
D	1.46	0.46	–	–	–	1.25	25.22	–	28.39	(2), (60), (417)
D	1.58	0.50	–	–	–	1.26	25.38	–	28.72	(2), (60), (416)
A	*	*	*	*	*	*	*	*	*34.30*	(1), (78), (432)
STEEL (LIQUID) from pig iron and scrap (MJ/kg)										
(a) Basic Oxygen Furnace										
A	0.34	0.11	–	–	–	–	–	–	0.45	(1), (69), (592)
D	0.63	0.20	–	–	–	–	–	–	0.83	(2), (74), (421)
F	*	*	*	*	*	*	*	*	*1.40*	(1), (43), (384)
A	*	*	*	*	*	*	*	*	*1.57*	(1), (36), (384)
(b) Open-Hearth Furnace										
A	0.16	0.05	0.47	2.30	–	0.18	1.19	–	4.35	(1), (31), (391)
A	–	–	0.83	3.80	–	–	–	–	4.63	(1), (96), (388)
F	*	*	*	*	*	*	*	*	*5.22*	(1), (43), (385)
A	*	*	*	*	*	*	*	*	*5.86*	(1), (36), (385)
(c) Electric Arc Furnace										
D	3.77	1.19	–	–	–	0.18	1.25	–	6.39	(2), (61), (420)
D	4.65	1.47	0.11	0.62	–	0.02	0.10	–	6.97	(2), (61), (602)
D	5.26	1.66	–	–	–	0.03	0.20	–	7.15	(1), (98), (429)
D	5.16	1.63	–	–	–	0.07	0.44	–	7.30	(2), (60), (431)
D	5.13	1.62	–	–	–	0.08	0.50	–	7.33	(2), (61), (419)
A	5.64	1.78	–	–	–	–	–	–	7.42	(1), (72), (434)
A	5.70	1.80	–	–	–	–	–	–	7.50	(2), (101), (383)
A	5.83	1.84	–	–	–	–	–	–	7.67	(2), (83), (413)

System Type	Electricity: Fuel production energy	Electricity: Energy content of fuel	Oil Fuels: Fuel production energy	Oil Fuels: Energy content of fuel	Oil Fuels: Feedstock	Other Fuels: Fuel production energy	Other Fuels: Energy content of fuel	Other Fuels: Feedstock	Total System Energy Requirement	Notes (see page 376 *et seq.*)
D	6.30	1.99	–	–	–	0.05	0.35	–	8.69	(2), (74), (425)
F	*	*	*	*	*	*	*	*	*8.85*	(1), (43), (383)
D	6.14	1.94	–	–	–	0.11	0.80	–	8.99	(2), (62), (427)
D	7.16	2.26	0.02	0.09	–	0.04	0.58	–	10.15	(2), (97), (433)
D	(9.06)	(2.86)	–	–	–	–	–	–	(11.92)	(1), (99), (383)
(d) Miscellaneous										
A	*	*	*	*	*	*	*	*	*5.46*	(1), (36), (382)
H	*	*	*	*	*	*	*	*	*6.31*	(1), (90)
A	*	*	*	*	*	*	*	*	*7.88*	(1), (36), (389)
A	8.27	2.61	–	–	–	–	–	–	10.88	(1), (95), (387)
STEEL LIQUID from sponge iron and scrap (MJ/kg)										
A	6.13	1.93	–	–	–	–	–	–	8.06	(1), (72), (435)
A	6.40	2.02	–	–	–	–	–	–	8.42	(1), (69), (390)
A	7.28	2.30	–	–	–	–	–	–	9.58	(1), (94), (386)
STEEL PIPE from ore (MJ/kg)										
F	*	*	*	*	*	*	*	*	*54.73*	(1), (43)
STEEL PREHEATING FURNACES (MJ/kg output)										
A	–	–	–	–	–	0.08	0.60	–	0.68	(2), (61), (436)
STEEL REHEATING FURNACES (MJ/kg output)										
A	–	–	–	–	–	0.22	1.52	–	1.74	(2), (102), (439)
A	0.16	0.05	–	–	–	0.21	1.48	–	1.90	(2), (83), (443)
A	–	–	–	–	–	0.26	1.85	–	2.11	(2), (104), (441)
A	–	–	–	–	–	0.27	1.90	–	2.17	(2), (103), (441)
A	–	–	–	–	–	0.28	1.93	–	2.21	(2), (104), (441)
A	–	–	–	–	–	0.29	2.00	–	2.29	(2), (103), (440)

System Type	Electricity: Fuel production energy	Electricity: Energy content of fuel	Oil Fuels: Fuel production energy	Oil Fuels: Energy content of fuel	Oil Fuels: Feedstock	Other Fuels: Fuel production energy	Other Fuels: Energy content of fuel	Other Fuels: Feedstock	Total System Energy Requirement	Notes (see page 376 *et seq.*)
A	–	–	–	–	–	0.30	2.08	–	2.38	(2), (102), (447)
A	–	–	–	–	–	0.35	2.42	–	2.77	(2), (102), (446)
A	0.28	0.09	–	–	–	0.37	2.63	–	3.37	(2), (97), (460)
A	–	–	–	–	–	0.43	3.01	–	3.44	(2), (102), (445)
A	0.29	0.09	–	–	–	0.49	3.40	–	4.27	(2), (97), (444)
A	3.07	0.97	–	–	–	0.04	3.29	–	4.37	(2), (102), (437)
A	3.42	1.08	–	–	–	–	–	–	4.50	(2), (102), (438)
A	3.99	1.26	–	–	–	–	–	–	5.25	(2), (105), (442)
STEEL (ROLLED-BAR) from ingot or billet (MJ/kg)										
A	1.17	0.37	–	–	–	–	–	–	1.54	(2), (62), (449)
STEEL SCRAP from domestic refuse (MJ/kg)										
A	0.50	0.16	0.05	0.24	–	0.03	0.22	–	1.20	(1), (16), (451)
STEEL BUNDLING PRESS (MJ/kg)										
A	0.06	0.02	–	–	–	–	–	–	0.08	(2), (83), (452)
STEEL (SCRAP DRYING PLANT) (MJ/kg output)										
A	–	–	–	–	–	0.03	0.24	–	0.27	(2), (83), (453)
STEEL (SCRAP HANDLING) (MJ/kg)										
A	0.09	0.03	–	–	–	–	–	–	0.12	(2), (83), (606)
STEEL (SCRAP PREHEATING PLANT) (MJ/kg steel output)										
A	–	–	–	–	–	0.07	0.49	–	0.56	(2), (83), (454)
STEEL (SCRAP-SHREDDING PLANT) (MJ/kg)										
A	0.13	0.04	–	–	–	–	–	–	0.17	(2), (83), (455)

System Type	Electricity		Oil Fuels			Other Fuels			Total System Energy Requirement	Notes (see page 376 *et seq.*)
	Fuel production energy	Energy content of fuel	Fuel production energy	Energy content of fuel	Feedstock	Fuel production energy	Energy content of fuel	Feedstock		
STEEL SHEET (COLD-ROLLED) from hot rolled sheet (MJ/kg)										
_D_	_(20.96)_	_(6.62)_	–	–	–	–	–	–	_27.58_	(1), (36), (607)
STEEL SHEET from ore in the ground (MJ/kg)										
_J_	*	*	*	*	*	*	*	*	_38.55_	(1), (42)
_A_	5.39	1.70	0.53	2.54	–	2.47	34.65	–	47.28	(1), (16), (457)
_F_	*	*	*	*	*	*	*	*	_48.50_	(1), (44), (412)
_H_	*	*	*	*	*	*	*	*	_49.96_	(1), (55), (127), (448)
_H_	*	*	*	*	*	*	*	*	_58.59_	(1), (58)
_D_	*	*	*	*	*	*	*	*	_61.60_	(1), (36)
STEEL SHEET from ore at the blast furnace (MJ/kg)										
_A_	_(5.13)_	_(1.62)_	–	–	–	_(2.66)_	_(17.18)_	–	_(26.59)_	(4), (45), (608)
_D_	5.25	1.66	0.42	2.05	–	3.94	33.37	0.42	47.11	(1), (26), (456)
STEEL SLAB from steel ingot (MJ/kg)										
_A_	0.36	0.12	–	–	–	neg	0.02	–	0.50	(2), (610), (611)
STEEL SOAKING PIT (MJ/kg output)										
_A_	0.08	0.03	–	–	–	0.16	1.10	–	1.37	(2), (610), (612)
_A_	–	–	–	–	–	0.28	1.98	–	2.26	(2), (102), (458)
_A_	0.29	0.09	–	–	–	0.27	1.86	–	2.51	(2), (97), (459)
STEEL (STAINLESS COLD-ROLLED) from ore (MJ/kg)										
_D_	*	*	*	*	*	*	*	*	_79.03_	(1), (36), (609)
_F_	*	*	*	*	*	*	*	*	_81.77_	(1), (43)
STEEL (STAINLESS HOT-ROLLED) from ore (MJ/kg)										
_F_	*	*	*	*	*	*	*	*	_79.03_	(1), (43)

System Type	Electricity: Fuel production energy	Electricity: Energy content of fuel	Oil Fuels: Fuel production energy	Oil Fuels: Energy content of fuel	Oil Fuels: Feedstock	Other Fuels: Fuel production energy	Other Fuels: Energy content of fuel	Other Fuels: Feedstock	Total System Energy Requirement	Notes (see page 376 *et seq.*)
STEEL WIRE from ore (MJ/kg)										
F	*	*	*	*	*	*	*	*	*63.42*	(1), (43)
STEEL (WIRE ROD) from steel casting										
A	*	*	*	*	*	*	*	*	*19.10*	(11), (98)
STYRENE from crude oil (MJ/kg)										
H	2.80	0.89	4.09	19.94	54.36	1.56	10.89	–	94.55	(1), (15), (178)
H	4.97	1.57	11.20	54.70	57.68	–	–	–	130.12	(2), (15), (20), (461)
STYRENE from ethylene and benzene										
A	0.75	0.24	–	–	–	1.56	10.89	–	13.44	(1), (16), (462)
STYRENE BUTADIENE RUBBER – see under S.B.R.										
SULPHUR from sulphur dioxide										
A	–	–	–	–	–	2.66	18.61	–	21.27	(1), (675), (676)
SULPHUR from underground deposits (MJ/kg)										
H	–	–	0.35	1.71	–	0.57	3.98	–	6.61	(2), (20), (464)
A	0.03	0.01	0.01	0.10	–	0.82	5.73	–	6.70	(1), (704), (730)
SULPHURIC ACID from sulphur dioxide (MJ/kg)										
A	1.88	0.60	–	–	–	–	–	–	2.48	(1), (31), (465)
J	*	*	*	*	*	*	*	*	*3.25*	(1), (57)
SULPHURIC ACID from sulphur in the ground (MJ/kg)										
A	0.29	0.09	neg	0.03	–	0.28	1.93	–	2.62	(1), (704), (731)

System Type	Electricity: Fuel production energy	Electricity: Energy content of fuel	Oil Fuels: Fuel production energy	Oil Fuels: Energy content of fuel	Oil Fuels: Feedstock	Other Fuels: Fuel production energy	Other Fuels: Energy content of fuel	Other Fuels: Feedstock	Total System Energy Requirement	Notes (see page 376 *et seq.*)
TELLURIUM from anode slimes in copper refining (MJ/kg)										
D	297.41	93.92	249.11	1193.63	–	–	–	–	1834.07	(1), (31), (466)
TEREPHTHALIC ACID from crude oil (MJ/kg)										
H	36.51	11.53	8.94	43.67	36.78	–	–	–	137.43	(2), (15), (348)
TILES (ROOF) from raw materials (MJ/sq.yd)										
H	*	*	*	*	*	*	*	*	*154.80*	(2), (50)
TIMBER – see under WOOD										
TIN from ore concentrate (MJ/kg)										
A	14.57	4.60	–	–	–	–	–	–	19.17	(12), (107), (468)
A	–	–	–	–	–	3.93	25.42	–	29.50	(2), (107), (469)
A	–	–	–	–	–	5.98	41.84	–	47.82	(2), (107), (470)
A	–	–	8.57	41.84	–	–	–	–	50.41	(2), (107), (467)
TINPLATE from sheet steel (MJ/kg)										
A	1.27	0.40	–	–	–	–	–	–	1.67	(1), (31), (472)
TINPLATE from scrap (MJ/kg)										
A	9.31	2.94	0.37	1.83	–	0.80	5.60	–	20.85	(1), (16), (471)
TINPLATE from steel casting (MJ/kg)										
J	*	*	*	*	*	*	*	*	*24.60*	(2), (48)
TINPLATE from raw materials (MJ/kg)										
A	6.65	2.09	0.04	0.21	–	5.14	31.68	–	45.81	(2), (15)
H	*	*	*	*	*	*	*	*	*54.70*	(2), (48)

System Type	Electricity: Fuel production energy	Electricity: Energy content of fuel	Oil Fuels: Fuel production energy	Oil Fuels: Energy content of fuel	Oil Fuels: Feedstock	Other Fuels: Fuel production energy	Other Fuels: Energy content of fuel	Other Fuels: Feedstock	Total System Energy Requirement	Notes (see page 376 *et seq.*)
TITANIUM from scrap (MJ/kg)										
A	195.96	61.88	–	–	–	–	–	–	257.84	(1), (31), (479)
TITANIUM from ore in the ground (MJ/kg)										
A	418.51	132.16	–	–	–	24.25	169.78	–	744.70	(1), (31), (473)
A	535.45	169.09	–	–	–	24.31	170.21	–	899.06	(1), (31), (475)
A	552.44	174.46	–	–	–	24.03	168.03	–	919.16	(1), (31), (476)
A	553.57	174.82	–	–	–	24.31	170.21	–	922.91	(1), (31), (474)
A	559.84	176.80	–	–	–	25.50	178.56	–	940.70	(1), (31), (477)
A	807.57	255.02	–	–	–	25.73	180.14	–	1268.46	(1), (31), (478)
TITANIUM SHEET from ore in the ground (MJ/kg)										
H	*	*	*	*	*	*	*	*	*559.82*	(1), (55), (127), (480)
TOLUENE from crude oil (MJ/kg)										
H	2.50	0.79	4.19	20.47	51.83	–	–	–	79.78	(2),(15),(20),(163),(481)
TRUCK TRANSPORT – see under ROAD TRANSPORT and Chapter 9										
TYRES from raw materials in the ground (M/kg) (see under ROAD TRANSPORT and Chapter 9)										
D	*	*	*	*	*	*	*	*	*85.77*	(1), (36)
UREA from raw materials in the ground (MJ/kg)										
D	4.58	1.45	0.63	3.06	–	0.99	6.96	15.08	32.75	(2), (56), (681)
D	5.01	1.58	0.88	4.30	–	0.98	6.90	14.38	34.03	(2), (56), (682)
D	4.75	1.51	1.01	4.93	–	0.97	6.84	14.25	34.26	(2), (56), (685)
D	4.04	1.58	1.01	4.93	–	0.99	6.96	15.08	34.59	(2), (56), (683)
D	5.06	1.60	0.85	4.15	–	0.99	6.96	15.08	34.69	(2), (56), (684)

System Type	Electricity: Fuel production energy	Electricity: Energy content of fuel	Oil Fuels: Fuel production energy	Oil Fuels: Energy content of fuel	Oil Fuels: Feedstock	Other Fuels: Fuel production energy	Other Fuels: Energy content of fuel	Other Fuels: Feedstock	Total System Energy Requirement	Notes (see page 376 *et seq.*)
VEHICLE TRANSPORT see under ROAD TRANSPORT and Chapter 9										
VINYL CHLORIDE from acetylene and hydrogen chloride (MJ/kg)										
A	0.75	0.24	0.78	3.81	–	–	–	–	5.58	(1), (19), (485)
VINYL CHLORIDE from ethylene and chlorine (MJ/kg)										
A	22.80	7.20	1.71	8.37	–	–	–	–	40.08	(4), (45), (486)
VINYL CHLORIDE from 1.2–dichloroethane (MJ/kg)										
A	25.70	8.16	89.54	435.50	–	–	–	–	558.90	(1), (19), (483)
A	26.29	8.35	89.54	435.50	–	–	–	–	559.68	(1), (19), (484)
VINYL CHLORIDE from crude oil (MJ/kg)										
D	*	*	*	*	*	*	*	*	*74.38*	(1), (36)
H	–	–	8.47	41.33	24.66	–	–	–	74.66	(2), (15), (481)
H	15.96	5.04	5.46	27.17	24.00	–	–	–	77.73	(2), (20), (482)
WASTE DISPOSAL – refuse collection and landfill (MJ/kg waste)										
A	–	–	0.01	0.07	–	–	–	–	0.08	(2), (15), (531)
A	–	–	0.02	0.10	–	–	–	–	0.12	(2), (15), (531)
A	–	–	0.02	0.12	–	–	–	–	0.14	(2), (15), (531)
A	–	–	0.03	0.13	–	–	–	–	0.16	(1), (16)
A	–	–	0.03	0.14	–	–	–	–	0.17	(2), (15), (531)
A	*	*	*	*	*	*	*	*	*0.23*	(1), (58)
A	–	–	0.04	0.19	–	–	–	–	0.23	(2), (15), (531)
A	*	*	*	*	*	*	*	*	*0.31*	(1), (44)
A	–	–	0.06	0.28	–	–	–	–	0.34	(2), (15), (531)
H	*	*	*	*	*	*	*	*	*0.40*	(1), (44), (487)

System Type	Electricity: Fuel production energy	Electricity: Energy content of fuel	Oil Fuels: Fuel production energy	Oil Fuels: Energy content of fuel	Oil Fuels: Feedstock	Other Fuels: Fuel production energy	Other Fuels: Energy content of fuel	Other Fuels: Feedstock	Total System Energy Requirement	Notes (see page 376 *et seq.*)
WASTE DISPOSAL – incineration (MJ/kg waste)										
A	*	*	*	*	*	*	*	*	*0.34*	(1), (44), (686)
A	0.57	0.18	–	–	–	–	–	–	0.75	(4), (45)
WATER from raw water (MJ/imperial gallon)										
J	*	*	*	*	*	*	*	*	*0.004*	(6), (22)
A	0.006	0.002	–	–	–	–	–	–	0.008	(2), (56), (204)
H	*	*	*	*	*	*	*	*	*0.040*	(2), (38)
A	0.032	0.010	–	–	–	–	–	–	0.042	(2), (15)
H	0.035	0.011	–	–	–	–	–	–	0.046	(2), (20)
WATER from raw water (MJ/1000 litres)										
A	1.80	0.57	–	–	–	–	–	–	2.37	(2), (56), (204)
H	7.03	2.22	–	–	–	–	–	–	9.25	(2), (20), (203)
WEAVING (FIBRES) into cloth (MJ/kg)										
A	34.42	10.87	–	–	–	–	–	–	45.29	(2), (15), (349)
WIRE – see under the relevant metal										
WOOD from standing timber (MJ/kg)										
J	*	*	*	*	*	*	*	*	*3.33*	(1), (40)
F	*	*	*	*	*	*	*	*	*5.20*	(2), (59), (127)
H	*(4.89)*	*(1.54)*	*	*	*	*	*	*	*(6.43)*	(2), (50), (127), (687)
H	*	*	*	*	*	*	*	*	*7.20*	(2), (52), (127)
WOOD POLES from standing timber (MJ/kg)										
A	*(9.57)*	*(3.02)*	–	–	–	–	–	–	*(12.59)*	(1), (92), (688)

System Type	Electricity: Fuel production energy	Electricity: Energy content of fuel	Oil Fuels: Fuel production energy	Oil Fuels: Energy content of fuel	Oil Fuels: Feedstock	Other Fuels: Fuel production energy	Other Fuels: Energy content of fuel	Other Fuels: Feedstock	Total System Energy Requirement	Notes (see page 376 *et seq.*)
WOOD BOARD from standing timber (MJ/board metre)										
H	*	*	*	*	*	*	*	*	*17.84*	(1), (55), (127), (490)
WOOD PULP from standing timber (MJ/kg)										
A	5.60	1.77	–	–	–	–	–	–	7.37	(5), (15), (488)
A	2.91	0.93	2.66	12.92	–	–	–	–	19.42	(2), (521), (523)
A	22.46	7.09	–	–	–	–	–	–	29.55	(2), (521), (522)
A	9.69	3.06	–	–	–	0.88	16.80	–	30.43	(5), (524), (525)
A	28.02	8.85	–	–	–	–	–	–	36.87	(5), (15), (489)
H	9.76	3.06	–	–	–	0.50	16.30	18.38	48.00	(2), (15), (500)
XYLENES (o & m) from crude oil (MJ/kg)										
H	2.50	0.79	4.19	20.47	51.83	–	–	–	79.78	(2), (15), (20), (491)
XYLENE (p) from crude oil (MJ/kg)										
H	12.83	4.05	7.27	35.51	52.75	–	–	–	111.91	(2), (15), (20), (492)
ZINC from ore (MJ/kg)										
J	*	*	*	*	*	*	*	*	*50.50*	(6), (22)
A	2.90	0.92	–	–	–	6.60	46.21	–	56.63	(1), (63), (494)
H	*	*	*	*	*	*	*	*	*58.28*	(1), (55), (127), (493)
D	*	*	*	*	*	*	*	*	*63.92*	(1), (36)
A	32.75	10.34	–	–	–	2.96	19.16	–	65.21	(1), (63), (495)
H	*	*	*	*	*	*	*	*	*69.67*	(1), (46)
H	*	*	*	*	*	*	*	*	*72.00*	(2), (50), (127)
A	14.99	4.73	0.47	2.34	–	3.76	48.10	–	74.39	(1), (704), (733)
J	*	*	*	*	*	*	*	*	*79.18*	(1), (42)
J	*	*	*	*	*	*	*	*	*80.00*	(2), (52)

System Type	Electricity: Fuel production energy	Electricity: Energy content of fuel	Oil Fuels: Fuel production energy	Oil Fuels: Energy content of fuel	Oil Fuels: Feedstock	Other Fuels: Fuel production energy	Other Fuels: Energy content of fuel	Other Fuels: Feedstock	Total System Energy Requirement	Notes (see page 376 *et seq.*)
A	45.17	14.26	0.20	1.02	–	3.02	23.86	–	87.53	(1), (704), (732)
A	55.04	17.38	0.22	1.12	–	0.89	6.10	–	80.75	(1), (704), (734)
ZINC from scrap (MJ/kg)										
H	*	*	*	*	*	*	*	*	*8.46*	(1), (43)
H	*	*	*	*	*	*	*	*	*9.00*	(2), (50), (127)
J	*	*	*	*	*	*	*	*	*10.57*	(1), (42)
ZINC from solution (MJ/kg)										
A	38.76	12.24	–	–	–	–	–	–	51.00	(1), (63), (496)
ZINC CASTINGS from ore in the ground (MJ/kg)										
A	*	*	*	*	*	*	*	*	*90.99*	(1), (43)
ZINC DIE-CASTINGS from ore in the ground (MJ/kg)										
D	*	*	*	*	*	*	*	*	*105.76*	(1), (36)
ZINC (ELECTROLYTIC) from sulphide concentrate (MJ/kg)										
A	42.02	13.27	–	–	–	–	–	–	55.29	(2), (66)
ZINC SHEET from ore (MJ/kg)										
H	*	*	*	*	*	*	*	*	*82.22*	(1), (43)
D	*	*	*	*	*	*	*	*	*96.00*	(1), (36)
ZINC SHEET from ingot (MJ/kg)										
D	*	*	*	*	*	*	*	*	*14.18*	(1), (36)

NOTES ON THE TABLE

General Notes

Values printed in Roman type (for example 214.39) are derived directly from the quoted reference when sufficient information is available to allow a complete breakdown of the energy requirement by fuel type and hence allow energies to be calculated in accordance with Table 6.13.

Values printed in italics (for example *214.39*) indicate that this value is quoted in the reference but the information given is insufficient to permit a complete breakdown by fuel type. Such values are not corrected to the fuel production efficiencies of Table 6.13 and hence cannot be compared with similar italicised values nor with entries printed in Roman type.

Values surrounded by brackets (for example (214.39)) represent suggested breakdown values. Such entries have usually been derived by analogy with more complete entries for the same system or by comparison with similar systems.

A hyphen (–) is used to denote a zero entry (that is 0.00). An entry 'neg' indicates a value less than 0.01.

An asterisk (*) denotes that a contribution may exist in a column but the information given in the reference is inadequate to allow its determination.

The horizontal completeness of a system (see Chapter 3) may be determined by reference to the input and output materials stated for the system.

The vertical completeness (see Chapter 3) of a systen may be determined by reference to the letter *A* to *J* of the system type. These letters refer as follows:

Letter	System
A	Main process only
B	Main process plus transport
C	Main process plus capital
D	Main process plus services
E	Main process plus transport plus capital
F	Main process plus transport plus services
G	Main process plus capital plus services
H	Main process plus transport plus capital plus services
J	Unspecified system

In general when there is some uncertainty about the system reported in the reference, the letter which most closely approximates is used and further details are given in the supplementary notes referred to by the entry.

In most instances the supplementary notes will give as much information concerning the basis of the calculation as is available in the reference together with any assumptions or omissions that have been made in the calculation of the values shown in the Table.

SUPPLEMENTARY NOTES REFERRED TO IN THE MAIN TABLE

(1) U.S. data.
(2) U.K. data.
(3) Canadian data.
(4) Swedish data.
(5) Finnish data.
(6) Dutch data.
(7) Swiss data.
(8) German data.
(9) Italian data.
(10) Mexican data.
(11) Belgian data.
(12) West African data.
(13) Japanese data.
(14) Indian data.
(15) Author's data.
(16) U.S. Environmental Protection Agency (1974) Report No. EPA/530/SW-91c, 'Resource and environmental process analysis of nine beverage container alternatives'.
(17) Based on the direct energy to convert acrylonitrile, styrene and polybutadiene monomers into ABS resin (0.454 kWh of

electricity and 5.25 cu.ft. of natural gas). The process requires an input of 0.761 kg of acrylonitrile, 0.127 kg of styrene and 0.12 kg of polybutadiene (all per kg of output ABS). Other materials inputs are 0.0005 kg of catalysts and 0.0115 kg of additives as well as 5.22 gallons of water (all per kg of output ABS). These ancilliary materials and services have not been included in the quoted energy requirement.

(18) Based on a direct electrical energy requirement of 0.082 kWh/kg to clean and grind scrap ABS bottles to produce a re-usable resin. The process requires 1.053 kg of scrap to produce 1 kg of output resin. Other materials used are 0.005 kg of cleaning compounds and 0.18 gallons of water per kg of output resin but the energy associated with these has not been included in the values quoted.

(19) Ayres, R., Saxton, J. and Stern, M., *Materials - Process - Product Model,* International Research and Technology Corporation, Arlington, Va. July 1974.

(20) Smith, H., 'Transactions of the World Energy Conference', 18, Section E (1969).

(21) Based on the direct electrical consumption of 0.154 kWh/kg to manufacture acrylonitrile from ammonia and propylene. The process requires 0.879 kg of propylene and 0.285 kg of ammonia per kg of output acrylonitrile and allowance has been made for usable by-products. In addition, the process requires 25.3 gallons of water, 0.0015 kg of catalyst, 0.0005 kg of oxalic acid and 0.0035 kg of sulphuric acid per kg of output but these ancilliary materials have not been included in the quoted energy requirements.

(22) Krejger, P. C., Proceeding of the 9th International TNO Conference, *The Energy Accounting of Materials, Products, Processes and Services,* p 141, February 1976, Rotterdam.

(23) No fuel breakdown is quoted but it is stated that capital energy has not been included. The fuel production efficiencies used are not known.

(24) Dite, R. V. and Rayner, C. J., 'Current practice and future trends in the manufacture of bituminous mixtures for roads', Soc. Chem. Ind., March 1975.

(25) Altenpohl, D., Daughterty, T. S. and Blum, W., Proceedings of the 9th International TNO Conference, February 1976, Rotterdam.

(26) Maddox, K. P., *Mineral Industries Bulletin,* **18**, Colarado School of Mines, July 1975.

(27) Value has been calculated from the requirements for 0.102 lb of alumina of 103 Btu of coal, 85 Btu of oil fuels, 180 Btu of natural gas and 248 Btu of electricity by converting from the following fuel production efficiencies: coal 93.8%, oil fuels 89.3%, natural gas 75.5% and electricity 26.7%.

(28) Sheldon, A, C., in *Energy use and conservation in the metals industry,* Met. Soc. AIME, p. 157 (1975).

(29) Rotary kiln for calcining aluminium hydroxide coming from the Bayer process. Energy requirement is 4.64 MJ/kg Al_2O_3, assumed to be natural gas.

(30) Flash calciner for converting aluminium hydroxide from the Bayer process. Energy requirement 3.25 MJ/kg Al_2O_3, assumed to be natural gas.

(31) Bravard, J. C., Flora, H. B. and Portal, C., 'Energy expenditures associated with the production and recycle of metals', Oak Ridge National Laboratory Report, ORNL. NSF. EP24 (1972).

(32) Based on data for the extraction of aluminium from anorthosite, a feldspar ore of composition $(CaO.Al_2O_3.SiO_2)_X.(Na_2O.Al_2O_3.6SiO_2)_Y$. The energy quoted is based on the stated direct energy requirements per ton of alumina of 770 kWh electricity, 26.11 Btu gas (assumed natural gas) and 13.35×10^6 Btu steam. Steam has been converted at the rate of 3.50 MJ/kg (*q.v.*). In addition to these fuel requirements the process also uses limestone and soda ash, but the energy associated with the provision of these inputs has not been included.

(33) Based on the stated direct energy requirements per ton of alumina of 102 kWh electricity, 24.28×10^6 Btu natural gas and 16.97×10^6 Btu steam. Steam has been converted at the rate of 3.50 MJ/kg (*q.v.*). The process also requires a supply of nitric acid but the energy associated with this has not been included.

(34) Based on the total energy for the production of alumina from 50% bauxite ore in the ground. Calculation of energy based on a

mining, crushing and grinding requirement of 250 kWh/ton of Al_2O_3; digestion of alumina using steam raised by gas of 15.7×10^6 Btu/ton Al_2O_3; clarification and filtering of 25 kWh electricity, calcination of the trihydrate by gas firing of 3.88×10^6 Btu/ton Al_2O_3.

(35) Calculation based on the direct energy requirements for the production of alumina from bauxite by the Bayer Process (that is 0.14 kg coal, 0.017 gall oil, 5.94 cu.ft. natural gas and 0.77 kWh, all per kg of output). The process requires 1.523 kg bauxite per kg alumina plus 0.07 kg of additives and 0.44 gallons of water. The energy associated with the production of ancilliary materials is not included in the quoted values.

(36) Samples, D. K., Seminar at the Institute of Sciences and Technology, University of Michigan, 28 August 1974, 'Energy in the automobile'.

(37) Bayer process based on the production of 1.88 lb of alumina from 4.47 lb of bauxite for which the energy requirement is 5500 Btu.

(38) Chapman, P. F., Open University Energy Research Group Report ERG 001 (1973), 'Energy costs of producing copper and aluminium from primary sources'.

(39) The Bayer process. Based on a requirement per ton of alumina of 3.5 ton of steam, 240 kWh of electricity and 110 kg of fuel oil (assumed medium fuel oil). Steam has been converted at the rate of 4.49 MJ/kg (*q.v.*). The ancilliary requirement of 120 kg of NaOH per ton alumina has not been included in the calculation.

(40) Pick, H. J. and Becker, P. E., *Applied Energy,* **1**, 31 (1975).

(41) MacKillop, A., *Ecologist,* **2**, No. 12, 1, (1972).

(42) Tien, J, K., Clark, R. W. and Malu, M. K., *Technology Review,* p. 39, February 1975.

(43) Berry, R. S. and Fels, M. F., *Science and Public Affairs,* p. 11, December 1973.

(44) Berry, R. S. and Makino, H., *Technology Review,* p. 33, February 1974.

(45) Sundstrom, G. Investigation of the energy requirements from raw materials to garbage treatment for four Swedish beer packaging alternatives. Report for Rigello Pak AB (1973).

(46) Agarwal, J. C. and Sinek, J. R., in 'Energy use and conservation in the metals industry', Met. Soc. AIME, p. 121 (1975).

(47) Mangalick, M. C., in 'Energy use and conservation in the metals industry', Met. Soc. AIME, p. 101 (1975).

(48) 'The competitiveness of LDPE, PP and PVC after the 1973 oil crisis', Imperial Chemical Industries, London (1974).

(49) Long II, T. V., Proceedings of the 9th International TNO Conference (1976) Rotterdam.

(50) Chapman, P. F., *Energy Policy,* p. 47, March 1975.

(51) Roberts, F., *Metals and Materials,* p. 167, March 1974.

(52) Hinde, P. T. and Probert, S. D., *Applied Energy,* **2**, p. 17 (1976).

(53) Odok, A., Thym, G. and Wurgler, R., 'The potential for energy conservation by strip casting of aluminium', 4th International BNF Conference on Energy and Non-ferrous Metals, Barcelona (1975).

(54) Schipper, L. and Lichtenberg, A. J., *Science,* **194**, 1001 (1976).

(55) Makhijani, A. B. and Lichtenberg, A. J., *Environment,* **14**, No. 5, p. 10 (1972).

(56) Leach, G. and Slesser, M., 'Energy equivalent of network inputs in food production', University of Strathclyde (1973).

(57) Kellogg, A. J., *J. Metals,* p. 25, June 1974.

(58) Hannon, B. M., 'System energy and recycling', ASME (1973).

(59) Gartner, E. M. and Smith, M. A., *Energy Policy,* p. 144, June 1976.

(60) Thornton, D. S. and Williams, D. I. T., *Ironmaking and Steelmaking,* p. 241, No. 4 (1975).

(61) Sheridan, A. T., *Ironmaking and Steelmaking,* p. 262, No. 4 (1975).

(62) Smart, J. D., *Ironmaking and Steelmaking,* p. 320, No. 4 (1975).

(63) PoramReswaren, K. and Nadkarni, R., in *Energy use and conservation in the metals industry,* Met. Soc. AIME, p. 271 (1975).

(64) Laurila, A. and Andersson, B., 'Modern flash smelting technology', 4th International BNF Conference on Energy and Non-Ferrous Metals, Barcelona (1976).

(65) Owen, M. and Jacobi, J. S., *Journal of Metals,* p. 10, April

1975.
(66) Gray, P. M. J., 'Conservation in primary metal extraction', 4th International BNF Conference on Energy and Non-Ferrous Metals, Barcelona (1976).
(67) Gregory, P. and Day, P., 'Energy consumption and waste products in the conversion of high electrical conductivity copper wire bar to wire', 4th International BNF Conference on Energy and Non-Ferrous Metals, Barcelona (1976).
(68) Barnes, R. S., *Proc. Inst. Mech. Eng. London,* **190**, 153 (1976).
(69) Brown, J. W. and Riddy, R. L., *Iron and Steel Engineer,* p. 37, June 1976.
(70) Schroer, C. A. and Clark, D. W., *Iron and Steel Engineer,* p. 37, June 1976.
(71) Sigworth, G. K., in *Energy use and conservation in the metals industry,* Met. Soc. AIME, p. 19 (1975).
(72) Larwood, G. M. and Leary, R. J., in *Energy use and conservation in the metals industry,* Met. Soc. AIME, p. 259 (1975).
(73) Ban, T. E., in *Energy use and conservation in the metals industry,* Met. Soc. AIME, p. 173 (1975).
(74) Barnes, R. S., *Ironmaking and Steelmaking,* p. 271, No. 4 (1975).
(75) Wild, R., *Chemical and Process Engineering,* p. 55, February 1969.
(76) Anon., *Ironmaking and Steelmaking,* p. 327, No. 4 (1975).
(77) Pitt, R. S., *Ironmaking and Steelmaking,* p. 242, No. 4 (1975).
(78) Elliot, J. F., in *Energy use and conservation in the metals industry,* Met. Soc. AIME, p. 4 (1975).
(79) Koenig, H., *Canadian Mining and Met. Bulletin,* p. 174, February 1970.
(80) Colautti, F. and Barbi, A., *Chemical Engineering,* April 25, (1977).
(81) Clay, H., *Proc. Inst. Mech. Eng. London,* **184**, 1279 (1970).
(82) Miller, J. R., *Steel Furnace Monthly,* 169, March 1973.
(83) Sharp, J. D., *Ironmaking and Steelmaking,* p. 320, No. 4 (1975).
(84) Peklke, R. D., *Unit processes of extractive metallurgy,* Elsevier, 1975.
(85) Ydstie, B., in *Energy use and conservation in the metals industry,* Met. Soc. AIME, p. 131 (1975).
(86) Smithson, D. J. and Sheridan, A. T., *Ironmaking and Steelmaking,* p. 286, No. 4 (1975).
(87) Anon., *Ironmaking and Steelmaking,* p. 279–280, No. 4 (1975).
(88) Jacobi, J. S., 'The secondary metals industry as a conservation agent', 4th International BNF Conference on Energy and Non-Ferrous Metals, Barcelona (1975).
(89) *Pulp and Paper International,* Jan. 1974, pp 31, 32 and 36.
(90) Kakela, P., *Environment,* **17**, No. 2, p. 27 (1975).
(91) Montgomery, W., *Ironmaking and Steelmaking,* p. 231, No. 4 (1975).
(92) Kellogg, H. H., in *Energy use and conservation in the metals industry,* Met. Soc. AIME, p. 87 (1975).
(93) Russell, C. S. and Vaughan, W. J., *Steel production – processes, products and residuals,* Johns Hopkins University Press)1976).
(94) Rodriguez, F. A., *Iron and Steel Engineer,* p. 57, January 1977.
(95) Barraclough, K. C., 'The specialist steelmaking processes', *Metals and Materials,* p. 192, March 1974.
(96) Stone, J. K., *Iron and Steel Engineer,* p. 38, March 1976.
(97) Montgomery, W., *Ironmaking and Steelmaking,* **2**, p. 231, No. 4 (1975).
(98) Decker, A., 'Energy accounting and steel', Proceedings of the 9th International TNO Conference, Rotterdam, 1976.
(99) Barnhart, T. F., in *Energy use and conservation in the metals industry,* Met. Soc. AIME, p. 41 (1975).
(100) Cleathero, B., *Ironmaking and Steelmaking,* p. 313, No. 4 (1975).
(101) Langman, R. D. and Harris, T. H., *Ironmaking and Steelmaking,* p. 253, No. 4 (1975).
(102) Smithson, D. J. and Sheridan, A. T., *Ironmaking and Steelmaking,* p. 286, No. 4 (1975).
(103) Hopkins, N. and Gandhi, K. C., *Ironmaking and Steelmaking,* p. 295, No. 4 (1975).
(104) Harrison, G., *Ironmaking and Steelmaking,* p. 299, No. 4 (1975).
(105) Anon., *Ironmaking and Steelmaking,* p. 299, No. 4 (1975).
(106) Shreve, R. N., *Chemical Process Industries,* McGraw Hill (1967).

(107) Wright, P. A., *Extractive metallurgy of tin,* Elsevier, London (1966).

(108) No detailed fuel breakdown is available.

(109) Energy for raising steam at 250 psig in a coal-fired boiler with an efficiency of 80%. The available energy in the steam is 2.67 MJ/kg. No allowance has been made for the provision of water.

(110) Energy for raising steam of 300 psig in a coal-fired boiler with an 80% efficiency. The available energy in the steam is 1.8 MJ/kg. No allowance has been made for the provision of water.

(111) Boiler with an efficiency of 85% assumed to be oil-fired. No allowance has been made for the provision of water.

(112) Fuel type is unspecified in the reference so it has been assumed that an oil-fired boiler is used. No allowance has been made for the supply of water.

(113) Calculation based on quoted inputs of 1.61×10^6 Btu/1000 lb steam (assumed oil fuels) and 120 gallons (U.S.) of water per 1000 lb steam. No allowance has been made for the provision of water.

(114) Based on a fuel requirement of 60 l oil (assumed medium fuel oil) per 1000 kg steam. Data relates to a steam raising plant in a Swedish polyethylene plant.

(115) Calculation based on practice with U.K. petrochemical plants. The authors' values have been converted to the efficiencies used in this book.

(116) Calculated from a quoted value of 361 kcal/lb steam with an oil production efficiency of 82.7%. No allowance has been made for the provision of water.

(117) Values are for methane partial oxidation. Inputs are 3598 lb methane and 1941 lb of oxygen to produce 497 lb of acetylene and 5533 lb of fuel gas. To treat these quantities, the direct fuel inputs are 139×10^6 Btu of oil fuel, 5176 kWh of electricity and 176,000 lb of steam. Steam has been converted using an energy requirement of 4.49 MJ/kg (*q.v.*) and the input fuels have been apportioned on a mass basis over the two outputs. Feedstock has been included as shown by Table 7.7.

(118) Based on a total direct energy requirement of 40 MJ for light-weight aggregate. It is assumed that this is all oil fuels.

(119) Based on the partial oxidation of naphtha for which an input of 722 lb of naphtha and 2504.99 lb of fuel oil will yield 495.7 lb of acetylene, 188.6 lb of C_3/C_4 fraction, 466 lb of ethylene and 4733 lb of fuel gas. The direct fuel requirements to treat these materials are 2134 kWh of electricity, 54×10^6 Btu of oil fuel and 228,000 lb of steam. Fuel requirements have been partitioned over all outputs on a mass basis and steam has been converted at the rate of 4.49 MJ/kg (*q.v.*). No feedstock is included.

(120) Essentially the same calculations as in Note 119 above, but feedstock has been included as in Table 7.6 by assuming that fuel gas and naphtha have equal feedstock values.

(121) Wulff process data in which in input of 643 lb of naphtha is converted to 510 lb acetylene, 447 lb of ethylene and 1507 lb of fuel gas for the direct consumption of 2660 kWh of electricity, 349×10^6 Btu of oil fuels and 216,000 lb of steam. All inputs have been apportioned on a mass basis over all outputs and steam has been converted at the rate of 4.49 MJ/kg (*q.v.*). No feedstock is included.

(122) Essentially the same calculation as Note 121 above, but feedstock values based on 0.261 kg naphtha per kg acetylene using Table 7.6 have been included.

(123) Based on the partial oxidation of naphtha for which a total cumulative energy requirement excluding feedstock is given as 16,530 kcal/kg, of which 870 kcal/kg are electricity calculated with an efficiency of 26.5%. The remainder is oil with an efficiency of 87.5%. Values have been converted to the final production energies of Table 6.13 and feedstock has been included.

(124) Based on the route via propylene and ammonia.

(125) The references give no details of the fuel use or of the efficiencies associated with these fuels. Unlikely however that any fuel efficiency has been used for the stated requirement of 0.095 MJ/kg.

(126) A direct total energy requirement of 16.2 MJ/kg alumina with no fuel breakdown given. The process also requires 0.17 kg of

sodium hydroxide per kg alumina, but this has not been added into the calculation.

(127) Value based on statistical information.

(128) Based on the direct fuel required to transport, shred and melt aluminium scrap to produce recycled aluminium. These requirements per kg of aluminium are 0.01 gall of oil, 5.12 cu.ft. of natural gas and 0.396 kWh of electricity, of which 0.275 kWh is for shredding. This value may also be used as a representative average.

(129) No detailed fuel breakdown is given in this reference but electricity provision is stated as 32.5% efficient. Stated requirement is 280×10^6 Btu/ton.

(130) This value is based on the melting of secondary aluminium in a reverberatory furnace with a recirculation pump which improves heat transfer in the bath and with the metal charge preheated in the flue gas outlet.

(131) As in Note 130 but with no scrap preheating.

(132) Based on the energy to run a Hall cell-room and associated electrode production facility. The process also requires 31.9 gallons of water per kilogramme of aluminium but this has not been included in the calculation.

(133) Electricity to run Hall cells only.

(134) Based on the total energy to produce aluminium by the electrolysis of alumina in the Hall cell. Includes electrode consumption based on 0.5 kg electrode/kg Al.

(135) Based on total energy to produce aluminium by the Bayer process and Hall cell starting with a 50% bauxite ore in the ground.

(136) As in Note 135 but starting with a 30% ore.

(137) Electricity to Hall cell only.

(138) Gas supply to reverberatory furnace used in scrap melting.

(139) Aluminium production by the Alcoa chloride process.

(140) Electric melting furnace.

(141) Average value for washed aggregate suitable for most building purposes.

(142) Data for a smelting plant based on a consumption of 7 Btu of oil fuel, 1165 Btu of natural gas and 3958 Btu of electricity for the production of 0.052 lb of product, and converting from energy production efficiencies for the fuels of 89.3% for oil fuels, 85.5% for natural gas and 76.7% for electricity.

(143) Does not include bauxite processing.

(144) Based on the direct fuel requirement to manufacture aluminium beverage can ends from sheet aluminium, that is 1.1 cu.ft. of natural gas, 0.094 kWh of electricity (both per kg of can ends).

(145) Thought to be representative average values for hot rolling 300 to 500 mm aluminium billets on a reversing stand. The exact starting and finishing sizes are not given.

(146) Based on a quoted value of 0.96 kWh direct electrical consumption by the rolling mill.

(147) Based on the direct energy to produce aluminium sheet from ingots by rolling (7.72 cu.ft. of natural gas/kg and 0.43 kWh of electricity/kg). The process also requires 12 gallons of water per kg aluminium processed but the energy associated with the provision of this has not been included.

(148) Based on a production of 0.051 lb of product for which the total fuel consumption is 25 Btu of oil fuels, 80 Btu of natural gas and 66 Btu of electricity, measured at fuel production efficiencies of 89.3% for oil fuels, 85.5% for natural gas and 26.7% for electricity.

(149) Based on a total fuel consumption for the production of 0.059 lb of output of 523 Btu of natural gas and 306 Btu of electricity at energy efficiencies of production of 85.5% and 26.7% respectively.

(150) Value represents electric power to continuous-casting machine for aluminium strip. No water consumption is quoted and so has been excluded from the calculation.

(151) Thought to be typical values for the oil route to ammonia. No fuel breakdown is available.

(152) Based on steam reformation of natural gas for which the stated requirements per kg of ammonia are 0.0242 kWh electricity, 8.17 cu.ft. of natural gas and 2.05 kg of steam. We have converted 1 cu.ft. of natural gas as 1.054 MJ and steam raising has been converted at the rate of 4.49 MJ/kg (*q.v.*). Feedstock has been

included in accordance with Table 7.7.

(153) Based on direct process requirements of 60 kWh electricity/ton ammonia and 19,200 cu.ft. of natural gas per ton ammonia. Feedstock has been calculated in accordance with Table 7.7 assuming that all natural gas input is feedstock. However, if the values are compared with Note 152 it seems likely that a proportion of the natural gas requirement in this instance will be used in steam raising, hence the need to bracket three of the terms.

(154) Based on a total cumulative energy requirement for steam reforming of naphtha of 4540 kcal/kg, of which 10 kcal/kg represents electricity measured at an efficiency of 26.5%. The remainder has been assumed to be oil calculated at an efficiency of 87.5%. Feedstock is excluded from these values but has been included as an average of a number of published and private values. The values shown in the Table have been converted to the fuel production efficiencies of Table 6.13.

(155) Based on the difference between the energy of formation of gas and solution of 232 kcal/lb. We have assumed this to be supplied by natural gas.

(156) Based on the steam reformation of natural gas for which the feedstock will be 10,500 Btu/lb ammonia and the natural gas needed for fuel will be 4000 Btu/lb ammonia. Electricity is based on the use of 850–950 hp compressors.

(157) Based on what is thought to be a representative 1966 plant for which the requirements to produce 2000 lb of ammonia solution are 21,900 cu.ft. of natural gas as feedstock, 12,400 cu.ft. of natural gas as fuel and 780 kWh of electricity. The process also requires 79,000 imp. gallons of water, but the energy associated with provision of this service has not been included. Feedstock has been calculated as in Table 7.7.

(158) Based on a 1970 plant design incorporating centrifugal steam turbines consuming 850 MBtu as feedstock and 575 MBtu as fuel (both natural gas) for the production of 1 lb of ammonia as solution. Feedstock has been calculated as in Table 7.7.

(159) The energy required to produce paper and board from timber varies considerably depending upon whether or not the pulp and paper plants are integrated. For integrated plants, waste products from the pulping process may be used as fuels in the paper or board-making process and this reduces the dependence of the plant on alternative fuels. Such plants predominate in Canada and Scandinavia. The values quoted here assume a non-integrated plant.

(160) Based on average U.S. mining data per kg of product of 0.00036 gallons of oil fuel, 0.44 cu.ft. of natural gas, and 0.00773 kWh of electricity. The process requires on average 0.014 gallons of water but the energy associated with this has not been included in the calculation.

(161) Based on mining 4.47 lb of bauxite at 1205 Btu.

(162) Based on the distillation of crude oil, catalytic reforming and separation of aromatics. The process requires 1030 lb of crude oil per 1000 lb of benzene as well as 49 kWh of electricity and 6000 cu.ft. of natural gas. The calculation excludes the delivery of crude oil. The process also requires 5 lb of catalyst and 4200 gallons of water per 1000 lb of benzene but the energy associated with these has been excluded from the calculation.

(163) Based on the catalytic reforming of naphtha for which the total cumulative energy requirement excluding feedstock is given as 6280 kcal/kg, of which 690 kcal/kg are electricity calculated with an efficiency of production of 27.5%. The remainder is oil fuel calculated with an efficiency of 87.5%. Values have been corrected to the fuel production efficiencies of Table 6.13 and feedstock has been included.

(164) Based on a quoted energy breakdown of an electrical consumption of 0.1 kWh/kg and a non-electrical consumption of 1.5 kWh/kg. Non-electrical energy assumed to be natural gas.

(165) Based on the consumption of 0.13 kWh electricity and 5626 Btu of oil fuels to produce 1 lb of formed glass. The value given in the Table assumes all raw materials at the glass factory. The raw materials requirements per lb of glass are stated as soda ash (0.2758 lb), limestone (0.2482 lb), sand (0.5408 lb), calcium silicate (0.0179 lb), potassium aluminate (0.0322 lb) and cullet (0.102 lb) as well as 0.71 gallons of water for cooling. The

production energy of these material inputs are not included in the quoted value.

(166) Thought to be an average for all containers. No fuel breakdown is given but electrical efficiency of 29.4% is assumed by the reference.

(167) Based on a typical grinding machine which processes 400 kg of waste plastic for an electrical consumption of 40 kWh.

(168) Compare this with Note 165 where the main source of melting energy was oil – here it is natural gas.

(169) Based on an average direct energy to manufacture glass containers from raw materials (0.00616 gallons oil fuel, 11.77 cu.ft. natural gas, 0.289 kWh of electricity, all per kg of product). the process requires 0.66 kg of sand, 0.167 kg of limestone, 0.077 kg soda ash, 0.075 kg feldspar, 0.02 kg of unspecified minor ingredients and 0.8 gallons of cooling water. The energy associated with these materials inputs have not been added into the value in the table.

(170) Based on the direct energy to manufacture corrugated cardboard boxes from wood-fibre (30.39 MJ natural gas, 0.0124 gallons diesel, and 0.352 kWh electricity, all per kg of product). Process requires 0.996 kg of wood fibre plus 0.07 kg of additives and 26.95 gallons of water per kg of output. Feedstock energy not included because the first processing stage (that is wood fibre) is excluded.

(171) Based on a stated requirement of 10,300 kWh/ton. No detailed fuel breakdown given but the electrical efficiency used is 32.8%.

(172) Based on the direct energy to manufacture bleached kraft paperboard cartons from wood fibre (33.87 MJ natural gas, 0.013 gallons diesel and 0.352 kWh electricity – all per kg of output). The process requires 1.02 kg of wood fibre and 47.3 gallons of water per kg of output but the energy associated with the provision of these has not been included in the calculation. No feedstock energy is shown because the initial conversion from wood to fibre is excluded.

(173) Based on direct manufacturing energy of 1.3 butadiene by dehydrogenation of n-butenes (0.48 kWh electricity and 5.90 cu.ft. of natural gas per kg of product). The process also requires 8.33 gallons of water per kg of butadiene but energy associated with water has been excluded from the calculation.

(174) Based on the direct energy to manufacture 1.3 butadiene from n-butanes by dehydrogenation (0.72 kWh electricity and 19.04 cu.ft. natural gas per kg of product). The process requires 1.901 kg of n-butanes to produce 1 kg product and 8.33 gallons of water.

(175) Based on the direct energy to manufacture 1.3 butadiene as a by-product from naphtha cracking to produce olefine (0.15 kWh electricity and 4.49 cu.ft. of natural gas per kg of product). The process also requires 8.33 gallons of water but the energy associated with water provision has been excluded from the calculation.

(176) Based on the route via extraction from C_4 olefin stream in cracking naphtha, for which a total cumulative energy requirement of 10,680 kcal/kg is given. Of this 850 kcal/kg is electricity calculated at an efficiency of 26.5%. The remainder is oil at an efficiency of 87.5%. All values have been corrected to the production efficiencies of Table 6.13 and feedstock has been included.

(177) From para-xylene and ethylene glycol via dimethyl terephthalate, for which the total cumulative energy excluding feedstock is given as 39,020 kcal/kg. Of this 16,340 kcal/kg is electricity calculated with an efficiency of 26.5%. The remainder is oil at an efficiency of 87.5%. Values have been corrected to the efficiencies shown in Table 6.13 and feedstock has been included.

(178) Based on requirements of 1730 lb benzene and 640 lb ethylene per ton styrene (from Note 106). Energies evaluated as ethylene 68.12 MJ/kg (*q.v.*), benzene as 79.78 MJ/kg (*q.v.*), and the production of styrene from these reactants as 13.44 MJ/kg (*q.v.*).

(179) Calcination of limestone requires 0.09 kg coal, 0.00091 gallons oil fuels, 1.837 cu.ft. natural gas and 0.0308 kWh electricity, all per kg of output. Process also requires 0.25 gallons of water per kg of calcium oxide. System starts with limestone at the calcining plant.

(180) System starts with limestone at kiln and is essentially the natural gas used to fire the kiln. No electricity is listed.
(181) Calcination of limestone at calciner. Based on 12 Btu of oil, (89.3% production efficiency), 40 Btu of natural gas (85.5%) and 5 Btu of electricity (26.7%) all per 0.028 lb of calcium oxide.
(182) Quarrying, crushing and drying of natural gypsum.
(183) Calumite Brand Slag is a registered trade mark of the Calumite Corporation of America and this preparation is used to assist melting in glass furnaces. It is made by treating selected slags using a patented process.
(184) Detailed fuel breakdown not available, but the calculation is based on an electricity efficiency of 30% and an oil efficiency of 87.5%.
(185) Value is quoted in the reference as 10 kWh_e + 10 kWh_t per kg of anode. Using the electrical energy efficiency of 24% in Table 6.13, the electrical component is converted to a gross value of 41.67 kWh_t. The non-electric component is probably the fuel value of the carbon in the electrode.
(186) A detailed fuel breakdown is not available but 50.4 MJ represents the feedstock value of the carbon. Fuel production efficiencies are not known. The stated production energy is 25.2 MJ/kg.
(187) Fuel production efficiencies are not known nor is the extent of the system. The stated fuel energy requirement is 22.90 MJ/kg.
(188) Based on a direct consumption of 6000 kWh electricity/ton of electrode and 3.4×10^6 Btu coal per ton electrode. System starts with raw materials at the electrode-making plant.
(189) Based on the total energy for the production of carbon electrodes from raw materials in the ground. No fuel breakdown is given and efficiencies for fuel production are not known.
(190) Based on a quoted value of 1520 Btu per 0.6 lb carbon electrode. This energy will probably be almost all thermal fuel for baking the electrode with a small electrical requirement for compacting the coke.
(191) Carbon dioxide is a by-product of a number of processes of which ammonia plant is one of the most significant. All of the energy is usually 'charged' to the ammonia because the carbon dioxide is not always collected. The energy therefore represents that needed for compression and liquefaction of the gas.
(192) Based on a heat of sublimation for carbon dioxide of 70 kcal/lb.
(193) Stated as the energy of all processing operations but excluding feedstock. Calculation based on an electrical efficiency of 30% and an oil efficiency of 87.5%, but further detail is not given.
(194) Based on direct energy consumption per tonne of cement of 367 MJ electricity, 45.6 MJ coal, 343 MJ oil and 857 MJ natural gas. The system is not specified but it is known that capital energy is not included.
(195) Quoted value of 2.2 kWh_t derived from statistical sources. Detailed fuel breakdown is not known but an electrical efficiency of 23.8% has been used in the calculation.
(196) Based on energies per 2000 lb cement of 2200 kWh for production, 50 kWh for capital and 50 kWh for transport. An electrical efficiency of 30% is used, although the detailed fuel breakdown is not known.
(197) Based on the manufacture of h.c. cement by the dry route, for which the quoted energy requirements per kg of cement are 0.06 MJ for mining and transport, 1.41 MJ for kiln fuel, 0.77 MJ as electricity and 0.14 MJ for transport to final consumer. We have assumed that kiln firing is with natural gas and that mining and transport are all oil fuel powered.
(198) Manufacture of h.c. cement by the wet route. Quoted energy requirements are as in Note 197 except that kiln firing requires 2.32 MJ/kg cement.
(199) Manufacture of Portland cement by the dry route. Quoted fuel requirements per kg of finished cement are 0.05 MJ for mining and transport (assumed oil), 3.60 MJ for kiln firing (assumed natural gas), 1.11 MJ as electricity and 0.15 MJ for delivery to final consumer (assumed oil). The 1.11 MJ for electricity includes allowance for electricity production efficiency of 31.5% so that the electrical energy is 0.35 MJ/kg. In the entry this has been converted to authors' efficiency.
(200) Manufacture of Portland cement by the wet route. Fuel requirements per kg of finished cement are 0.03 MJ for mining and

transport (assumed oil), 5.93 for kiln firing (assumed natural gas), 1.11 MJ as electrical energy and 0.15 MJ for final delivery (assumed oil). Electricity has been converted at 31.5% efficiency in original reference and amended to authors' efficiency in this reference.

(201) Based on the conversion of HCl gas to chlorine such that 1.02921 lb of HCl yields 1 lb of chlorine and 0.02828 lb of hydrogen, for an electrical consumption of 0.80946 kWh. Electricity has been apportioned on a mass basis over both products. The process also requires 0.13751 lb of water for the reaction and 4.87807 gallons (U.S.) of cooling water. These are not included in the quoted energy value.

(202) Manufacture by mercury cells with brine recirculation. The values are based on a quoted total energy consumption of 27.15 MJ/kg of which 93% is stated to be electricity at an efficiency of 26.5%. The remainder has been assumed oil at an efficiency of 87.5%.

(203) All fuel consumption is assumed to be electricity. Value calculated from a quoted value of 2000 kcal/m^3 with an electrical efficiency of 26.5%.

(204) All fuel consumption is assumed to be electricity. Values calculated on the basis of 2250 kcal/imperial gallon with an electrical efficiency of 27.5%.

(205) Process route not stated but the calculation is based on a production of 1 lb of chlorine from 1.0285 lb of hydrogen chloride and 0.25495 lb of oxygen as air. The requirements are electricity (0.04945 kWh/lb), steam (1.31321 lb/lb), cooling water (11.84774 U.S. gallons) and process water of 5.79076 lb. Cooling water is assumed to be recirculated. Steam has been converted at the rate of 4.49 MJ/kg (*q.v.*) and process water at the rate of 9.25 MJ/1000 l (*q.v.*).

(206) Assumed derived from the liquefaction of air and cryogenic separation.

(207) Based on the production of 0.11 lb of oxygen for an energy requirement of 1 Btu of oil (efficiency 89.3%), 10 Btu natural gas (85.5%) and 30 Btu electricity (26.7%).

(208) Based on direct fuel requirements of 0.0012 gallons oil fuels, 1.68 cu.ft. of natural gas and 0.4565 kWh, all per kg oxygen. The process also requires 5.13 gallons of water per kg of oxygen.

(209) Process design not given but the detailed requirements per lb of chlorine are 0.07598 kWh of electricity, 3.9659 lb of steam, 0.24374 lb of pure oxygen and 1.04508 lb of HCl. In addition 6.19394 gallons of cooling water are required but this has been assumed to be recirculated. Steam has been converted at the rate of 4.49 MJ/kg (*q.v.*) and oxygen at the rate of 9.33 MJ/kg (*q.v.*).

(210) Electrolysis of brine for which an input of 6.20084 lb of brine yields 1 lb of chlorine and 0.02816 lb of hydrogen. Process requirements are 0.0179 kWh of electricity and 2.05009 lb of steam. Inputs have been apportioned over both outputs on a mass basis and steam has been converted at the rate of 4.49 MJ/kg (*q.v.*).

(211) Based on the reaction between ammonium chloride, oxygen and hydrogen. Inputs are 0.10253 kWh electricity, 11.04121 MBtu (assumed oil), 0.57526 lb of steam, 14.28863 U.S. gallons cooling water (assumed recirculating), 0.54646 lb reaction water, 0.43051 lb oxygen as air, 0.03752 lb hydrogen and 1.55370 lb of NH_4Cl. The corresponding outputs are 1 lb chlorine, 0.47793 lb of ammonia and 0.11227 lb of carbon dioxide. Inputs have been apportioned over all outputs on a mass basis and steam has been converted at the rate of 4.49 MJ/kg (*q.v.*).

(212) Electrolysis of brine. System starts with brine in the ground. Electrical efficiency of 33% used.

(213) Electrolysis of saturated sodium chloride solution starting with solid sodium chloride. Inputs 1.46032 kWh electricity, 6.20083 lb steam, 36.6897 lb of reaction water, 1.70538 lb of NaCl. Outputs 1 lb of chlorine, 0.02909 lb of hydrogen and 1.12616 lb of sodium hydroxide. Steam has been converted at the rate of 4.49 MJ/kg (*q.v.*) and water at 0.046 MJ/gallon (*q.v.*). Cooling water at the rate of 23.40306 gallons is also needed but this has been assumed to be recirculated. All inputs apportioned over all outputs on mass basis.

(214) Mercury cell electrolysis with inputs of 1.65028 kWh electricity, 1.02041 lb steam, 3.70037 gallons cooling water, 4.49072 lb of reaction water and 1.67347 lb of sodium chloride to give outputs of 1 lb of chlorine, 0.02825 lb of hydrogen and 0.95993 lb of sodium hydroxide. Steam has been converted at the rate of 4.49 MJ/kg (*q.v.*) and water at the rate of 9.25 MJ/1000 l. Cooling water is assumed to be recirculated and all inputs are apportioned over the three outputs on a mass basis.

(215) Operation of diaphragm cell only.

(216) Based on a requirement that 1500 m^3/hour of compressed air requires a 165 kW motor.

(217) Primary smelting and refining. Presumably includes both fire refining and electrolytic refining, although this is not stated in the reference.

(218) Includes mining and processing but does not state an ore grade. Probably an average value for finished copper products.

(219) Secondary copper production by an unspecified route requireing 14.7×10^6 Btu/ton. This will be a combination of oil or natural gas for melting and fire refining and electricity for electro-refining.

(220) Smelting of calcined copper concentrate of unspecified copper content in an electric furnace. Excludes capital energy and electrode consumption.

(221) Smelting of calcined copper concentrate of unspecified copper content in a gas-fired reverberatory furnace.

(222) Smelting of green copper concentrate of unspecified copper content in a gas-fired reverberatory furnace. The furnace charge is taken straight from the flotation cells and therefore contains water.

(223) Fire refining of blister copper in a gas-fired reverberatory furnace. Based on a natural gas consumption of 1.008 kWh_t/kg copper.

(224) High intensity electrolytic refining of impure copper anodes for which the requirements per ton of refined copper are 25 lb steam and 500 kWh of electricity.

(225) Based on the following requirements per ton of copper produced. Mining with a stripping ratio of 2.3:1 demands 6×10^6 Btu of diesel plus 550 kWh of electricity. Concentrating based on 200 tons of rock per ton of copper requires 2.75×10^6 Btu of natural gas plus 2625 kWh of electricity. Smelting consumes 21×10^6 Btu of natural gas plus 300 kWh of electricity and refining needs 0.31×10^6 Btu of natural gas plus 720 kWh of electricity.

(226) Based on electrolysis energy of 280 kWh/ton Cu directly consumed plus a gross capital energy requirement of 30 kWh/ton Cu. The electrical consumption has been converted to a gross value using an efficiency of 24%. Table entry has changed electrical consumption to authors' efficiency and assumes that electrical component of capital energy requirement is 35%.

(227) Based on the total energy to produce electrolytic copper from impure scrap by fire refining and electrolytic refining. 1.037 kWh natural gas/kg and 0.197 kWh electricity/kg. It includes coke limestone and soda ash used in fire refining.

(228) Based on the total energy to produce electrolytic copper from 0.3% ore. Includes the energy of flotation agent and silicate and limestone consumption. Fuel requirements 9684 kWh of gas or oil per ton of copper and 14,622 kWh of electricity per ton of copper.

(229) Based on total energy to produce electrolytic copper from 1% ore. Includes the energy of the flotation agent, and silicate and limestone consumption. Fuel requirements are 4572 kWh of electricity and 8545 kWh of gas, both per ton of copper produced.

(230) Electrical energy to produce copper from ore by the hydrometallurgical route. Treatment up to electrolyte production consumes approximately 50% of the total.

(231) Fuel consumption of 1000 kWh electricity/ton copper and 300 kg oil/ton copper used in the production of cathode copper from an ore of unspecified copper content by the conventional smelting route.

(232) Smelting copper concentrate containing 25% copper in a reverberatory furnace. Fuel assumed to be natural gas.

(233) Use of a reverberatory furnace with waste heat boiler recovering 50% of the energy consumed. The figure quoted is not the fuel consumption but that attributable to the copper after energy credits have been made. A 25% copper content is assumed in the

concentrate feed.

(234) Complete smelting in a reverberatory furnace using waste heat boiler as in Note 233. Includes contribution from fluxes, furnace linings, air blower, direct power and capital. Stated values are based on 4307 kWh_t + 282 kWh_e. Electrical component has been converted using a 24% efficiency for electricity. The values assume a 25% copper content in the concentrate feed.

(235) Based on requirements of fuel and anode carbon for electric furnace smelting of a 25% copper content concentrate. Basic data from reference is 1956 kWh_e + 249 kWh_t per ton of copper. The electrical component has been converted using an electrical efficiency of 24%.

(236) Oil fired reverberatory furnace for melting copper cathodes prior to casting into wire bar. Based on fuel consumption of 24 gallons of fuel oil per tonne copper melted.

(237) Melting cathode copper in electric furnace prior to casting into bar. Based on an electricity consumption of 300 kWh/tonne.

(238) Melting of clean copper scrap in electric furnace prior to casting. Based on an electricity consumption of 0.25 kWh/kg.

(239) Based on the energy to produce 8 mm copper rod from wire bar. Includes energy to reheat the wire bar prior to hot rolling (6 gallons oil/tonne copper) and lubricant consumption of 0.155 gallons/tonne of copper rolled. The electrical energy consumption for hot rolling is 70 kWh/tonne.

(240) Pickling hot rolled copper in 8–10% w/v sulphuric acid at 40–60°C. Steam heating of the pickling solution requires 15.8 MJ/tonne copper and electricity for pumping and electrolysis of used acid is 6.3 kWh/tonne copper.

(241) Energy of shaving machine used to remove surface defects from copper rod (55 kWh/tonne Cu). Typically some 8–10% of the input mass is removed and the shavings are remelted and used again.

(242) Based on 20,000 kWh/short ton for production, from unspecified ore grade, 800 kWh/short ton for capital energy and 200 kWh/short ton for transport. Electrical efficiency of 30% is used in the calculations but a detailed fuel breakdown is not available.

(243) Production of sponge copper by dump leaching followed by cementation. The energy is consumed almost entirely in the pumping of acidified water to the dumps. The process uses 3 tons of steel per ton of copper precipitated but the energy associated with this is not included in the calculation.

(244) Electric furnace smelting of calcined copper concentrate. No account has been taken of the electrode consumption. Furnace requires 300 kWh/ton of calcine. This entry has been calculated assuming 22% copper in the calcine.

(245) Energy to anneal cold-drawn copper wire in an electrical pit pot vacuum annealing furnace (56 kWh/tonne copper). About 2% of the copper input is lost due to 'stickiness' of the spooled wire.

(246) Average electrical energy (379 kWh/tonne) for the breaking down and drawing of 8 mm copper rod to a range of sizes down to 0.025 mm. 4% scrap is produced in the operation.

(247) Based on 20,000 kWh/short ton for production, 800 kWh/short ton for capital energy and 200 kWh/short ton for transport. A fuel breakdown is not given but an electrical efficiency of 30% is used in the calculation.

(248) Based on a stated energy requirement of 53,700 Btu/lb from unspecified ore. No fuel breakdown is given but an electrical efficiency of 33% is used.

(249) Based on a stated energy requirement of 20,000 Btu/lb. Process is approximately 98% efficient in materials.

(250) Based on a metallic yield of 0.63% of ore. System includes all processes (unspecified) up to a concentrate suitable for smelting and requires 0.39 MJ/kg of ore. Table entry calculated assuming electrical:non-electrical consumption as in entry associated with Note 251, changed to authors' efficiencies.

(251) Based on stated requirement per ton ore of 15 kWh_e + 18.5 kWh_t. Values include energy associated with all ancilliary materials as well as making allowance for metal lost through abrasion.

(252) Energy expended in blasting, crushing, grinding and flotation of copper ore to produce concentrate. Includes the energy of the explosives.

(253) Stated energy requirement is 230,000 Btu/ton ore. Includes energy of crushing, grinding, flotation, de-watering and mineral transport as well as an equivalent fuel value for steel consumed as mill-liners, balls and rods, plus fuel value of flotation reagents. No fuel breakdown is given but an electrical efficiency of 33% is used.

(254) Injection moulding of high-density polyethylene beer crates of mass 1.5 kg. Value in Table is *per kg* of polymer moulded.

(255) Vapour phase direct chlorination of ethylene. Inputs of 94 kWh of electricity, 2650 lb steam, 138,000 gallons of cooling water, 210 lb chlorine, 208 lb of ethylene and 2230 lb of reaction water yield 207 lb of 1.2 dichloroethane. Cooling water has been assumed recirculated thus attracting no energy. Steam has been converted at the rate of 4.94 MJ/kg (*q.v.*) and reaction water at 9.25 MJ/1000 l. The energies associated with the production of chlorine and ethylene are *not* included in this system.

(256) Vapour phase oxychlorination of ethylene. Inputs of 479 kWh of electricity, 19,450 lb of steam, cooling water of 80,160 gallons (assuming recirculating), 118 lb air, 450 lb of HCl and 219 lb of ethylene yield 211 lb of dichloroethane. Steam has been converted at 4.45 MJ/kg (*q.v.*). The energy associated with provision of materials inputs has *not* been included.

(257) Gas phase combined chlorination/oxychlorination of ethylene. Inputs of 533 kWh electricity, 153,395 lb steam, 225,749 gallons of cooling water (assumed recirculated), 111 lb air, 423 lb of HCl, 229 lb chlorine and 441 lb of ethylene yield 424 lb of dichloroethane. Steam has been converted at the rate of 4.49 MJ/kg (*q.v.*). The energy associated with provision of materials inputs other than services has *not* been included.

(258) Liquid phase chlorination of ethylene. Inputs of 93.2 kWh electricity, 20,000 lb steam, 60,660 gallons cooling water (assumed recirculated), 213 lb of chlorine and 216.6 lb of ethylene yield 212.6 lb of dichloroethane. Steam has been converted at the rate of 4.94 MJ/kg (*q.v.*) and the energy associated with the provision of chlorine and ethylene has *not* been included.

(259) Liquid phase oxychlorination of ethylene. Inputs of 685 kWh of electricity, 230 lb steam, 378,599.9 gallons of cooling water (assumed recirculating), 934 lb hydrogen chloride, 480 lb ethylene yields 465 lb of dichloroethane. Steam has been converted at the rate of 4.94 MJ/kg (*q.v.*). The energy associated with the provision of HCl and ethylene has *not* been included.

(260) Combined liquid phase chlorination/oxychlorination of ethylene. Inputs of 4000 kWh electricity, 2220 lb steam, 230,880 gallons cooling water (assumed recirculating), 112.6 lb oxygen, 400 lb of hydrogen chloride, 225.8 lb chlorine and 440 lb ethylene yield 430 lb of dichloroethane. Steam has been converted at 4.94 MJ/kg (*q.v.*) and oxygen at 9.33 MJ/kg (*q.v.*). Energy of other material inputs are *not* included.

(261) Thermal cracking of naphtha, for which the total cumulative energy requirement excluding feedstock is given as 4070 kcal/kg of which 120 kcal/kg is electricity calculated at an efficiency of 26.5%. The remainder is oil fuels at an efficiency of 87.5%. All values have been corrected to the fuel production efficiencies of Table 6.13 and feedstock has been included.

(262) Based on the direct energy to manufacture ethylene from L.P.G. (0.132 kWh electricity and 14.19 cu.ft. of natural gas – both per kg of product). The process requires 1.01 kg of L.P.G., 0.0002 kg of catalysts and 2.2 gallons of water per kg of ethylene. Note feedstock is *not* included in the calculation.

(263) Based on electrical efficiency of 30% and oil efficiency of 87.5%.

(264) Pyrolysis of ethane in fired-tube reactor. Inputs of 2124 kWh electricity, 461,000 MBtu oil fuel, 274,000 lb steam, 2,660,000 gallons cooling water (assumed recirculated) and 78,939 lb ethane yield 63,421 lb ethylene, 3687 lb of C_3/C_4 fraction, 2634 lb gasoline and 9158 lb residue gas. Steam has been con-converted at 4.94 MJ/kg (*q.v.*). No feedstock has been included.

(265) Autothermic cracking of crude oil at refinery. Inputs of 5000 kWh electricity, 293,000 MBtu oil fuel, 803,000 lb steam, 6.7×10^6 gallons cooling water (assumed recirculating), 113,176 lb air and 282,936 lb crude oil yield 63,421 lb ethylene, 12,732 lb coke, 17,202 lb butanes, 15,922 lb ethane, 3767 lb propane, 34,314 lb propylene, 2247 lb fuel oil, 68,849 lb fuel gas and

11,318 lb naphtha. Steam has been converted at 4.94 MJ/kg (*q.v.*). Inputs apportioned over all outputs. Includes crude oil feedstock but *excludes* winning and delivery of crude oil to refinery.

(266) Thermal cracking of naphtha at a Swedish refinery.

(267) Based on a 50:50 mix of ammonium nitrate and fuel oil.

(268) 0.3 to 0.6 kg ammonium nitrate plus fuel oil to give an energy content of 5553 kcal/kg for explosive.

(269) No detailed fuel breakdown given but electricity used is 29.4% efficient. Stated energy requirement is 406.62 Btu/lb.

(270) Calculated from the stated requirements of 9 Btu oil (at 89.3% efficient), 1 Btu natural gas (85.5% efficient) and 10 Btu electricity (26.7% efficient).

(271) Based on the direct energy to produce feldspar by open-pit mining (0.0037 gallons oil fuels, 0.0626 kWh electricity and 0.066 cu.ft. of natural gas). The process requires 4.12 gallons of water per kg of feldspar (not included in the calculation). Oil fuels assumed as medium fuel oil.

(272) Quarrying, crushing, separation, drying and grading of feldspar in Finnish operation.

(273) No fuel breakdown but electricity used at 33% efficiency. Stated energy requirement is 20,000 Btu/lb.

(274) Based on the assumption that any electricity used is generated on-site from oil fuels.

(275) No fuel breakdown but electricity efficiency is 23.85%.

(276) Based on the direct energy to operate a hypothetical scaled-up plant similar to the wet separation plant designed by Black Clawson at Franklin, Ohio (0.0000462 gallons oil fuels, 0.0088 cu.ft. natural gas and 0.44 kWh electricity).

(277) No fuel breakdown available but electrical efficiency is 29.4%.

(278) Based on the direct energy to extract gold from anode slimes in copper refining. Assumes that there is 4.69 kg gold per tonne of slime. Includes the energy to manufacture sulphuric acid needed for leaching (0.81 kg/kg slime).

(279) Based on 18 kWh/short ton for production, 1 kWh/short ton for capital and 2 kWh/short ton for transport. 30% electrical efficiency used.

(280) All operations within foundry only. Based on a stated consumption of 17.94 MJ/kg. In the entry it has been assumed that the whole energy consumption is natural gas for foundry furnaces. There will, however, be a small amount of electricity consumption.

(281) Foundry operations plus the energy to supply pig iron, scrap and ferro-alloys. Does not specify the ratios pig iron:scrap:alloy and does not specify in detail the precise starting point.

(282) Based on average energy for all operations within Chrysler foundry producing engine blocks for automobiles, heads and crankshafts. Stated requirement is 21.15 MJ per kg, which is assumed to be all natural gas. There will in fact be a small electrical component.

(283) Based on average energy for all operations in Chrysler foundry plus the energy of ferro-alloy and non-metallics used as additives in iron-making. No details of energies attributed to these additives are given.

(284) Another estimate of energy consumption in all foundry operations plus energy associated with inputs. No details of energies attributed to additives are given.

(285) Based on foundry energy plus energy of pig iron (30%) and additives. No details of energies attributed to pig iron and other additives are given.

(286) Blast furnace operation with double injection of oil and natural gas.

(287) Based on the energy to operate the blast furnace.

(288) Almost all fuel is coke, although a small amount of electricity and oil are used. Based on a stated energy requirement of 23.47 MJ/kg.

(289) Fuel energy (electricity and coke) to produce pig iron in an electric smelting furnace from concentrate containing 63% iron and 7.8% silica. Inputs are 0.303 kg coke and 2.163 kWh electricity per kg iron. The furnace charge is 908 kg of iron pellets, 90.8 kg of scrap and 135.8 kg of limestone. The energy associated with the provision of these inputs is not included.

(290) Based on the fuel energy for the mining and the blast furnace production of pig iron from high grade (unspecified) haematite

ore.

(291) Based on fuel energy to mine and produce pig iron in the blast furnace from magnetic taconite ore containing 32.5% iron and 43% silica. Excludes transport.

(292) Based on the fuel energy for the mining and the blast furnace production of pig iron from specular haematite ores containing 38% iron and 50% silica. Excludes transport.

(293) Based on the fuel energy for the mining and the blast furnace production of pig iron from non-magnetic taconite ores containing 28% iron and 43% silica. Excludes transport.

(294) Based on the fuel energy for the mining and blast furnace production of pig iron from iron laterite ore containing 48% iron, 0.8% nickel. Transport excluded.

(295) Energy to operate a blast furnace/coke oven complex which is attributable to iron smelting in the blast furnace. Energy credits from coke-making, blast furnace gas, and so on have been made. No account has been taken of the energy needed to produce the materials feeds to the complex.

(296) Blast furnace operation coal input only.

(297) Blast furnace operation only based on best U.K. practice requiring 470 kg coke/tonne iron and 91.25 kWh electricity/tonne iron. Energy associated with provision of iron ore and limestone has *not* been included.

(298) Operation of the MIDREX shaft furnace for direct reduction. Fuel requirements per tonne of sponge iron are 146.73 kWh electricity and 414.3 m^3 of natural gas. Energy associated with materials feeds to the furnace are not included. 1 tonne of sponge iron contains 0.912 t of iron.

(299) Operation of SLRN process for which the fuel requirements per tonne of sponge iron are 640 kg coal and 68 kWh of electricity.

(300) Operation of the MIDREX process for which fuel requirements per tonne of sponge iron are 3.3 Gcal natural gas and 120 kWh electricity.

(301) Direct reduction of iron ore using natural gas. Note that this represents the fuel requirement for the reduction itself. It does not include electrical energy associated with operation of ancilliary plant (see Note 300).

(302) Based on the direct fuel to mine iron ore (0.396 cu.ft. natural gas, 0.000726 gallon oil fuels and 0.0308 kWh of electricity, all per kg ore). It assumes that 3.5 kg of waste is removed per kg ore produced. The process requires 3.21 gallons of water per kg product but the energy associated with the provision of water has not been included.

(303) Operation of pelletisation plant (fuels only) based on fuel requirement per tonne of pellets of 45.7 kWh electricity and 1465 MJ blast furnace gas.

(304) Operation of pelletisation plant to produce pellets containing 65% iron.

(305) Based on a total direct energy requirement of 124.98 MJ/kg and a total indirect energy of 47.23 MJ/kg. The fuels involved are assumed to be oil and coal at production efficiencies of 89.2% and 30% respectively. Values shown in the Table have been corrected to the fuel production efficiencies of Table 6.13.

(306) Based on a fuel consumption of 1900 t cal/t. All fuels assumed to be electricity.

(307) Based on 12,000 kWh/short ton for production, 700 kWh/short ton for capital and 200 kWh/short ton for transport. Electrical efficiency of 30% is used in the calculation.

(308) Sinter production and blast furnace operation for which the fuel requirements per ton of lead are 5.14×10^6 Btu natural gas and 469 kWh electricity.

(309) No fuel breakdown given but 32.5% electrical efficiency is used in the calculations. Stated energy requirement is 31.09 MJ/kg.

(310) Based on the direct fuel requirements to mine limestone (1.2×10^{-5} kg coal, 0.000204 gallons oil fuels, 0.01 cu.ft. natural gas and 0.022 kWh electricity, all per kg of product). The process also consumes 0.083 gallons water per kg product but the energy associated with this has not been added.

(311) Quarrying, loading and crushing.

(312) Quarrying, crushing, drying and grading.

(313) Mining, crushing, pulverizing, drying and grading.

(314) Based on the fuel consumption to produce magnesium by

electrolysis of sea water. Excludes limestone mining energy.

(315) Operation of Dow cell, which electrolyses magnesium chloride.

(316) Direct energy for electrowinning of manganese (99.98%) from leach liquor.

(317) Based on a stated requirement of 1956 kWh electricity/ton and 249 kWh thermal/ton for smelting in an electric furnace. Thermal fuel assumed to be natural gas for electrode baking.

(318) Energy to produce an aluminium can holding 16 fl.oz.

(319) Based on the direct energy to manufacture aluminium can bodies by drawing and ironing from aluminium sheet (23.76 cu.ft. natural gas and 2.07 kWh electricity, both per kg of can). The process also uses 0.52 gallons of water per kg of product, which is not included in the calculation of energy.

(320) Based on the average direct fuel consumption to fabricate open-top steel beverage cans from sheet steel (2.4 cu.ft. natural gas, and 0.3 kWh electricity). The process also requires the following materials inputs for which the energy associated with their provision has not been included in the calculation; 0.016 kg solder, 0.01 kg paint and lacquer, 0.026 kg solvents, 0.049 kg of corrugated packaging material and 0.29 gallons of water, all per kg of product.

(321) Based on fuels required to fill one 12 oz soft drinks can (0.015 cu.ft. natural gas and 0.0017 kWh electricity).

(322) Based on fuels required to fill with beer and pasteurise one 12 oz can (0.009 kg coal, 0.00013 gallons oil. 0.055 cu.ft. natural gas and 0.002 kWh electricity).

(323) Based on the steam reforming of naphtha for which the total cumulative energy requirement excluding feedstock is 6110 kcal/kg, of which 1280 kcal/kg are electricity calculated with an efficiency of 26.5%. The remainder is oil fuels calculated with a production efficiency of 87.5%. The values in the Table have been adjusted to the fuel production efficiencies of Table 6.13 and feedstock has been included.

(324) Based on a more recent plant than that reported in Note 323 above.

(325) Operation of electric-powered shovel. Values based on a fuel energy consumption of 2.4 kWh electricity/ton ore, and gross energies of 1.652 kWh for the materials and assembly of the machine and 0.6 kWh for repairs and maintenance.

(326) Electric-powered iron ore crusher operating at 3.5 kWh electricity per ton of ore.

(327) Average value including the energy equivalent of explosives, truck tyres, drill bits and other supplies.

(328) Grinding in rod and ball mills.

(329) Based on a cumulative energy of 1240 kcal/kg *excluding* feedstock. Electrical efficiency of 26.5%, oil efficiency 87.5%; values shown are corrected to the production efficiencies of Table 6.13 and feedstock has been included.

(330) Based on thermal cracking in oil-fired furnace with fuel requirement of 1400 kWh per tonne naphtha. Feedstock included as in Table 7.5. Electricity consumption for compressors and so on not included.

(331) Power to the electric smelting stage only in the production of nickel from lateritic ores. Excludes all treatment up to the smelting stage and other fuels used as reducing agents.

(332) All stages from lateritic ore to ingot production. No detailed fuel breakdown available but electricity efficiency used is 32.5%.

(333) All stages from sulphide ore to ingot production. No detailed fuel breakdown but electrical efficiency used is 32.5%.

(334) Based on the route via ammonia oxidation for which a total cumulative energy requirement of 700 kcal/kg is required as electricity, calculated using an efficiency of production of 26.5%. Efficiency has been corrected to the value of Table 6.13.

(335) Based on the following requirements for the production of 0.273 lb of paper; 283 Btu oil (at 89.3% efficiency), 2506 Btu natural gas (85.5%), 610 Btu electricity (26.7%), other fuels 1401 Btu.

(336) Based on energy per short ton of 5900 Btu for production, 300 Btu for capital and 200 Btu for transport. Electrical efficiency 30%. Includes no feedstock.

(337) Based on electrical consumption of 1.5 kWh/kg and non-electrical consumption of 5.1 kWh/kg. Excludes feedstock.

(338) Based on electrical consumption of 1.5 kWh/kg and non-electrical

consumption of 8.0 kWh/kg. Excludes feedstock.

(339) No detailed fuel breakdown available, but electricity converted at 32.8% efficient.

(340) Excludes feedstock.

(341) Excludes feedstock. Data relates to production from wood at the paper-making plant. Based on consumption of 1950 Mcal/t assumed oil, and 420 kWh/t electricity.

(342) Based on a cumulative energy of 860 kcal/kg with an electrical efficiency of 26.5% and an oil efficiency of 87.5%.

(343) Based on a cumulative energy of 55,500 kcal/kg, which is 88% electricity calculated as 26.5% efficient. The remainder is oil fuels at 87.5% efficiency.

(344) Based on the direct fuel required to convert 1.3 butadiene to polybutadiene (5.13 cu.ft. natural gas per kg product). The process requires 1.015 kg of 1.3 butadiene, 0.025 kg solvent (hexane), 0.0005 kg catalyst, 0.005 kg modifier, 0.0023 kg antioxidant and 6.1 gallons of water, all per kg of product. The energy associated with the provision of these materials inputs is not included.

(345) The polymerization of ethylene, for which the fuel requirements are 690 kWh electricity per tonne polyethylene and 0.2 t steam (for which the fuel requirement is 60 l oil per tonne).

(346) Ziegler catalytic polymerization of ethylene for which inputs of 2202 kWh electricity, 73,080 lb steam, 339,960 gallons cooling water (assumed recirculating) 7853 lb ethylene and 29,534 lb reaction water yield 7648 lb of high density polyethylene. Steam has been converted at the rate of 4.94 MJ/kg and water at 9.25 MJ per 1000 l.

(347) Based on a total direct consumption of 71.48 MJ/kg and a total indirect consumption of 43.89 MJ/kg, measured with production efficiencies for electricity and oil of 30% and 89.2% respectively.

(348) Based on a direct energy requirement of 55.2 MJ/kg and a total indirect energy requirement of 32.19 MJ/kg, calculated with fuel production efficiencies for oil and electricity of 89.2% and 30% respectively.

(349) Based on a direct energy requirement of 2600 kg cal/kg assumed all electricity.

(350) Based on a cumulative energy of 10,510 kcal/kg *excluding* feedstock, calculated using an electrical efficiency of 26.5% and an oil efficiency of 87.5%. It is also stated that 45% of this total energy is due to electricity. Efficiencies have been corrected to Table 6.13 and feedstock has been included at 45 MJ/kg of polymer.

(351) Metal oxide catalytic polymerization of ethylene. Inputs of 2626 kWh electricity, 169,500 lb steam, 825,660 gallons of cooling water (assumed recirculating), 7974 lb ethylene and 91,073 lb of reaction water yield 7610 lb of low density polyethylene. Steam has been converted at the rate of 4.94 MJ/kg (*q.v.*) and water at the rate of 9.25 MJ/1000 litres (*q.v.*).

(352) Based on a total direct consumption of 165.20 MJ/kg and a total indirect consumption of 71.48 MJ/kg, with oil and electrical fuel production efficiencies of 89.2% and 30% respectively.

(353) Based on the direct fuel requirements to manufacture polystyrene foam from crude oil and natural gas (46.42 cu.ft. of natural gas and 2.16 kWh electricity both per kg of product). In addition to these fuels the process also requires 0.966 kg crude oil, 0.316 kg natural gas, 0.009 kg of unspecified additives and 15.23 gallons of water, all per kg polystyrene. The energy of feedstock is not included in the Table.

(354) As in Note 353 above, but including feedstock.

(355) Based on feedstock of 1.3 toe and processing energy of 1.88 toe, both per tonne of finished product.

(356) Production of PVC from crude oil. polymerization by the suspension process. Based on a feedstock of 0.55 toe/t polymer and processing energy of 1.5 toe/t polymer. Electrical efficiency 30%, oil efficiency 87.5%.

(357) Polymerization of vinyl chloride monomer. Inputs of 0.0867 kWh electricity, 0.142 lb steam, 4.5 gallons of water (assuming recirculating) and 1.002 lb of monomer yield 1 lb of PVC. Steam has been converted at the rate of 4.94 MJ/kg (*q.v.*).

(358) Suspension polymerization of vinyl chloride monomer. Inputs of 0.1461 kWh electricity. 1.29586 lb steam, 8.57551 gallons of

cooling water (assume recirculating) and 1.05066 lb of monomer yield 1.00 lb of polymer. Steam has been converted at the rate of 4.94 MJ/kg (*q.v.*).

(359) Bulk polymerisation of vinyl chloride monomer. Inputs of 0.19462 kWh electricity, 2773.25 Btu natural gas fuel, 0.33605 lb steam, 15.58238 gallons cooling water (assumed recirculated) and 1.04617 lb monomer yield 1.00 lb PVC. Steam has been converted at the rate of 4.94 MJ/kg (*q.v.*).

(360) Based on a cumulative energy of 16,620 kcal/kg excluding feedstock for the chlorination of acetylene using an electrical efficiency of 26.5% and an oil efficiency of 87.5%. Electricity comprises 38% of this cumulative energy. Efficiencies have been corrected to Table 6.13 and feedstock added.

(361) Based on requirements for polymerization of 3.14 t steam, 0.64 MWh electricity and 70 m^3 cooling water, all per tonne of PVC. Steam is stated to need 188.4 l of fuel oil per tonne of PVC.

(362) Sylvinite is an ore containing a 50:50 mixture of Na:K as chloride. To produce 1 kg of potassium chloride from ore requires 2 kg of sylvinite, 1.136 kg of steam and 0.05 kWh electricity. Steam is converted at the rate of 4.49 MJ/kg.

(363) Based on 92 kWh electricity to print 5520 m^2.

(364) Electrical input to operate a 400 mm steel rolling mill. Based on a total energy of 0.37 MJ per kg of steel input of which 94.2% appears as useful products.

(365) Electrical input to operate an 850 mm steel rolling mill. Based on a total energy of 0.37 MJ per kg of steel input of which 96.4% appears as useful product.

(366) Electrical input to 5 stand cold reduction mill.

(367) Glass-making quality.

(368) Based on 18 kWh/short ton for production, 1 kWh/short ton for capital and 2 kWh/short ton for transport with electricity assumed to be 30% efficient.

(369) Based on fuel requirements to produce glass-making sand (0.0058 kg coal, 0.0004546 gallons oil fuel, 0.4741 cu.ft. natural gas and 0.0153 kWh electricity, all per kg sand). The process also requires 1.65 gallons of water per kg sand.

(370) Styrene-butadiene co-polymer production for which a total cumulative energy excluding feedstock is given as 16,900 kcal/kg. Of this, 2470 kcal/kg is electricity calculated at an efficiency of 26.5%. The remainder is oil at 87.5% efficiency. Values have been corrected to the efficiencies of Table 6.13 and feedstock has been added.

(371) Based on an energy of 1.9 toe per tonne for compound used to seal end joints in metal cans. Based on an oil production efficiency of 84%.

(372) Based on an electrical energy requirement of 2.1 toe/tonne film. Shrinkwrap is an oriented crosslinked polyethylene film.

(373) Based on the direct fuel energy required to extract selenium from anode slime from copper refining. 4400 kWh electricity and 60,160 kWh oil fuels, both per ton of selenium. The calculation assumes 75 kg of selenium per tonne of slime and includes the energy to manufacture sulphuric acid used for leaching (0.81 kg H_2SO_4 per kg slime).

(374) Based on the direct fuel energy to extract silver from anode slimes from copper refining (1406.1 kWh electricity, 2.876 kWh gas oil, both per ton of slime). Calculation also assumes 206.3 kg of silver per tonne of slime and includes the energy to manufacture the sulphuric acid used in the leaching (0.82 kg, H_2SO_4 per kg slime).

(375) Direct electricity used in Down's Cell of 5 kWh/lb sodium.

(376) Solvay process from brine in the ground. Based on a total cumulative energy requirement of 2870 kcal/kg, of which 160 kcal/kg was electricity calculated at an efficiency of 26.5%. The remainder is assumed to be coal. Values shown have been converted to the fuel production efficiencies of Table 6.13.

(377) Based on the direct fuel energy to manufacture synthetic sodium carbonate by the Solvay process (0.00026 kg coal, 0.00011 kg coke, 0.0153 gallons oil and 1.32 cu.ft. natural gas, all per kg of product). The process requires 1.65 kg salt, 1.25 kg limestone, 0.0045 kg minor ingredients and 14.7 gallons of water, all per kg of product.

(378) Based on the direct fuel energy to mine rock salt (0.00011 gallons

oil fuels, 0.185 cu.ft. natural gas, and 0.0935 kWh electricity, all per kg of product). The process also consumes 0.48 gallons water per kg of salt.

(379) Based on the direct fuel energy to produce brine from rock salt (0.0001 kWh electricity, 0.19 cu.ft. natural gas and 0.0001 gallons of oil, all per kg of product). The process also requires 1.1 gallons of water per kg of brine.

(380) Vacuum crystallisation of saturated brine.

(381) Based on the direct energy requirement to produce sodium hydroxide by electrolysis of brine (1.59 kWh electricity per kg of sodium hydroxide). The process requires 1.017 kg sodium chloride, 0.016 kg process additives (unspecified) and 6.38 gallons of water, all per kg of product. The energy associated with these materials inputs are not included in the calculation.

(382) Average value for U.S. steel industry in 1967 for all steel converters. Excludes oxygen supplies and other materials inputs.

(383) Steel making in the electric arc furnace.

(384) Steel making in the basic oxygen furnace. Not specified whether the electrical consumption is for rotating the converter or whether it also includes oxygen energy.

(385) Steel making in the open-hearth furnace.

(386) Melting of 70% sponge iron, 30% scrap in electric arc furnace. Based on 640 kWh electricity to furnace per tonne of liquid steel, of which 50 kWh is used during refining after melt-down.

(387) Power to vacuum induction furnace. Based on middle of range 700 to 750 kWh electrical consumption per tonne liquid steel.

(388) Production of steel in the open-hearth furnace using preheated air. (25 gallons oil/ton steel.)

(389) 1967 average of all U.S. steel converters (basic oxygen furnace, electric arc and open-hearth). Of the total energy, 5.46 MJ was consumed as furnace energy and 2.39 MJ as oxygen and other materials.

(390) Electric arc furnace melting of 70% sponge iron and 30% scrap. Based on a direct fuel energy of 570 kWh electricity per ton steel. Furnace also requires 0.0065 ton electrode and 0.75 ton flux per ton steel, which are not charged in the entry.

(391) Operation of open-hearth furnace. For the production of 1 ton of steel the requirements are 246 kWh heavy fuel oil, 186 kWh petroleum tar, 219 kWh coal tar pitch, 302 kWh coke-oven gas, 35 kWh mixed gases, 120 kWh steam, 15 kWh electricity. In addition 0.004 tons of lime and 0.13 ton oxygen are needed per ton of steel; these last two have not been added into the calculation.

(392) Based on a steel production energy of 48.81 MJ/kg plus 3.02 MJ/kg for alloying additives.

(393) Based on the electrical energy to operate a cogging mill. Cogging mills give typically 87% of the input material as finished product.

(394) Operation of a 3-stand continuous caster producing 120 mm × 120 mm billet at 30 tonne per hour. Handling between furnace and caster adds an additional 0.01 MJ/kg to the overall energy.

(395) Operation of continuous-bloom casting machine. 94% of the materials input emerges as finished product.

(396) Energy calculated assuming an electrical efficiency of 33%. System starts with raw materials at blast furnace. Value is U.S. industry average.

(397) U.K. average for 1968 based on statistical sources with an electrical generation efficiency of 23.85% and an assumed scrap usage in steelmaking of 50%.

(398) Process route from ore preparation via direct reduction and electric arc furnace to continuous-caster and finishing. Electrical efficiency is 33% but no fuel breakdown is given. The suggested energy apportionment between the different operations is; ore preparation = 6%, direct reduction = 39%, electric arc furnace = 24%, continuous-casting = 12%, secondary reduction and final finishing = 19%.

(399) Process route fron ore processing, via blast furnace and basic oxygen furnace to continuous-casting and finishing. Electrical efficiency is 33% but no fuel breakdown is given. Suggest energy apportionment between the different operations are; agglomerating = 6%, blast furnace and coke ovens = 55%, oxygen furnace = 5%, continuous-casting and secondary conversion = 13%, final finishing = 21%.

(400) Route is electric arc furnace and continuous-casting followed by finishing. Input materials are not specified. Electrical efficiency is 33%. Energy apportioned between operations as electric arc furnace = 39%, casting and conversion = 23%, finishing = 38%.

(401) Production of unspecified steel product via the blast furnace. System inputs are ore to the blast furnace and coal to coke ovens. The system is said to be average for the whole industry although no electricity is quoted. Based on requirements of coal = 16.95 MJ/kg steel, natural gas = 4.88 MJ/kg steel and oil = 2.09 MJ/kg steel.

(402) Based on 30 kWh/ton raw steel input for sheet mill galvanizing unit. Energy of zinc excluded.

(403) Average of the processes operated by one company adjusted to include mining and transportation.

(404) Average 1969 data for U.S. steel industry includes all processes and all products with raw materials at the blast furnace.

(405) As Note 404 but with an allowance for mining.

(406) As Note 403 but excluding mining.

(407) Based on a quoted value for all steel products of 14,400 kcal/kg.

(408) Based on direct fuel energies of 1.62 MJ natural gas and 0.10 MJ electricity, both per kg of *input*. 99% of the input ends up as finished product.

(409) Operation of hot rolling mill. Electricity will be the dominant fuel type.

(410) Energy for handling between furnace and casting.

(411) Based on a steelmaking energy (via blast furnace and open hearth) of 34.3 MJ/kg (see note 432) and assuming a casting operation which has a materials efficiency of 69%.

(412) No fuel breakdown given but electrical component converted at an electrical efficiency of 32.5%.

(413) Electric arc melting of 100% scrap. Does not include any allowance for electrodes or additives.

(414) Production of steel by direct reduction of iron ore using natural gas followed by electric arc melting of 20% scrap and 80% direct reduced iron. Energy apportioned as 12.94 MJ to natural gas, 0.16 MJ to carbon electrode, 0.40 MJ to lime and 1.53 MJ to electricity.

(415) Based on an average for the production of 1.6 m lengths of straight vitrified clay pipes.

(416) Blast furnace to produce liquid pig iron followed by conversion of this pig iron to steel by the basic oxygen process using additives of 8.3% iron ore, oxygen, lime and acidic gangue. The direct energy is apportioned as 25.38 MJ to coal, 0.21 MJ to oxygen and 0.29 MJ to lime. Process requires 1.051 kg pig iron, 0.1 kg ore per kg liquid steel.

(417) Blast furnace to produce liquid pig iron followed by conversion to steel by the basic oxygen process using as additives 8.3% iron ore, oxygen, lime and basic gangue. The energy is apportioned as 25.22 MJ to coal, 0.20 MJ to oxygen and 0.26 MJ to lime. Process requires 1.051 kg pig iron and 0.1 kg ore per kg of liquid steel.

(418) Blast furnace operation followed by conversion of the liquid pig iron in a basic oxygen furnace. The blast furnace has a double injection of oil and natural gas. The input to the basic oxygen furnace in addition to pig iron is 27% scrap. The direct energy is apportioned as 9.84 MJ to coal, 2.72 MJ to oil, 8.84 MJ to natural gas, 0.23 MJ to oxygen and 0.20 MJ to lime.

(419) Electric arc furnace for which the direct energy requirements are apportioned as 1.62 MJ to electricity, 0.39 MJ to coke and 0.11 MJ to electrodes. No allowance has been made for oxygen or scrap.

(420) Electric arc furnace with scrap preheater. Direct energy apportioned as 1.19 MJ to electricity for the furnace, 1.17 MJ to natural gas for the preheater and 0.08 MJ for electrodes.

(421) Operation of basic oxygen furnace only. Furnace charged with 800 kg liquid iron, 300 kg scrap and 57 kg flux. The electrical consumption is made up of 0.06 MJ to the furnace and 0.14 MJ to the oxygen.

(422) Blast furnace followed by basic oxygen furnace with hot hydrogen injected at the tuyeres at the rate of 656 m^3 per tonne of steel. No energy has been included for the provision of hydrogen. The energy is composed of 5.38 MJ for coke, 0.45 MJ for process

electricity and 0.15 MJ for oxygen production. No contribution has been include for ore, limestone or scrap.

(423) Steel making by blast furnace; basic oxygen process with cold hydrogen injected at the tuyeres (277 m^3 per tonne steel). No energy allowance has been made for hydrogen, ore, limestone or scrap. The energy has been apportioned as 9.34 MJ to coke, 0.54 MJ to direct electrical use and 0.15 MJ to oxygen.

(424) The conventional blast furnace; basic oxygen process. No allowance has been made for the energy associated with ore, limestone and scrap. The energy is apportioned as 10.76 MJ to coke, 0.55 MJ to direct electricity use and 0.15 MJ to the production of oxygen.

(425) Electric arc furnace using a mixture of sponge iron and scrap. The furnace is charged with 811 kg sponge iron, 300 kg of scrap and 100 kg of flux. 200 kg of slag is produced and 40 kg of iron are lost. The energy may be apportioned as 1.98 MJ to direct electricity consumption, 0.35 MJ to electrodes and coke and 0.01 MJ to oxygen. The energy associated with the provision of the materials inputs other than oxygen and electrodes has not been included.

(426) Direct reduction of iron ore by natural gas using MIDREX process followed by electric arc furnace. System starts with ore concentrate. The energy is apportioned as 10.69 MJ natural gas, 2.62 MJ direct plant electricity consumption, 0.01 MJ for oxygen production and 1.23 MJ for pellet drying and firing.

(427) Electric arc furnace melting 100% scrap. Includes scrap bundling, shredding, drying, preheating, handling and the operation of the electric arc furnace itself. Assumes 1.1 t scrap yields 1 t steel.

(428) Based on a total energy requirement of 0.27 toe per tonne, assumed electrical energy at 30% production efficiency.

(429) Electric arc furnace melting 100% scrap with consumption per tonne of steel of 460 kWh electricity and 8 kg coke. No allowance has been made for the energy associated with the provision of scrap.

(430) Blast furnace followed by basic oxygen furnace for which consumption per tonne of steel is 194 kWh electricity and 378 kg coke.

(431) Electric arc furnace with 100% scrap input. 1.026 kg scrap yields 1 kg steel. No allowance has been made for the energy associated with the provision of scrap. The energy is apportioned as 0.34 MJ to coke, 0.16 MJ to electrodes, 0.10 MJ to lime and 1.47 MJ to direct electricity consumption.

(432) Blast furnace followed by conversion to steel with either basic oxygen furnace of open-hearth furnace. Includes electricity converted at 33% but no fuel breakdown is given.

(433) Operation of electric arc furnace melting 100% scrap. Energy consumption apportioned as 2.08 MJ to direct electricity consumption, 0.45 MJ to anthracite, 0.13 MJ to natural gas, 0.18 MJ to electrodes and 0.09 MJ to the provision of oxidised scrap.

(434) Electric arc furnace with inputs per short ton of steel of 1960 lb scrap, 40 lb pig iron, 57 lb lime, 9 lb fluorspar and 450 kWh electricity. No charge has been made for any of the materials inputs.

(435) Electric arc furnace with inputs per short ton of steel of 800 lb of metallized pellets, 1200 lb scrap, 9 lb fluorspar, 101 lb lime and 488 kWh electricity. No charge has been made for any of the materials inputs including the sponge iron.

(436) Natural gas fired preheating furnace to achieve a temperature of 700°C.

(437) Furnace is initially gas-fired to raise the temperature of billets to 300°C and then electrical resistance heating is used to achieve 1060°C.

(438) Electrical induction slab heating furnace dealing with 400 t per hour.

(439) One zone reheating furnace. Values refer to normal operation which involves one light-up per week followed by near continuous operation. With intermittent operation the total energy requirement rises to 3.21 MJ/kg steel.

(440) 230 tonne per hour pusher furnace fuelled by natural gas. No electrical consumption given.

(441) 230 tonne per hour walking beam furnace. No electrical

consumption given.

(442) Electric induction reheating furnace achieving a temperature of 1200°C with a consumption of 350 kWh/t.

(443) Unspecified 30 to 40 tonne per hour reheating furnace using 410 kWh/tonne of gas and 14 kWh/tonne electricity.

(444) Energy to operate reheating furnace taking steel from cogging mill and feeding to a hot rolling mill. Such furnaces yield 98% of the input materials to the rolling mill.

(445) 5-zone reheating furnace fuelled by natural gas. Quoted values refer to normal operation, that is one light-up per week followed by near continuous operation. With intermittent operation the total energy requirement rises to 3.86 MJ/kg steel. No electricity given for driving slabs through the furnace.

(446) Fuel supply to a 3-zone gas reheating furnace. No electricity given for driving slabs through the furnace.

(447) Fuel supply to a natural gas 2-zone reheating furnace. Normal operation is 1 light-up per week followed by near continuous operation. With restricted operation the fuel energy requirement rises to 2.42 MJ/kg steel. No electricity is quoted for driving the slabs through the furnace.

(448) Based on a total production energy of 11,700 kWh/short ton and an electrical efficiency of 30%. Included is 700 kWh/short ton for capital and 200 kWh/short ton for transport.

(449) Electric power to rolling mill reducing 140 mm × 140 mm ingot or 100 mm × 100 mm billet to 6–32 mm round bar (103.5 kWh/t).

(450) Mini-steel works, includes fuels for scrap treatment, melting, casting, reheating and rolling. The fuel energy per kg product is apportioned as 1.84 MJ to melting, 0.04 MJ to scrap handling, 0.02 MJ to continuous-casting, 1.53 MJ (gas) to reheating, 0.35 MJ to rolling, 0.01 MJ to mill cranes. 0.02 MJ to lighting, 0.09 MJ to fume treatment, 0.03 MJ to water circulation and treatment and 0.02 MJ to compressed air production.

(451) Based on the direct fuel energy for the magnetic separation of steel scrap from domestic refuse (that is 0.44 kWh electricity, 0.209 cu.ft. natural gas and 0.00128 gallons oil, all per kg of scrap).

(452) 30 to 35 tonne per hour processing rate at 6 kWh/t.

(453) Gas-fired plant operating at 120°C.

(454) Gas-fired preheating furnace operating at 400°C.

(455) 16 tonne per hour electric scrap shredder (12 kWh/t).

(456) Precise route not specified, but for the production of 0.104 lb of sheet steel the fuel requirements are 1282 btu coal (93.8% efficient), 103 Btu oil (89.3%), 424 Btu gas (85.5%), 278 Btu electricity (26.7%) and 19 Btu of other fuels.

(457) Based on the direct fuel energy per kg of steel strip for iron-making, steelmaking and steel rolling (0.9 kg coal, 0.116 gallons oil, 0.00183 gallons of tar and pitch, 0.000165 gallons L.P.G., 7.623 cu.ft. natural gas and 0.473 kWh electricity). Calculation made by aggregating oil, tar, pitch and L.P.G. and converting as if medium fuel oil. The process also requires 0.255 kg limestone, 1.37 kg of iron ore (grade not specified), 0.041 kg of other unspecified materials and 42.16 gallons of water, all per kg of product. The energy associated with the provision of these ancilliary materials has not been included in the calculation.

(458) Soaking pit fed with approximately 46% cold ingots; the remainder are fed hot. Temperature maintained at 1600 K. The reference gives a range of energy requirements based on differing inputs of cold ingots.

(459) Fuel for soaking pits used prior to hot rolling. Thermal fuel assumed to be natural gas. About 2.5% of the steel is lost by oxidation in the furnace.

(460) Approximately 1% of the input mass is lost in this process.

(461) Based on the route via the dehydrogenation of ethylbenzene for which the total cumulative energy requirement excluding feed-stock is given as 16,360 kcal/kg, of which 1420 kcal/kg are electricity calculated at an efficiency of 26.5%. The remainder is oil at a production efficiency of 87.5%. Values have been corrected to the efficiencies given in Table 6.13 and feedstock has been included.

(462) Based on the direct fuel energy requirements for the production of 1 kg of styrene from ethylene and benzene (0.066 kWh

electricity and 10.33 cu.ft. natural gas). The process requires 0.862 kg benzene, 0.308 kg ethylene, 0.074 kg catalyst and 103.9 gallons of water. The energy associated with the provision of these materials inputs has not been included. The catalyst is either aluminium chloride or boron trifluoride/alumina.

(463) Based on the route via propylene and ammonia for which the requirements per kg acrylonitrile are 1.049 kg propylene, 0.424 kg ammonia gas (see Note 106). The energies associated with provision of these inputs have been taken as propylene = 68.12 MJ kg (*q.v.*) and ammonia = 44.02 MJ/kg (*q.v.*). The conversion to acrylonitrile has been taken as 2.30 MJ/kg (*q.v.*).

(464) Frasch process using electrical efficiency of 26.5% and oil efficiency of 87.5%.

(465) Production of sulphuric acid from SO_2 in a copper plant. No energy contribution is assigned to the by-product SO_2 used as starting material

(466) Extraction of tellurium from anode slime from copper refining. Assume 30 lb tellerium per ton of slime. Direct fuel energy requirements are 23,667 kWh electricity and 300,800 kWh of oil fuels. The calculation includes the provision of sulphuric acid used to leach the slime (0.81 kg per kg slime).

(467) Rotary furnace with no heat exchangers.

(468) Electric furnace with 70% tin in concentrate.

(469) Blast furnace with 20% tin in concentrate.

(470) Natural gas reverberatory furnace with 20% tin in concentrate.

(471) Scrap to electric arc furnace followed by all processes up to production of finished tinplate. Based on the direct fuel energy requirements per ton of tinplate of 500 kWh for the furnace, 330 kWh for rolling and plating, 5400 cu.ft. of natural gas and 10 gallons of oil. In addition the materials requirements are 2140 lb scrap, 30 lb oxygen, 14 lb iron ore, 92 lb limestone, 10 lb of carbon electrode, 10 lb of plating metal and 11,000 gallons of water. The energy associated with the provision of these materials inputs are *not* included.

(472) Based on direct fuel energy requirement of 100 kWh electricity per ton of tinplate for an electrolytic tinplating unit. Energy associated with provision of tin excluded.

(473) Based on the direct fuel energy required to produce titanium by the Kroll process from rutile obtained from beach sand containing 24% TiO_2 (33,300 kWh electricity and 42,790 kWh gas, both per ton titanium).

(474) Based on the direct fuel energy required to produce 1 ton of titanium from ilmenite slag produced from beach sand containing 1.4% TiO_2 (44,062 kWh electricity and 42,900 kWh gas).

(475) Based on the direct fuel energy required to produce 1 ton of titanium from ilmenite slag obtained from titanium bearing rock containing 20% TiO_2 (42,616 kWh electricity and 42,900 kWh gas).

(476) Based on the direct fuel energy required to produce 1 ton of titanium from ilmenite slag obtained from ferruginous rock containing 35% TiO_2 (43,965 kWh electricity and 42,400 kWh gas).

(477) Based on the direct fuel energy required to produce 1 ton of titanium from ilmenite slag obtained from high alumina clay containing 5% TiO_2 (44,560 kWh electricity and 45,000 kWh gas).

(478) Based on the direct fuel energy required to produce 1 ton of titanium ilmenite slag obtained from soil containing 0.2% TiO_2 (64,270 kWh electricity and 45,400 kWh gas).

(479) Titanium recycling using electrolysis in a fused salt bath (11.02 MJ/kg) followed by melting the sponge titanium in an electric arc furnace (6.17 MJ/kg).

(480) Based on a production energy of 140,000 kWh/short ton, capital energy 1000 kWh/short ton and transport 200 kWh/short ton. The calculation is based on an electrical efficiency of 30%.

(481) Based on a plant in which all electricity used is generated on the site from oil fuels.

(482) Based on a total cumulative energy requirement excluding feedstock of 11,970 kcal/kg, of which 4550 kcal/kg are electricity calculated with an efficiency of 26.5%. The remainder is oil with an efficiency of 87.5%. Values have been corrected to the final production efficiencies of Table 6.13 and feedstock has been

included.

(483) Pyrolysis of 1.2-dichloroethane for which inputs of 762 kWh electricity, 47,500 MBtu of oil fuel, 67,250 lb steam, 460,000 gallons of cooling water, 415 lb of dichloroethane, and 170 lb of reaction water produce 408 lb of vinyl chloride and 416 lb of hydrogen chloride. Steam has been converted at the rate of 4.94 MJ/kg (*q.v.*) and the cooling water is assumed to be recirculated. The process produces waste heat of 111,606 MBtu and it has been assumed that none of this is reclaimed. Inputs have been apportioned over both outputs on a mass basis. Note that no allowance has been made for steam raised elsewhere on the plant and used here; all steam is assumed raised by oil fuels.

(484) As for Note 483 except that electricity consumption is increased to 781 kWh.

(485) Hydrochlorination of acetylene for which 1 lb of vinyl chloride requires inputs of 0.0288 kWh electricity, 1.0325 lb steam, 10.7586 gallons of cooling water (assume recirculated), 0.61 lb hydrogen chloride and 0.4262 lb acetylene. Steam has been converted at the rate of 4.94 MJ/kg (*q.v.*).

(486) Combined chlorination/oxychlorination/pyrolysis plant.

(487) Energy to transport, shred and dump refuse including landfill maintenance energy.

(488) Chemical pulping with no allowance for chemicals or water.

(489) Mechanical pulping with no allowance for water used.

(490) Based on a production energy of 1.47 kWh/board ft. capital energy of 0.02 kWh/board ft. and 0.02 kWh/board ft. for transport with electricity converted at 30% efficiency.

(491) Catalytic reforming of naphtha with benzene as co-product. See Note 163.

(492) Based on the freezing and isomerisation of mixed xylenes for which the total cumulative energy requirement excluding feedstock is given as 13,350 kcal/kg. Of these 3650 kcal/kg are electricity calculated at an efficiency of 26.5%. The remainder are oil fuels calculated at an efficiency of 87.5%. The values shown here have been corrected to the fuel production efficiencies of Table 6.13 and feedstock has been included.

(493) Based on a production energy of 13,800 kWh/short ton, capital energy of 700 kWh/short ton and transport of 200 kWh/short ton, using an electrical efficiency of 30%.

(494) Direct fuel energy to operate vertical retort method.

(495) Direct fuel energy to operate the electrothermic method.

(496) Direct fuel energy for electrolysis from solution.

(497) Based on a direct fuel energy of 3.18 MJ/kg. This has been assumed all natural gas.

(498) Based on the production of carbon black from crude oil. Two processes are in use, one reclaims some of the heat of combustion while the other does not. The value given is thought to represent a reasonable average of the two processes. There will also be some variation from grade to grade of carbon black.

(499) An average value based on a variety of published and private information.

(500) An average value based on published data for mechanical and chemical production of wood pulp. Value includes wood feedstock energy.

(501) Typical value for unbleached paper.

(502) Average value for the production of unbleached paper in the U.K. from wood grown and pulped in Scandinavia.

(503) Based on the energy requirement of ammonia of 44.02 MJ/kg (*q.v.*) and an energy to produce nitric acid from ammonia of 3.25 MJ/kg (*q.v.*). Ammonia requirement 0.1165 kg NH_3/kg HNO_3 (see Note 106).

(504) Water gas route for which 1000 standard cu.ft. of hydrogen (19.04 kg) requires 42 lb coke, 445 lb steam and 3 kWh electricity. Steam has been converted at the rate of 4.49 MJ/kg (*q.v.*). There is by-product CO_2 which has been assumed wasted.

(505) As Note 504 except that CO_2 is reclaimed and energy apportioned on a mass basis.

(506) Reaction of natural gas with steam. Inputs for 1000 cu.ft. of hydrogen (19.04 kg) are 250 cu.ft. natural gas, 360 lb steam, 350,000 Btu natural gas as fuel and 0.9 kWh electricity. Steam has been converted at 4.49 MJ/kg (*q.v.*).

(507) The esterification and hydrolysis of ethylene. The production of

1000 U.S. gallons of 95% ethanol requires 4000 lb ethylene, 600 lb of sulphuric acid, 130 lb of sodium hydroxide, 240,000 gallons of water, 20,000 lb steam, 20 kWh electricity and 2,800,000 Btu of fuel (assumed oil). Steam has been converted at 4.49 MJ/kg (*q.v.*) and water as 0.046 MJ/imp gallon.

(508) As Note 507 except that the energy associated with the provision of the raw materials has been included at the rate of 29.88 MJ/kg for sodium hydroxide (*q.v.*), 2.48 MJ/kg for sulphuric acid (*q.v.*) and 68.12 MJ/kg for ethylene (*q.v.*).

(509) Route via ethylene chlorohydrin. The production of 2240 lb of ethylene oxide requires 1600 lb ethylene, 400 lb of chlorine, 3200 lb of calcium oxide, 180 kWh of electricity, 24,000 lb of steam and 60,000 gallons of water, The materials inputs have been converted to energy as follows: ethylene = 68.12 MJ/kg (*q.v.*) chlorine = 29.88 MJ/kg (*q.v.*), calcium oxide = 5.52 MJ/kg (*q.v.*), steam = 4.49 MJ/kg (*q.v.*) and water = 0.046 MJ/gallon (*q.v.*).

(510) The catalytic oxidation of ethylene for which the production of 2240 lb ethylene oxide requires inputs of 2200 lb ethylene, 1700 kWh electricity, 200 lb steam and 50,000 gallons water. Materials inputs have been converted to energy on the basis of ethylene = 68.12 MJ/kg (*q.v.*), steam = 4.49 MJ/kg (*q.v.*) and water = 0.046 MJ/gallon (*q.v.*).

(511) Based on the combustion of natural gas in sulphur for which inputs of 11,000 cu.ft. of natural gas and 1850 lb sulphur yields 2240 lb of carbon disulphide. Sulphur has been converted at the rate of 6.61 MJ/kg (*q.v.*). The value does *not* include the energy to melt the sulphur prior to the reaction.

(512) Based on a direct fuel gas energy requirement of 22×10^6 Btu per ton of ammonia liquified. This fuel is used to drive compressors.

(513) Based on requirements of 26,000 cu.ft. natural gas and 108 kWh electricity per ton ammonia.

(514) A supply of 1000 m^3 air per hour requires a 55 kw motor. Air density assumed 1.25 kg/m^3.

(515) Based on inputs of 1800 lb ethylene and 19,000 lb air.

(516) Requirements per kg nitric acid are 0.287 kg ammonia gas, 0.0099 kWh electricity, 0.875 kg steam and 59.5 gallons water. Ammonia gas has been converted at the rate of 44.02 MJ/kg (*q.v.*) steam at 4.49 MJ/kg (*q.v.*) and water at 0.04 MJ/gallon (*q.v.*).

(517) Requirements are 9 kWh electricity and 1.57×10^6 Btu fuel (assumed natural gas), both per ton of ammonia.

(518) Based on the conversion requirements of crude oil to 1.3 butadiene from Note 176 and from 1.3 butadiene to polybutadiene from Note 344.

(519) Based on the requirements given in Note 17. Monomer production has been used as 92.43 MJ/kg for acrylonitrile (*q.v.*), 130.12 MJ/kg for styrene (*q.v.*) and 108.46 MJ/kg for polybutadiene (*q.v.*).

(520) Based on a requirement of 110 kg of fuel oil per ton of alumina for calcination. The fuel is assumed to be light fuel oil.

(521) Newton, G. E. H., Fuel Power Conference (1973), Paper 11, 'Fuel and productivity in the pulp and paper industry'.

(522) Based on a direct electrical requirement of 2000 kWh electricity for mechanical pulping. The lowest value expected corresponds to a value of 17.7 MJ/kg for the total energy requirement.

(523) Based on the direct requirement of 3.5 ton of steam and 250 kWh of electricity per ton of pulp produced. Steam has been converted at the rate of 4.49 MJ/kg (*q.v.*). These values relate to chemical pulping.

(524) Wahlman, M., *Pulp and Paper International*, 60, November 1974.

(525) Average values based on requirements of 16.8 MJ steam per kg pulp and 0.85 kWh electricity per kg pulp, thought to represent a balance between mechanical and chemical pulping.

(526) Based on average typical requirements per ton of paper of 0.5 ton of steam (assumed oil-raised) and 300 kWh electricity.

(527) Based on average typical requirements of 25.6 MJ/kg of steam (assumed coal-based) and 1.77 kWh electricity per kg for a non-integrated pulp-paper processing plant. In practice only a proportion of the energy requirement will be satisfied by fuels bought externally. Typically the purchased fuel would be 10.5

MJ steam/kg and 0.82 kWh electricity per kg. The remainder of the requirement will be satisfied by waste wood products internally produced. Feedstock energy of wood has not been included.

(528) Based on average typical requirements of 22.2 GJ steam per ton paper and 1560 kWh electricity per ton paper for an integrated pulp-paper mill. In practice only a proportion of this requirement is satisfied by purchased fuels because internally produced waste wood products are used. Typically the purchased fuels will be 7.1 GJ steam per ton paper and 760 kWh electricity per ton.

(529) Based on a stated total coal equivalent fuel energy requirement. It is assumed that one third of this energy is consumed as electricity.

(530) Based on the production of 0.019 kg bottles produced at the rate of 660 per hour. The total electrical consumption of the moulding machine is 30 kW, of which 0.5 kW operates the heaters. These are switched on for only 60% of the time that the machine is in continuous operation. Air is the main service and this is produced by a compressor driven by a 15 hp motor which is on-load for only 80% of the time.

(531) Data for different local authorities in the U.K. representing collection from urban, mixed urban/rural and rural areas. There is no simple pattern between rural or urban collection.

(532) Based on a direct electrical energy requirement of 2760 kWh/ton of film.

(533) Based on a direct electrical energy requirement of 940 kWh/ton of bottles produced.

(534) Based on a direct electrical requirement of 5.24 kWh per 1000 bottles each of mass 20 g.

(535) Requirements of 2.8 ton steam (250 psig) and 49.5 kWh electricity convert 0.22 ton ammonia gas and 0.815 ton nitric acid to 1 ton of ammonium nitrate. Steam has been converted at the rate of 4.49 MJ/kg (*q.v.*).

(536) As in Note 535 except that steam has been assumed raised from an 80% efficient coal-fired boiler for which the direct energy input is 1437.5 Btu per lb of steam produced.

(537) Based on the requirements stated in Note 535 above, with ammonia converted at the average rate of 28.82 MJ/kg (*q.v.*) and nitric acid converted at the rate of 19.30 MJ/kg (*q.v.*).

(538) Based on the requirements of Notes 536 and 537 above.

(539) The process operated by the Girdler Corporation, for which 1 kg of ammonium nitrate requires inputs of 0.239 kg ammonia solution, 0.8832 kg nitric acid, 1.429 kg steam and 0.0281 kWh electricity. Ammonia solution has been converted at the average rate of 45.50 MJ/kg (*q.v.*), nitric acid at the rate of 19.30 MJ/kg (*q.v.*) and steam at the rate of 4.49 MJ/kg (*q.v.*).

(540) Fuel and steam requirements only as given in Note 539 above.

(541) Based on the Stenge process, for which the inputs per kg of ammonium nitrate are 0.22 kg ammonia solution, 0.815 kg nitric acid, 0.0543 kWh electricity, 2.8 kg steam and 2.181 imperial gallons of water. Ammonia solution has been converted at the average rate of 45.50 MJ/kg (*q.v.*), nitric acid at the rate of 19.30 MJ/kg (*q.v.*), steam at the rate of 4.49 MJ/kg (*q.v.*) and water at the rate of 0.046 MJ/imperial gallon (*q.v.*).

(542) Fuel and service requirements only from Note 541 above.

(543) Based on the requirements per kg of ammonium sulphate of 0.26 kg ammonia solution, 0.8625 kg sulphuric acid, 0.0297 kWh electricity and 1.525 MJ oil fuels. Ammonia solution has been converted at the average rate of 45.50 MJ/kg (*q.v.*) and sulphuric acid at the rate of 2.48 MJ/kg (*q.v.*).

(544) Fuel and services only as given in Note 543 above.

(545) *Encyclopaedia of Chemical Technology,* Volume 1, page 947, 'Aluminium and Aluminium Alloys'.

(546) Electrorefining of aluminium from the Hall cell to produce very high purity aluminium. For 1 ton of purified aluminium the input requirements are 18,000 kWh electricity, 20 kg graphite and 60 kg of bath materials. These last two have not been taken into account.

(547) Lath, R. P. and Menon, P. M., *Trans. Ind. Institute of Metals,* December, 1969, p. 39.

(548) Based on requirements of 3.5 tons steam and 240 kWh electricity per ton of alumina produced. Steam has been converted at the

rate of 4.49 MJ/kg (*q.v.*). The process also requires 120 kg sodium hydroxide per ton alumina but this has not been taken into account in the calculation.

(549) Based on a total energy requirement of 0.21 toe per tonne, assumed to be all electricity produced with an efficiency of 30%.

(550) Based on the requirements of fuel for melting cathode copper in a reverberatory furnace prior to casting into wire bar. Furnace consumes 2.83 cu.ft. of natural gas, 0.0042 kWh electricity and 0.0036 kg coke per kg of copper melted. Other requirements not costed in the Table are 0.0223 kg of wood poles/kg copper and a fuel energy consumption of 0.026 MJ/kg for ancilliary services for which there is no fuel breakdown. When the furnace is operated with a waste heat boiler the reference estimates that 286 kg of steam can be generated/kg copper melted. This represents an energy credit.

(551) Based on the requirements of fuel for melting cathode copper in an Asarco shaft furnace and of fuel for the holding furnace. Shaft furnace consumes 1.25 cu.ft. of natural gas/kg copper melted. Holding furnace requires 0.041 cu.ft. natural gas/kg copper. Electricity consumption is 0.006 kWh/kg copper melted. Not costed in the Table is a fuel energy consumption of 0.026 MJ/kg for ancilliary services for which there is no fuel breakdown.

(552) Based on the fuel requirements for melting cathode copper in an electric arc furnace. Furnace power is 0.216 kWh/kg, with an additional 0.0032 kWh/kg for tilting the furnace. Process consumes 0.002 kg of graphite electrode/kg of copper melted which is not costed in the Table. A fuel energy consumption of 0.026 MJ/kg for ancilliary services for which there is no breakdown is also excluded.

(553) Based on a stated energy requirement of 1.2 toe/tonne for the production of copper billet by an unspecified route and from an unspecified ore grade.

(554) Based on a stated energy requirement of 36,485 kWh_t/ton by an unspecified route and from an unspecified ore grade.

(555) Production of secondary copper for a stated energy requirement of 5.70 kWh_t/kg. The route is not specified so that a breakdown is not possible. Will be a mixture of thermal fuels for melting and fire refining and electricity for electrorefining.

(556) Based on a fuel energy requirement of 0.288 kWh_t/kg for a furnace fire refining copper. Fuel assumed to be all medium fuel oil.

(557) Energy to produce copper from copper matte in an oxygen blown converter. Process requires 10 m^3 oxygen and 15 kWh electricity per ton *concentrate* fed to the previous matte smelting stage. Table is constructed assuming a 22% copper concentrate feed so that the fuel requirement is 0.067 kWh electricity/kg copper output. The oxygen requirement of 0.045 m^3 O_2/kg copper is not included.

(558) Probably an average energy for rolled copper products. Electricity converted at an unspecified efficiency and no starting ore grade given.

(559) An average energy for rolled copper and copper alloy products.

(560) Probably an average energy for copper wire taken from values for a range of diameters. Unspecified starting ore grade.

(561) Entry calculated from entries covered by Notes 239, 240, 245 and 246 using the appropriate materials conversion efficiencies. Value quoted is energy requirement to produce fine annealed copper wire from wire bar.

(562) Based on a requirement of 0.197 kWh/kg for the electrorefining of copper.

(563) Davies, D., *Metals and Materials*, p. 176, March 1974.

(564) Based on the energy to produce electrolytic copper from scrap by secondary smelting and refining. Fuels consumed are 922 kWh electricity, 62 gallons oil and 0.76 tonnes coke per tonne of cathode copper.

(565) Cofer, D. B., 'SCR – The Southwire Continuous Rod System', *Metals and Materials*, p. 74, February 1972.

(566) Based on the energy to manufacture copper rod by melting cathode copper in an Asarco furnace and continuous-casting in a 30 ton per hour plant.

(567) Snelgrove, W. R. N. and Barber, M. J., 'The Continuous-Casting of Anodes at British Copper Refiners Ltd'.

(568) Based on the energy to produce copper strip by melting blister copper and scrap in an Asarco shaft furnace and continuous-casting. Fuel requirements are 0.037 kg butane and 0.0226 kWh electricity per kg strip.

(569) Based on the fuel energy to operate a blast furnace (9240 Btu/lb). No electricity consumption is charged but credits have been made for blast furnace gas and coke breeze. Entry has been calculated assuming energy source 100% coke.

(570) Based on the energy to produce pig iron in a submerged-arc electric smelting furnace. Process requires 0.3125 kg coke and 1.967 kWh per kg of hot metal.

(571) Based on the fuel energy to operate a blast furnace. Fuel inputs are 0.55 kg coke and 0.0983 kWh electricity per kg iron.

(572) Based on the fuel energy to operate a blast furnace. Fuel inputs are 0.491 kg coke and 0.01475 gallons fuel oil per kg iron. No electricity consumption given.

(573) Based on the fuel energy to operate a blast furnace. Fuel inputs are 0.6 kg coke, 0.358 cu.ft. natural gas and 0.0043 gallons oil per kg iron. No electricity consumption given. Data relate to average U.S. practice in 1973.

(574) Based on the fuel energy to operate a blast furnace. Fuel requirement is 0.55 kg coke/kg iron. Materials inputs are 1.45 kg iron ore pellets (66% Fe 5% SiO_2), 0.0925 kg limestone and 0.125 kg dolomite/kg iron. These inputs have not been costed in the Table.

(575) Based on natural gas consumption to operate HyL process for direct reduction. No electricity consumption given.

(576) Direct reduction of iron ore using natural gas in an unidentified process. No electricity consumption given.

(577) Direct reduction of iron ore using natural gas (0.373 m^3/kg) in the Purofer process. No electricity consumption given.

(578) Direct reduction of iron ore in the MIDREX process. No electricity consumption given.

(579) Direct reduction of iron ore in the MIDREX process. Reduction is by a mixture of CO and H_2 produced by reforming methane with O_2 and recirculated top gas. Production energy of fuel therefore taken as negligible. No electricity consumption given.

(580) Direct reduction of iron ore using coal in the SLRN process. No electricity consumption given.

(581) Direct reduction of iron ore in the Kinglor-Metor shaft furnace. Fuel inputs are 0.364 kg coal, 7.067 cu.ft. natural gas and 0.068 kWh electricity/kg sponge iron.

(582) Direct reduction of iron ore in the HyL process using 16.72 cu.ft. natural gas/kg sponge iron. No electricity consumption given.

(583) Direct reduction of iron ore in the HyL process using 19.96 cu.ft. of natural gas/kg sponge iron. No electricity consumption given.

(584) Direct reduction of iron ore in the MIDREX process using 11.80 cu.ft. natural gas per kg sponge iron. No electricity consumption given.

(585) Direct reduction of iron ore in an unspecified plant. Fuel assumed to be natural gas. No electricity consumption given.

(586) Electrical energy (0.0095 kWh/kg ore) to operate an ore blending plant.

(587) Based on the natural gas consumption of a pelletisation plant handling taconite iron ores (0.5×10^6 Btu/ton pellets). The oxidation of Fe_3O_4 to Fe_2O_3 supplies part of the overall heat requirements.

(588) Based on the natural gas consumption of a pelletisation plant handling non-taconite iron ores. There is no Fe_3O_4 to be oxidised so that the fuel energy requirement is higher than that described under Note 587.

(589) Production of unspecified steel product from ore.

(590) Production of unspecified steel product. Number quoted is fuel energy only with no fuel production energy taken into account.

(591) Production of unspecified steel product (probably sheet) from raw materials.

(592) Production of steel in the basic oxygen furnace. Electric power requirement (for tipping furnace) is 30 kWh/ton. Process consumes 0.075 tons O_2 and 0.075 tons flux/ton steel. These inputs not charged in the entry.

(593) Production of steel from recycled metal by an unspecified route.

(594) Production of steel from scrap by an unspecified route.

(595) Production of continuous-cast steel from unspecified inputs, probably raw materials. Based on a steel value of 34.3 MJ/kg with 79% yield from continuous-casting. Electricity component converted at 33% efficiency but no fuel breakdown given.

(596) Authors' estimate.

(597) Energy quoted will be a mix of electrical consumption for rolling and gas or oil for reheating furnaces. No fuel breakdown given.

(598) Based on requirements for one ton of finished steel of 1.63 tons ore, 0.28 tons limestone, 0.89 tons coal for coke. No other fuel breakdown given.

(599) Production of steel by direct reduction of iron ore with natural gas followed by electric arc furnace. Fuel requirements are 731 kWh electricity/tonne and 336 m^3/tonne natural gas.

(600) Production of steel via blast furnace; basic oxygen furnace route. Basic oxygen furnace charged with hot metal and 21% direct-reduced sponge iron. Takes account of coal (via coke) input to blast furnace but not other materials inputs. Natural gas consumption for production of 21% direct-reduced metal included. Takes account of raw materials to b.o.f. (oxygen and lime). Oxygen charged as in entry associated with Note 206 and lime charged as in entry associated with Note 497. Process operated with basic gangue. No electricity charged other than for O_2 manufacture.

(601) Production of steel via blast furnace; basic oxygen furnace route. Basic oxygen furnace charged with hot metal and 25% direct-reduced sponge iron. Takes account of coal (via coke) input to blast furnace but no other materials inputs. Natural gas consumption for the production of the 25% direct-reduced iron included. Takes account of oxygen and lime usage in b.o.f., oxygen charged as in the entry associated with Note 206 and lime as in entry associated with Note 497. Process operated with acid gangue and requires 0.86 kg hot metal and 0.286 kg sponge iron/kg steel. No electricity charged other than for O_2 manufacture.

(602) Electric arc furnace; average consumption for two adjacent furnaces with interchangeable roofs, one preheating with LPG while other melting down. Includes consumption of 0.0045 kg of electrode/kg steel.

(603) Production of steel via blast furnace; basic oxygen furnace route. Basic oxygen furnace charged with 4% ore, 14% scrap, with balance hot metal. Takes account of coal (via coke) inputs to blast furnace but not other materials inputs to this furnace. Takes account of oxygen and lime usage in b.o.f. (see Note 601). No charge made for scrap and ore to b.o.f. Process operated with basic gangue and requires 0.916 kg pig iron, 0.162 kg scrap and 0.046 kg ore/kg steel. No electricity charged other than for O_2 manufacture.

(604) Production of steel via blast furnace; basic oxygen route. Takes account of coal (via coke) input to blast furnace and oxygen and lime to b.o.f. (see Note 601). B.o.f. charged with 0.3 kg scrap and 0.8 kg pig iron/kg steel. No electricity consumption given other than for O_2 manufacture.

(605) Same ratio of ore/scrap/hot metal as in Note 603 but process operated with acid gangue. 1 kg of steel under these conditions requires 0.918 kg pig iron, 0.162 kg scrap and 0.044 kg ore.

(606) Electric power consumption (7.2 kWh/t) for scrap handling (cranes and so on) within a mini-steelworks.

(607) Direct energy to 6.62 MJ/kg assumed to be all electrical.

(608) Production of unspecified sheet product requiring 457.8 kWh electricity per ton and 4,176,306 kcal of unspecified thermal fuel/ton. Entry calculated assuming thermal fuel is coke.

(609) Based on an average steel production energy of 48.81 MJ/kg plus 30.22 MJ/kg for stainless production.

(610) Feltoe, F. J., 'Energy Requirements and the Problems of Energy Management in a Newly Commissioned Steelworks', *J. Inst. Fuel*, p. 66, February 1965.

(611) Based on the energy to roll ingot to slab. 32.56 kWh electricity/ton and 0.128 therms of steam/ton. Calculation assumes steam raised in coal-fired boiler 80% efficient.

(612) Based on the fuel energy to fire a steel-soaking pit; 0.00736 kWh electricity and 10.65 therms gas/ton steel.

(613) Based on the fuels and material inputs to an Outokumpu flash smelter. Process requires 110 kg steam, 110 m^3 oxygen, 86 kWh

electricity and 38 kg of oil per metric ton of concentrate smelted. Steam and oxygen have been charged for in the entry. A typical concentrate contains 22% copper. The values given are those with no waste heat recovery. With waste heat recovery 900 kg of steam is generated which provides all process steam, oxygen manufacture and electric power requirements and yields an overall surplus of 130 kg of steam.

(614) Based on an electrical consumption of 0.1 kWh/kg and a non-electrical consumption of 1.9 kWh/kg, which for the purposes of the Table entry has been assumed to be natural gas.

(615) Based on information from Laboratorium Meteriaalkunde, report No. ML/75/1, Technische Hogeschool, Eindhoven, Holland, in which average data for Dutch brick industry indicates that electricity forms 2.4% of fuel energy content.

(616) Reference quotes a value for glass making of 7740 Btu/lb. No fuel breakdown is given but the electricity component is converted with an efficiency of 29.4%.

(617) Reference quotes a value of 0.48 toe/tonne. No fuel breakdown is given but the electrical component assumes a value of 30% for efficiency and the oil fuels component a value of 87.5%.

(618) Based on a stated requirement of 360 kWh of electricity per tonne and 0.22 toe/tonne (assumed oil fuels).

(619) Based on a stated requirement of 0.22 kWh/bottle (assumed all electricity).

(620) Reference quotes fuel requirements of 0.112 kWh electricity, 2 cu.ft. natural gas, 0.01 gallons fuel oil and 2.1 gallons of gasoline for the bottling and transport to consumer of 1 U.S. gallon of beverage. Table entry calculated assuming that the gasoline is used for transport and the remaining fuels for bottling (converted to Imperial gallons).

(621) Based on a requirement of 17.12 kWh electricity and 36.98 kWh of thermal fuels/1000 boxes. The thermal fuel has been assumed to be medium fuel oil in the Table entry.

(622) Based on a fuel energy of 0.2 kWh/brick. No conversion efficiencies have been assumed in the reference when arriving at this figure. The Table entry has been calculated assuming that electricity is 2.4% of this total with the remainder natural gas.

(623) Energy requirement per m^2 for the materials inputs for the manufacture of a four storey block of flats with vertical brick load-bearing elements.

(624) Energy requirement per m^2 for the materials inputs for the manufacture of a nine storey block of flats with reinforced concrete load-bearing structure.

(625) Energy requirement per m^2 for the materials inputs for the manufacture of a conventional two storey house with a brick or block load-bearing structure.

(626) Energy requirement per m^2 for the materials inputs for the manufacture of a two storey house with load-bearing crosswalls and lightweight infills.

(627) Energy requirement per m^2 for industrial buildings.

(628) Based on a stated requirement of 2500 Btu/lb. No fuel breakdown is given but any electricity component converted at 33% efficiency.

(629) Based on a stated requirement of 6.3 MJ/kg. The electrical component in this value has been converted at 31.5% efficiency. No other fuel efficiencies have been used.

(630) Based on a natural gas consumption of 68.57 m^3/tonne of anode fabricated from petroleum coke for aluminium smelting. In addition to the natural gas usage there will be a small electrical contribution for mixing ingredients and compacting the anode prior to firing.

(631) Based on a stated fuel requirement of 0.45 kWh electricity, 0.23 m^3 natural gas and 0.084 l oil per 1000 crowns.

(632) Energy to manufacture tinplate used for crown-making represents 69% of the total energy requirement.

(633) Value quoted excludes transport and capital.

(634) System is unspecified but excludes transport and capital. The range of values quoted is for concrete made from cement, sand and aggregate in the proportions 1:3:6. Probably includes the energy to manufacture the concrete and the energy to manufacture the ingredients.

(635) As in Note 634, except the range of values quoted is for a 1:2:4

mix.

(636) Any electrical or oil fuel contributions to the overall total are converted at efficiencies of 26.5% and 87.5% respectively.

(637) Values quoted are for the same mix as given in 635, but with the use of lightweight aggregate.

(638) Value quoted is probably the energy requirement for the manufacture of concrete plus the energy requirement for the provision of sand, gravel, cement and water. Any electricity component will have been converted at 31.5% efficiency. No other fuel production efficiencies used. Excludes capital energy.

(639) As in 638 except that the value quoted makes allowance for 100 kg of steel reinforcement/m^3 concrete. Steel is charged at 23 MJ/kg.

(640) Laboratorium Materiaalkunde Report No. ML/75/4, Technische Hogeschool Eindhoven, Holland, April 1975.

(641) Based on a stated energy requirement of 0.011 MJ electricity (converted at 31.5% efficiency), 0.001 MJ natural gas, 0.006 MJ fuel oil, and 0.04 MJ diesel for transport.

(642) Based on a stated consumption of 281 kWh electricity/ton and 4307 kWh thermal fuel/ton. Latter assumed to be natural gas.

(643) Stated energy requirement is 40.44 MJ/kg. This will probably include electricity converted at an unspecified efficiency plus any fuels required for intermediate annealing treatments.

(644) Value based on table entries for 1% ore, 25% concentrate.

(645) Stated energy requirement is 0.25 MJ/kg but fuel type not specified.

(646) System is unspecified but excludes capital. Any electricity component converted at 32.8% efficiency.

(647) Based on a stated energy requirement of 1.5 toe/tonne. No fuel breakdown given.

(648) Stated energy requirement is 0.12 kWh/lb. No fuel breakdown given.

(649) Average values for U.K. industry.

(650) Calculated value by author, see Chapter 12.

(651) A stated energy requirement of 21 MJ/kg which contains electricity converted at 31.5% and no fuel conversion efficiencies applied to the other energy sources.

(652) Based on a stated requirement of 6700 kWh thermal/short ton for production, 300 kWh_t for capital, 200 kWh_t for transport. Electrical efficiency 30%. Breakdown has converted entries to authors' efficiencies.

(653) A stated requirement of 30.22 MJ/kg. No fuel breakdown given.

(654) Based on a stated requirement of 17.43 MJ/kg. No fuel breakdown given.

(655) The stated requirement is 28.59 MJ/kg. No fuel breakdown given.

(656) Any electrical component converted at 31.5% efficiency. No other fuel efficiencies used.

(657) Production energy for beer labels of mass 0.8 g.

(658) Stated energy consumption is 8×10^6 Btu/ton. Entry has been calculated assuming that this is all natural gas.

(659) Any electricity component has been converted at an efficiency of 29.4%.

(660) Value quoted is for tinplate can with easy-open aluminium end. Includes all the fuels used in the factory including factory heating.

(661) Value quoted is for a tinplate can with easy-open aluminium ends.

(662) The values have been calculated from data on the requirements to convert 107,800 tonne of steel into beer cans with an aluminium end. These requirements are 8 Gwh and 3897 toe for pressing and enamelling and 13.1 Gwh and 1026 toe for making-up. The entry assumes that all non-electrical fuel is medium fuel oil.

(663) Average value for electrically-powered walking dragline. Reference quotes a range of consumption for mining phosphate rock of 0.9–1.2 kWh/m^3.

(664) Reference quotes following data: 0.536 kWh of electricity per ton ore for drilling, 0.67 kWh thermal for drill bits, 0.003 kWh thermal for the drilling machine construction and 3.948 kWh_t/ton for explosives. Assumes that 3 tons of overburden removed for each ton of ore mined.

(665) Stated requirements per kg of concentrate are explosives 5 kcal,

mining 9 kcal, and beneficiation (flotation) 52 kcal.

(666) Production route is a wet process for which the requirements for 1 kg of orthophosphoric acid are 1.636 kg of concentrate, 0.591 kg steam, 0.011 gallons water and 0.105 kWh electricity. Entry takes account of electricity usage and steam usage for which the values associated with Note 111 have been employed. The water and concentrate have not been charged.

(667) Production of phosphoric acid as a 75% solution by a wet process for which the requirements per kg of acid are 3.06 kg concentrate, varying amounts of H_2SO_4, 4.55 kg steam and 0.248 kWh electricity. Entry takes account of electricity and steam usage only for which the value associated with Note 111 has been employed. The reaction is strongly exothermic and significant heat energy could be reclaimed.

(668) Production of phosphoric acid as a 35% solution by the Chemical and Industrial Corporation process. The requirements for 1 kg of acid are 2.433 kg concentrate, 1.861 kg H_2SO_4, 0.744 kg steam and 0.1103 kWh electricity. Entry takes account of electricity and steam usage only. The value associated with Note 111 has been assumed for steam.

(669) Production of 35% P_2O_5 concentrate requires 0.00041 kg of NaOH, 3.25 imperial gallons water, 0.025 kWh electricity, 0.0021 kg of fuel oil and 0.0027 kg of fatty acids. Electricity and fuel oil only are charged in the entry.

(670) Sealing compound assumed to be polybutadiene.

(671) Based on an energy requirement of 13.75 kWh/kg and 0.86 kg of carbon as coke per kg silicon.

(672) Based on a stated requirement of 0.5 ton coal/ton of sodium carbonate.

(673) Based on a fuel requirement of 0.07 kg of oil/kg steam.

(674) Based on a stated energy requirement of 1.0 toe/tonne.

(675) Biswas, A. K. and Davenport, W. G., *Extractive Metallurgy of Copper*, Pergamon Press, 1976.

(676) Based on a stated requirement of 500 m^3 natural gas/ton.

(677) Based on a stated requirement of 0.5×10^5 kcal/tonne for a *liquid* charge.

(678) Based on a stated requirement fo 300 kWh/tonne.

(679) Based on a stated requirement of 4×10^5 kcal/tonne.

(680) Based on a stated requirement of 25 kWh/tonne.

(681) Based on a stated requirement of 0.58 kg NH_3, 0.76 kg CO_2, 0.9 kg steam, 0.125 kWh electricity for the Snam Progetti Process. Values used for steam, ammonia and CO_2 are 3.75 MJ/kg, 45.50 MJ/kg and 0.70 MJ/kg (*q.v.*) respectively.

(682) Mitsui process for which the stated requirements are 0.575 kg NH_3, 0.76 kg CO_2, 1.3 kg steam and 0.165 kWh electricity. Assumed data as in Note 681.

(683) Montecatini process for which the requirements are 0.58 kg NH_3, 0.76 kg CO_2, 1.5 kg steam, 0.16 kWh electricity. Assumed data as in Note 681.

(684) Chemico process for which the requirements are 0.58 kg ammonia, 0.75 kg CO_2, 1.25 kg steam, 0.167 kWh electricity. Assumed data as in Note 681.

(685) Stannication process for which the requirements are 0.57 kg NH_3, 0.755 kg CO_2, 1.5 kg steam, 0.145 kWh electricity. Assumed data as in Note 681.

(686) The stated requirement is 24 kWh/ton for incineration and 73 kWh/ton for transportation of residue to landfill. No fuel breakdown given.

(687) Based on a stated requirement of 1.8 kWh thermal/kg. Assumed to be all electrical and was converted in the original reference at 23.85% efficiency. Table entry corrected to authors' efficiency.

(688) Based on a stated requirement of 9.30 MJ/kg. Assumed to be all electricity. Electrical efficiency has been converted to authors' efficiency from value of 32.5% in original reference.

(689) Based on a stated requirement of 62,770 kWh/ton. No fuel breakdown given.

(690) Based on a stated requirement of 73,345 kWh/ton. No fuel breakdown given.

(691) Based on a stated requirement of 23.24 MJ/kg for operations in the U.S. Chrysler foundries.

(692) Based on a stated requirement of 3105 kWh/ton. No fuel breakdown given.

(693) Based on a stated requirement of 1.67 kWh_t/kg; assumed to be natural gas.

(694) Based on a stated requirement of 1750 kWh_t/ton; assumed to be natural gas.

(695) Based on a stated requirement of 73,345 kWh/ton, no fuel breakdown given.

(696) Based on a stated requirement of 1500 kWh/ton, assumed to be natural gas.

(697) Based on a stated requirement of 790 kWh/ton; assumed to be medium fuel oil.

(698) Based on a stated requirement of 5300 kWh/ton. Probably mostly electrical for operating rolling mills.

(699) Based on a stated requirement of 73,440 kWh/ton.

(700) Based on a stated requirement of 66,000 kWh for production, 1000 kWh for capital, and 200 kWh/short ton for transport. No fuel breakdown given.

(701) Based on a stated requirement of 9850 kWh/ton. No fuel breakdown and no indication as to whether feedstock included.

(702) Based on a stated requirement of 1.4 toe/tonne. Value includes electricity at 30% and oil at a production efficiency of 87.5%. No feedstock included.

(703) Based on a stated requirement of 970 kWh/ton. No fuel breakdown is given nor are any details of the process.

(704) 'Energy Use Patterns in Metallurgical and Nonmetallic Mineral Processing (Phase 4 – Energy Data and Flow Sheets, High Priority Commodities)', U.S. Dept. of Commerce, National Tech. Information Service Report No. PB–245 759, June 1975.

(705) Based on a stated electrical requirement of 16,000 kWh/ton.

(706) Estimate by Batelle Columbus Laboratories, U.S.A.

(707) Limestone mining only. Based on a stated requirement of 0.439 gallons (U.S.) of diesel, 0.031×10^6 Btu unspecified other fuels, 0.857 kWh electricity and 1.067 lb explosives for 2.88 tons limestone. No charge in entry for explosives but reference quotes energy requirement of 0.03×10^6 Btu for explosives.

(708) Based on a stated requirement of 0.03×10^6 Btu/lb.

(709) Based on a stated requirement of 4.512 gallons (U.S.) of diesel and similar oils, 57.527 kWh electricity (of which 8.95 kWh used for dust collection) 0.167×10^6 Btu other unspecified fuels, 3015.3 cu.ft. natural gas, 0.1557 tons coal and 0.048 gallons gasoline for mining, crushing and calcining limestone to yield calcium oxide. 1.067 lb of explosives also used but not charged in the Table. Other fuels assumed to have same production efficiency as natural gas in Table entry.

(710) Based on a stated requirement of 0.07×10^6 Btu for the mining of 1 ton of clay or shale.

(711) Based on a stated requirement of 0.605×10^6 Btu/ton.

(712) Manufacture of Portland Cement from raw materials in the ground. Based on a stated requirement of 2956 cu.ft. natural gas, 119 kWh electricity, 6.8334 gallons (U.S.) diesel oil, 0.0908 tons coal, 1.37 tons limestone (charged as for Note 707), 0.144 tons clay and shale and 0.057 tons sand per ton of cement. In addition 0.01 ton refractories, 0.048 ton gypsum and 2 lb of steel balls are consumed per ton of cement. These have not been charged in the Table.

(713) Based on a stated requirement of 1.11 tons of clay, 35.5517 kWh electricity, 2712 ft^3 natural gas, 1.06 gallons (U.S.) diesel, 0.00564 tons coal and 0.017 gallons (U.S.) petrol per ton of bricks.

(714) Production of sponge copper by dump leaching followed by cementation with detinned steel scrap. Based on a stated requirement of 3904.8 kWh/ton. Process requires 2.5 tons of steel/ton of copper. No energy requirement charged for the scrap.

(715) Calculation for an ore of average grade 0.7% copper. Based on a stated requirement of 4506.58 kWh electricity, 53,667.3 cu.ft. natural gas, 136.752 U.S. gallons oil fuel and 0.0133 tons of coal per ton of electrolytic copper. Process described uses 0.1 ton cement copper/ton of product and this has been charged as in Entry 714. Process also consumes 155 lb explosives, 273 lb steel balls, 28.5 lb organic reagents, 541 lb inorganic reagents and 0.0165 tons of sulphuric acid per ton of copper. These entries not charged in the Table. Transport energies of waste rock to dumps (921.4 ton miles by truck), ore to mill (1261.3

ton miles by electric loco) and anodes to refinery (910 ton miles) have also been omitted from the Entry. In addition to 1 ton of copper, 2.3 tons of sulphuric acid also produced from associated acid plant and a surplus of 1305 kg of steam. No credits have been made for acid and steam production.

(716) Estimate for the production of all plate and sheet-glass in the U.S. Reference gives an approximate percentage distribution of the processing energy by process steps for sheet and float/plate glass.

(717) Mining and processing energy. Electricity converted at 32.5% efficiency.

(718) Based on a stated requirement of 9375.6 cu.ft. natural gas, 249.4 kWh electricity, 11.381 U.S. gallons oil and petrol, and 0.6615 gallons (U.S.) propane per ton of containers. This energy is for batch handling and cullet-crushing, fining and conditioning, forming, post-forming, product handling and space-heating. Materials inputs are 0.1 ton cullet, 0.5918 tons sand, 0.1817 tons limestone, 0.0838 tons feldspar, 0.1192 tons synthetic soda ash 0.0788 tons natural soda ash and 0.0788 tons of minor ingredients. No energy charge has been made for these inputs.

(719) Based on a stated requirement of 420 kWh/ton of oxygen.

(720) Production of refined lead from ore in the ground. Based on a stated requirement of 882.68 kWh electricity, 5.91 gallons (U.S.) of diesel, 0.44 gallons (U.S.) petrol, 6147.24 cu.ft. natural gas, 0.025 tons coke breeze and 0.23 tons coke per ton of lead. Materials inputs not costed in the energy Table are 12.35 lb explosives, 37.42 lb steel balls, 1.4616 lb organic reagents, 0.29 lb inorganic reagents, 0.002 tons sulphur, 0.12 tons limestone. 0.07 tons iron ore, 0.09 tons sand (silica), 0.01 tons soda ash, 0.0056 tons zinc for desilvering, 0.00066 tons calcium, 0.00173 tons magnesium, 0.001 tons solid caustic soda and 0.00025 tons sodium nitrate. 643.5 ton miles of rail transport for commodities also excluded from the Table.

(721) Energy for *solid* caustic soda.

(722) Estimated by authors of Reference 704 from *Industrial Chemicals* by Faith, W. L., Keyes D. B. and Clark, R. L., 3rd Ed., John Wiley & Sons, 1965.

(723) Manufacture of gaseous nitrogen (at 250 psi) from air by liquefaction, fractional distillation and pressurisation on a modified Linde-Frankl cycle. Oxygen and inert gases are also separated. Nitrogen forms 33.5% by weight of liquefied air and air-liquefaction and separation energy is apportioned in this basis (113 kWh(ton N_2). Pressurisation to 250 psi requires a further 165 kWh/ton N_2.

(724) Manufacture of liquid N_2 by process as in 723; energy is 113 kWh for air-liquefaction and separation (per ton N_2) plus 662 kWh/ton N_2 for liquefaction of N_2.

(725) Production of ammonia by the steam-reforming of natural gas. Based on a stated requirement of 22,310 cu.ft. natural gas feedstock, 15,940 cu.ft. natural gas as fuel and 28.0 kWh electricity per ton of ammonia. Materials inputs not costed in the Entry are 0.5 lb reforming catalyst, 0.3 lb shift reaction catalyst, 0.3 lb ethanolamine for CO_2 removal, 0.5 lb methanation catalyst and 0.5 lb of synthesis catalyst.

(726) Estimate by authors of Reference 704.

(727) Based on a stated requirement of 3.76 (U.S.) gallons oil, 0.163 (U.S.) gallons petrol, 1615.47 cu.ft. natural gas, 12,602.45 kWh electricity and 1.4456 tons coke breeze per ton phosphorus. Coke breeze charged at 21.77 MJ/kg fuel content; no production energy charged. Materials consumed and not charged in the Entry are 1.229 tons silica pebbles and 0.018 tons carbon electrodes per ton phosphorus. Transport by rail of 446 ton-miles also omitted from the Entry.

(728) Manufacture of phosphoric acid from Florida rock pebble (averaging 31.1% P_2O_5) obtained by open-pit mining. Based on a stated requirement of 3.72 (U.S.) gallons oil fuels, 318.2 kWh electricity, 702.7 cu.ft. natural gas and 2300 lb steam per ton of product. Steam according to 704 requires 3.25 MJ/kg. For calculation of Table entry this has been assumed raised from oil fuels. Materials consumed and not charged are 8.8 lb inorganic reagents, 31.6 lb organic reagents and 2.8 tons sulphuric acid.

(729) Based on a stated requirement of 73.786 kWh electricity, 4660

cu.ft. natural gas and 4.1 gallons (U.S.) of pitch-binder per ton of bricks. Fuel consumption is for sizing, mixing, pressing, curing, drying and firing and pitch inpregnation. One ton of bricks requires 0.068 ton dead-burned magnesia, 0.612 ton of synthetic magnesia, 0.32 ton chrome ore and 0.025 ton dead-burned dolomite. These inputs not charged for in the Entry. Similarly transport of 462 ton-miles by rail, 3666 ton-miles by sea and 1.25 ton-miles by truck not charged.

(730) Based on a stated requirement of 34.49 kWh electricity, 3956 cu.ft. natural gas and 1.425 gallons (U.S.) oil fuels per ton of bricks. Process requires an input of 1.11 tons of clay per ton of bricks (not charged). Clay transport of 3.33 ton-miles by truck also excluded from the Entry.

(731) Mining of sulphur by the Frasch process. Based on a stated requirement of 1864 cu.ft. natural gas, 0.031×10^6 Btu oil fuels and 0.94 kWh electricity for 0.337 tons of sulphur.

(732) Production of sulphuric acid by the burning of sulphur mined by the Frasch process. Process requires the mining of 0.337 tons of sulphur to produce one ton of 100% sulphuric acid. Stated requirement is 1864 cu.ft. natural gas, 0.031×10^6 Btu oil fuels and 25.44 kWh per ton sulphuric acid. Burning reaction produces 0.45 kg of steam/kg of sulphuric acid as a by-product. No energy credit has been given for the steam. 270 ton-miles of barge transport also excluded from the Entry.

(733) Production of zinc by the electrothermic process. Based on a stated requirement of 4029.72 kWh electricity, 6.74 gallons (U.S.) diesel, 0.54 gallons (U.S.) petrol, 2327 cu.ft. natural gas, 0.19 tons coke breeze and 0.681 tons of coke per ton of zinc. Materials inputs not charged in the Entry are 46.4 lb steel, 15.4 lb explosives, 2.63 lb organic reagents, 39 lb inorganic reagents, 0.044 tons sand. 0.01 tons refractories and 2.2 lb of electrode carbon per ton zinc. Transport by rail of 1716 ton-miles also excluded. Process described also produces 1.705 tons of sulphuric acid/ton Zn from the associated acid plant. No credit given for acid production.

(734) Production of zinc by the vertical retort process. Based on a stated requirement of 1336.92 kWh electricity, 14,205 cu.ft. natural gas, 7.0 gallons (U.S.) diesel, 0.56 gallons (U.S.) petrol, 8.37 gallons (U.S.) fuel oil, 0.4 tons anthracite and 0.792 tons coal per ton of zinc. Materials inputs not charged in the Entry are 47.85 lb steel, 15.87 lb explosives, 42 lb bentonite, 2.76 lb organic reagents, 41.0 lb inorganic reagents and 0.317 t clay binder per ton of zinc. Transport by rail of 3148 ton-miles and by water of 5100 ton-miles also excluded. Plant produces 1.59 net tons of sulphuric acid per ton zinc. No credit given for acid production.

(735) Production of zinc by the electrolytic process. Based on a stated requirement of 4909.7 kWh electricity, 5150 cu.ft. natural gas, 7.4 gallons (U.S.) diesel, 0.6 gallons (U.S.) petrol and 8.14 gallons (U.S.) propane per ton of zinc. Materials inputs not charged in the Entry are 50.34 lb steel, 16.87 lb explosives, 85 lb manganese dioxide, 2.89 lb organic reagents and 58.9 lb inorganic reagents per ton of zinc. Transport by rail of 1050 ton-miles also excluded. Plant produces 1.54 net tons of sulphuric acid per ton zinc. No credit given for acid production.

(736) Production of iron ore pellets from ore in the ground. Based on a stated requirement of 119.56 kWh electricity, 275 cu.ft. of natural gas, 0.631 gallons (U.S.) diesel, 0.039 gallons (U.S.) petrol and 1.5 gallons (U.S.) fuel oil per ton of pellets. Materials consumed but not charged are 3.5 lb explosives, 20 lb of steel balls, rods and liners and 16 lb of bentonite.

(737) Production of iron ore sinter. Based on a stated requirement of 150 cu.ft. natural gas, 30 kWh electricity and 0.074 tons coke breeze per ton sinter. Materials inputs not charged in the Entry are 0.61 tons ore fines. 0.267 tons of returns, 0.143 tons flue dust and fines. 0.122 tons mill scale and 0.049 tons limestone per ton of sinter.

(738) Based on a direct fuel requirement of 7030 kWh electricity and 25.6×10^6 Btu of hydrocarbon fuels (assumed natural gas) per ton of electrode. Process requires 1.5 tons of petroleum coke, 700 lb of binder-pitch and 300 lb of impregnating-pitch per ton of graphite. These inputs have not been charged to the Entry.

Appendix 1
Conversion factors for units

Conversion factors

To convert from:	to:	multiply by:
barrels (petroleum)	litres	158.98
Btu's	joules	1054
calories	joules	4.184
cubic feet	litres	28.316
cubic inches	litres	0.01639
cubic yards	litres	764.534
feet	metres	0.3048
gallons (British)	litres	4.546
gallons (U.S.)	litres	3.7853
horsepower	joules/sec	745.7
inches	metres	0.0254
kilowatts	joules/sec	1000
kilowatt-hours	joules	3.6×10^6
knots	km/hr	1.852
knots	miles/hr	1.1508
miles (statute)	kilometres	1.609
miles/hr	km/hr	1.609
ounces (avdp)	kilogrammes	0.02835
ounces (troy)	kilogrammes	0.03110
pounds	kilogrammes	0.4536
square feet	square metres	0.0929
ton oil equivalent (toe)	joules	4.375×10^{10}
ton coal equivalent (tce)	joules	2.88×10^{10}
tons (long)	kilogrammes	1016.05
tons (short)	kilogrammes	907.185
watts	joules/sec	1
yards	metres	0.9144

References

[1] Agarwal, J. C. and Sinek, J. R., in *Energy use and conservation in the metals industry,* Met. Soc. A.I.M.E., p. 121 (1975).

[2] Altenpohl, D., Daugherty, T. S. and Blum, W., Proceedings of the 9th International TNO Conference, Rotterdam, February 1976.

[3] Atkins, P. R., *Engineering and Mining Journal* (1973).

[4] Ayres, R., Saxon, J. and Stern, M., *Materials–Process–Product Model,* International Research and Technology Corp., Arlington, Va., July 1974.

[5] Ban, T. E. in *Energy use and conservation in the metals industry,* Met. Soc. A.I.M.E., p. 173 (1975).

[6] Barnes, R. S., *Ironmaking and Steelmaking Quarterly,* No. 4, p. 271 (1975).

[7] Barnes, R. S., *Proc. Inst. Mech. Eng. London,* **190**, p. 153 (1976).

[8] Barnhart, T. F. in *Energy use and conservation in the metals industry,* Met. Soc. A.I.M.E., p. 41 (1975).

[9] Barraclough, K. C. 'The specialist steelmaking process', *Metals and Materials,* p. 192, March 1974.

[10] Berry, R. S. and Fels, M. F., *Science and Public Affairs,* p. 11, December 1973.

[11] Berry, R. S. and Makino, H., *Technology Review,* p. 33, February 1974.

[12] Berry, R. W., 'Energy management in iron and steelworks', Iron and Steel Institute, London (1967).

[13] Biswas, A. K. and Davenport, W. G., *The extractive metallurgy of copper,* Pergamon Press (1976).

[14] Bodsworth, C. and Bell, H. B., *Physical chemistry of iron and steel manufacture,* Longmans (1972).

[15] Boustead, I. and Hancock, G. F., *The glass industry, volume 1,* Glass Manufacturers Federation, London (1975).

[16] Boustead, I. and Hancock, G. F., 'Merolite and other containers', Imperial Chemical Industries Ltd., Welwyn, U.K. (1976).

[17] Boustead, I. and Hancock, G. F., 'The Hepsleve system - Hepsleve couplings', Hepworth Iron Co. Ltd., Stocksbridge, Sheffield, U.K. (1977).

[18] Bravard, J. C., Flora, H. B. and Portal, C., 'Energy expenditures associated with the production and recycle of metals', Oak Ridge National Laboratory Report ORNL.NSF.EP24 (1972).

[19] British Petroleum Co. Ltd., 'Our Industry – Petroleum', London (1970).

[20] British Standards Institute, London, 'Definitions of the calorific values of fuels', BS526:1961.

[21] British Standards Institute, London, 'Specification for petroleum fuels for oil engines and burners', BS2869:1970.

[22] Brown, J. W. and Riddy, R. L., *Iron and Steel Eng.*, p. 37, June 1976.

[23] Bullard, C. W. and Herendeen, R. A., *Energy Policy,* p. 268, December 1975.

[24] Campbell, I. M., *Energy and the atmosphere,* Wiley-Interscience (1977).

[25] Chapman, P. F., *New Scientist,* p. 408, 17 May 1973.

[26] Chapman, P. F., 'Energy costs of producing copper and aluminium from primary sources', Energy Research Group Report No. 1, Open University (1973).

[27] Chapman, P. F., 'The energy costs of delivered energy: U.K. 1968', Energy Research Group Report No. 3, Open University (1973).

[28] Chapman, P. F., *Metals and Materials,* p. 107, February 1974.

[29] Chapman, P. F., *Energy Policy,* p. 91, June 1974.

[30] Chapman, P. F., *Energy Policy,* p. 47, March 1975.

[31] Chapman, P. F., *Fuels Paradise,* Penguin (1975).

[32] Chapman, P. F. and Hemming, D. F., Proceedings of the 9th International TNO Conference, Rotterdam, February 1976.

[33] Chapman, P. F., Leach, G. and Slesser, M., *Energy Policy,* p. 231, September 1974.

[34] Chapman, P. F. and Mortimer, N. D., 'Energy inputs and outputs for nuclear power stations', Energy Research Group Report No. 5, Open University (1974).

[35] Chemical Rubber Company, *Handbook of Chemistry and Physics,* 57th edition (1977).

[36] Clay, H., *Proc. Inst. Mech. Eng. London,* **184**, 1279 (1970).

[37] Cleathero, B., *Ironmaking and Steelmaking Qurarterly,* No. 4, p. 313 (1975).

[38] Clemow, C. J., 'Understanding energy systems', Paper presented to the joint conference of the Operational Research Society and the Institute of Fuel, London, April 1975.

[39] Cofer, D. B., *Metals and Materials,* p. 74, Februray 1972.

[40] Colautti, F. and Barbi, A., *Chem. Eng.,* April 1977.

[41] Commercial Motor, 'Tables of Operating Costs', I.P.C. Business Press (1977).

[42] Cottrell, A. H., *An introduction to metallurgy,* Arnold 1967.

[43] Davies, D., *Metals and Materials,* p. 196, March 1974.

[44] Dite, R. V. and Raynor, C. J., Soc. Chem. Ind., March 1975.

[45] Decker, A., Proceedings of the 9th International TNO Conference, Rotterdam (1976).

[46] Easetop, T. D. and McConkey, A., *Applied thermodynamics for engineering Technologists in SI units,* Longmans (1970).

[47] Ecologist (edited), *A blueprint for survival,* Penguin (1972).

[48] Edison Technical Services Report, 'The plastics beer bottle' (1974).

[49] Einstein, A., *Ann. Physik.,* **17**, 891, (1950). See also, Einstein, A., *Relativity – special and general theory,* Methuen (1960).

[50] Elliot, J. F. in *Energy use and conservation in the metals industry,* Met. Soc. A.I.M.E., p. 4 (1975).

[51] *Encyclopaedia of Chemical Technology,* Volume 1, p. 947, 'Aluminium and aluminium alloys'.

[52] U.S. Environmental Protection Agency Report EPA/530/SW-91c., 'Resource and environmental process analysis of nine beverage container alternatives' (1974).

[53] Everall, P. F., 'The effects of road traffic conditions on fuel consumption', Transport and Road Research Laboratory Report No. LR226, U.K. (1968).

[54] Feltoe, F. J., *J. Inst. Fuel,* p. 66, February 1965.

[55] Gartner, E. M. and Smith, M. A., *Energy Policy,* p. 144, June 1976.

[56] Gray, P. M. J., 'Conservation in primary metal extraction', Paper presented to the 4th International BNF Conference on energy and non-ferrous metals, Barcelona (1976).

[57] Gregory, P. and Day, P., 'Energy consumption and waste products in the conversion of high electrical conductivity copper wire bar to wire', paper presented to the 4th International BNF Conference on energy and non-ferrous metals, Barcelona (1976).

[58] Hannon, B. M., *Environment,* **14**, No. 2, p. 11 (1972).

[59] Hannon, B. M., 'System energy and recycling', A.S.M.E. (1973).

[60] Harrison, G., *Ironmaking and Steelmaking Quarterly,* No. 4, p. 299 (1975).

[61] Heal, D. W., *Ironmaking and Steelmaking Quarterly,* No. 4, p. 222 (1975).

[62] Hinde, P. T. and Probert, S. D., *Applied Energy,* 2, p. 17 (1976).

[63] Hirst, E., 'Energy consumption for transportation in the U.S.', Oak Ridge National Laboratory Report ORNL-NSF-EP-15.

[64] 'Report on the Census of Production – Introductory Notes', Her Majesty's Stationary Office, London (1972).

[65] 'Input-Output Tables for the United Kingdom 1968', Her Majesty's Stationary Office, London (1973).

[66] Holloman, J. H. and Grenon, M., *Energy Research and Development,* Ballinger, Cambridge, Mass. (1975).

[67] Hopkins, N. and Gandhi, K. C., *Ironmaking and Steelmaking Quarterly,*

No. 4, p. 295 (1975).
[68] Hubbert, M. K. in *Resources and Man,* W. H. Freeman & Co. (1969).
[69] 'Energy Analysis', Workshop Report Number 6, Internation Federation of Institutes of Advanced Study (IFIAS), Stockholm (1974).
[70] Imperial Chemical Industries Ltd., London (1974), 'The competitiveness of LDPE, PP and PVC after the 1963 oil crisis'.
[71] Jacobi, J. S., 'The secondary metals industry as a conservation agent', a paper presented to the 4th International BNF Conference on energy and non-ferrous metals, Barcelona (1976).
[72] Kakela, P., *Environment,* **17**, No. 2, p. 27 (1975).
[73] Kavrak, I., *Energy Communications,* **2**, 521 (1976).
[74] Kellogg, A. J., *J. Metals,* p. 25, June 1974.
[75] Kellogg, H. H., in 'Energy use and conservation in the metals industry', *Met. Soc. A.I.M.E.,* p. 87 (1975).
[76] Koenig, H., *Canadian Mining and Met. Bulletin,* p. 174, February 1970.
[77] Krejger, P. C., Proceedings of the 9th International TNO Conference, Rotterdam, February 1976.
[78] Kroger, C., *Glastechn. Ber.* 7, 202 (1953).
[79] Langman, R. D. and Harris, T. H., *Ironmaking and Steelmaking Quarterly,* No. 4, p. 253 (1975).
[80] Larwood, G. M. and Leary, R. J. in *Energy use and conservation in the metals industry,* Met. Soc. A.I.M.E., p. 259 (1975).
[81] Lath, R. P. and Menon, P. M., *Trans. Ind. Inst. of Metals,* p. 39, December 1969.
[82] Laurila, A. and Andersson, B., 'Modern Flash Smelting', paper presented to the 4th International BNF Conference on energy and non-ferrous metals, Barcelona (1976).
[83] Leach, G., *Energy and food production,* International Institute for Environment and Development, London (1975).
[84] Leach, G. and Slesser, M., *Energy equivalent of network inputs in food production,* University of Strathclyde, (1973).
[85] Long, T., V., Proceedings of the 9th International TNO Conference, Rotterdam (1976).
[86] Maadah, A. G. and Maddox, R. N., *Energy Communications,* **2**, 237, (1976).
[87] Maddox, K. P., *Mineral Industries Bulletin of the Colorado School of Mines,* **18**, July 1975.
[88] Makhijani, A. B. and Lichtenberg, A. J., *Environment,* **14**, No. 5, p. 10, June 1972.
[89] Mangalick, M. C. in *Energy use and conservation in the metals industry,* Met. Soc. A.I.M.E., p. 101 (1975).
[90] Markus, T. and Slesser, M., *Energy Policy,* p. 76, March 1977.
[91] Meadows, H. M., Meadows, L. M., Randers, J. and Behrens, W. W., *The*

limits to growth, Pan Books (1972).
[92] McKillop, A., *Ecologist,* 2, No. 12, p. 1 (1972).
[93] McMullen, J. T., Morgan, R. and Murray, R. B., *Energy resources and supply,* Wiley-Interscience (1976).
[94] McNabb, J. W., Statement to the Illinois Pollution Control Board, Chicago, 20 February 1973.
[95] Miller, J. R., *Steel Furnace Monthly,* p. 169, March 1973.
[96] Montgomery, W., *Ironmaking and Steelmaking Quarterly,* No. 4, p. 231 (1975).
[97] Moore, W., J., *Physical Chemistry,* Longmans (1959).
[98] Mortimer, N. D., 'The energy cost of road and rail freight transport in the U.K. in 1968', Energy Research Group Report No. 4, Open University (1974).
[99] Mortimer, N. D., 'Gross Energy requirements of marine transport', Energy Research Group Report No. 7, Open University (1974).
[100] Mortimer, N. D., 'Energy analysis of burner reactor power systems', Ph.D. thesis, The Open University (1977).
[101] National Economic Development Office, 'Energy conservation in the U.K.', Her Majesty's Stationary Office, London (1974).
[102] National Technical Information Service, U.S. Department of Commerce, Report No. PB-245-759 (June 1975), 'Energy use patterns in metallurgical and non-metallic mineral processing'.
[103] Newton, G. E. H., 'Fuel and productivity in the pulp and paper industry', Paper 11 to the Fuel and Power Conference, U.K. (1963).
[104] Odok, A., Thym, G. and Wurgler, R., 'The potential for energy conservation by strip casting of aluminium', paper presented to the 4th International BNF Conference on energy and non-ferrous metals, Barcelona (1975).
[105] Odum, E. P., *Fundamentals of Ecology,* W. B. Saunders Co. (1971).
[106] Owen, M. and Jacobi, J. S., *J. Metals,* p. 10, April 1975.
[107] Peklke, R. D., *Unit processes of extractive metallurgy,* Elsevier (1975).
[108] Pick, H. J. and Becker, P. E., *Applied Energy,* **1**, 31 (1975).
[109] Pitt, R. S., *Ironmaking and Steelmaking Quarterly,* No. 4, p. 242 (1975).
[110] Porameswaren, K. and Nadkarni, R. in *Energy use and conservation in the metals industry,* Met. Soc. A.I.M.E., p. 271 (1975).
[111] Roberts, F., *Metals and Materials,* p. 167, March 1974.
[112] Roberts, F., 'Conference on Energy Accountancy', Institute of Mechanical Engineers, London, December 1976.
[113] Rodriguez, F. A., *Iron and Steel Engineer,* p. 57, January 1977.
[114] Russell, C. S. and Vaughan, W. J., *Steel production – processes, products and residuals,* Johns Hopkins University Press (1976).
[115] Samples, D. K., 'Energy in the automobile', seminar at the Institute of Sciences and Technology, University of Michigan, Chicago, 23 August

1974.
[116] Schipper, L. and Lichtenberg, A. J., *Science,* **194**, 1001 (1976).
[117] Schroer, C. A. and Clark, D. W., *Iron and Steel Engineer,* p. 37, June 1976.
[118] Sharp, J. D., *Ironmaking and Steelmaking Quarterly,* No. 4, p. 320, 1975.
[119] Sheldon, A. C. in *Energy use and conservation in the metals industry,* Met. Soc. A.I.M.E., p. 157 (1975).
[120] Sheridan, A. T., *Ironmaking and Steelmaking Quarterly,* No. 4, p. 262 (1975).
[121] Shreve, R. N., *Chemical process industries,* McGraw Hill (1967).
[122] Sigworth, G. K., in *Energy use and conservation in the metals industry,* Met. Soc. A.I.M.E., p. 19 (1975).
[123] Smart, J. D., *Ironmaking and Steelmaking Quarterly,* No. 4, p. 320 (1975).
[124] Smith, H., *Trans. World Energy Conf.,* **18**, Section E (1969).
[125] Smithson, D. J. and Sheridan, A. T., *Ironmaking and Steelmaking Quarterly,* No. 4, p. 286 (1975).
[126] Snelgrove, W. R. N. and Barber, M. J., 'The continuous casting of anodes at British Copper Refiners Ltd.', paper presented to the 4th International BNF Conference, Barcelona, 1976.
[127] Stone, J. K., *Iron and Steel Engineer,* p. 38, March 1976.
[128] Stull, D. R., Westrum, E. F. and Sinke, G. C., *The chemical thermodynamics of organic compounds,* Wiley-Interscience (1969).
[129] Sundstrom, G., 'Investigation of the energy requirements from raw materials to garbage treatment for four Swedish beer packaging alternatives', report for Rigello Pak A.B. (1973).
[130] Technische Hogeschool, Eindhoven, Holland, April 1975, Laboratorium Materiaalkunde Report No. ML/75/4.
[131] Thornton, D. S. and Williams, D. I. T., *Ironmaking and Steelmaking Quarterly,* No. 4, p. 241 (1975).
[132] Thuman, A., *Plant engineers and managers guide to energy conservation,* Van Nostrand Reinhold Co. (1977).
[133] Tien, J. K., Clark, R. W. and Malu, M. K., *Tech. Rev.,* p. 39, February 1975.
[134] Transport and Road Research Laboratory Supplementary Report No. 251, 'Future Transport Fuels' (1976).
[135] Tuininga, E. J., 'Energy analysis of transportation systems', proceedings of the 9th International TNO Conference, Rotterdam (1976).
[136] United Nations, *World Energy Supplies 1950–1974,* New York (1976).
[137] U.S. Department of the Interior, *Minerals Yearbook 1971,* volume 1, p. 22, Government Print Office, Washington.
[138] Wahlman, M., *Pulp and Paper International,* p. 60, November 1974.
[139] Waller, R. F., 'Primary requirements for the production of iron and steel

in the U.K.', Energy Research Group Report No. 11, Open University (1976).
[140] Wild, R., *Chem. and Process Eng.*, p. 55, February 1969.
[141] Wright, D. J., *Energy Policy*, p. 307, December 1974.
[142] Wright, P. A., *Extractive metallurgy of tin*, Elsevier (1966).
[143] Ydstie, B. in *Energy use and conservation in the metals in industry*, Met. Soc. A.I.M.E., p.131 (1975).

Index